ATM Signalling

PROTOCOLS AND PRACTICE

ATM Signalling

PROTOCOLS AND PRACTICE

Hartmut Brandt
GMD Fokus, Berlin, Germany

Christian Hapke
vectos GmbH, Berlin, Germany

JOHN WILEY & SONS, LTD
Chichester • New York • Weinheim • Brisbane • Singapore • Toronto

Baffins Lane, Chichester,
West Sussex, PO19 1UD, England

National 01243 779777
International (+44) 1243 779777
e-mail (for orders and customer service enquiries): cs-books@wiley.co.uk

Visit our Home Page on http://www.wiley.co.uk or http://www.wiley.com

Other Wiley Editorial Offices

John Wiley & Sons, Inc., 605 Third Avenue,
New York, NY 10158-0012, USA

Wiley-VCH Verlag GmbH
Pappelallee 3, D-69469 Weinheim, Germany

Wiley Australia Ltd, 33 Park Road, Milton,
Queensland 4064, Australia

John Wiley & Sons (Canada) Ltd, 22 Worcester Road
Rexdale, Ontario, M9W 1L1, Canada

John Wiley & Sons (Asia) Pte Ltd, 2 Clementi Loop #02-01,
Jin Xing Distripark, Singapore 129809

Library of Congress Cataloging-in-Publication Data
Brandt, Hartmut, 1963-
ATM signalling: protocols and practice / Hartmut Brandt, Christian Hapke.
p. cm.
Includes bibliographical references and index.
ISBN 0 471 62382 2
1. Asychronous transfer mode. 2. Signals and signalling. I. Hapke, Christian, 1974- II. Title.

TK5105.35. B73 2001
621.382'16 – dc21 00-069345

British Library Cataloguing in Publication Data

A catalogue record for this book is available from the British Library

ISBN 0 471 62382 2

Produced from LaTeX files supplied by the authors
Printed and bound in Great Britain by Bookcraft (Bath) Ltd
This book is printed on acid-free paper responsibly manufactured from sustainable forestry, in which at least two trees are planted for each one used for paper production.

Contents

Preface

After more than 10 years of standardisation effort, the asynchronous transfer mode (ATM) technology has experienced a remarkable growth in backbone and large private networks during the last few years. ATM has not fulfilled all its expectations: it was seen as the network technology of the future ranging from desktop applications to large scale backbone networks. With the advent of cheap and fast Ethernet solutions, the "ATM to the desktop" tendencies have slowed down and it is not clear whether this will change in the near future. However, ATM is growing, especially for large networks—it proves to be scalable, reliable and able to provide Quality of Service (QoS) guarantees for applications. All three of these features of ATM are essential for the deployment of this technology in integrated networks, i.e. networks that are equally suited to transport classical telephony traffic as well as data traffic.

The features of ATM do not come cheaply for. ATM technology is complex. Two major organisations have driven the standardisation process and continue to produce new standards: the ITU-T[1] and the ATM-Forum. There are many more than one hundred standards (or recommendations) issued by the ITU-T, mainly in the I and Q-Series, and the ATM-Forum web-site lists 155 standards at the time of writing. The situation is further complicated by the fact that most of the standards are written in a style that is very different from the more informal style used in Internet RFCs. This makes it hard to read and understand for people not used to this style. As an example, the standard describing the transport protocol used in ATM signalling (Q.2110), consists of about 50 pages of SDL (Structured Description Language) and only about 10 pages prose description. For the beginner in ATM it is often hard even to find out, which standards are the most relevant ones.

This book presents a small part of the entire ATM technology in a more practical way than is done in standard documents. It is addressed to people doing practical work with ATM and to people who want to understand what ATM is and how it works.

The book focusses on the signalling protocols used in private ATM networks and protocols built on top of signalling. ATM, in contrast to the IP protocol used in the Internet, is connection-oriented. This means that before one can send data, a connection must be established to the intended receiver. To establish and control a connection, a specialised suite of signalling protocols is used. These protocols are quite complex but a thorough understanding of them is necessary to understand ATM networks.

ATM networks are generally divided into two classes: public networks and private networks. Public ATM networks are used in the backbone networks of the large telecommunication operators. The exceptionally strong reliability and performance requirements of these

---

[1] International Telecommunication Union, the former CCITT (Comitè consultatif international tèlègraphique et tèlèphonique).

networks and the necessity of interworking with existing equipment and the smooth introduction of new technology have lead to a complex architecture and a set of complex protocols for these networks. In private networks, on the other hand, performance requirements are not so high and the existing infrastructure can be changed more easily. For this reason the protocols used in private ATM networks are sometimes different from those in public networks. Because it is most likely that people are going to use private ATM network equipment, the book focuses on the appropriate protocols. Where appropriate public ATM protocols are discussed in short.

The book not only explains the protocols, but also shows traces of the most relevant protocol operations. To really understand protocol operation, it is not enough to read, how a protocol should work, but hands-on experience is needed in different situations. One way to get this experience is to trace an existing network during operation, another is to use a controllable protocol implementation to force the protocol into different states and observe how it behaves by means of communication traces. In both cases it is essential to see how the communication between protocol instances work. The tools needed to obtain communication traces for the ATM signalling protocols are publicly available and can be used to repeat the experiments. Experience has shown that tracing is also an excellent tool for finding problems in a network. In the ATM network used for the experiments in this book, optical splitting boxes are inserted permanent into most strategic links (for example, links between major switches). These boxes deliver a copy of the traffic on the given link to test systems. If problems arise in the network, tracing tools can be used on the test systems to locate the problems.

The material in this book is based on ITU-T standards and ATM-Forum standards as well as on more than five years of practical experience with ATM. GMD Fokus was one of the early users of ATM technology in Germany. In the ATM laboratory of GMD Fokus, ATM equipment of different vendors was used and tested. Numerous protocols and testing tools were implemented. GMD Fokus took also part in several European Union ACTS (Advanced Communications Technologies and Services) projects, like INSIGNIA and ELISA, which were dedicated to ATM. In these projects applications and signalling stacks were implemented and successfully demonstrated. In the last two years, work on ATM in Fokus has moved to more advanced topics like wireless and mobile ATM, ATM over satellites and security in ATM networks.

Acknowledgements

We wish to thank all our friends for their support during the writing of this book. Their patience and encouragement made this project possible.

GMD Fokus's ATM Laboratory provided a good working environment where we could run most of our experiments on the available ATM infrastructure.

It has been a pleasure working with the Tina protocol tracing tools. These tools are the basis of experiments described in this book. We especially thank Jörg Micheel, Robert H. Fomin and Harold I. Coy of the Begemot Computer Associates team who were developing these tools together with both authors. We also thank Robert H. Fomin for the development of the protocol software that we used in many experiments.

A camera-ready copy of this book was produced by the authors. The book was typeset with LaTeX, a macro package for TeX. The figures were drawn using the `xfig` UNIX program. The bibliography was prepared using `bibtex`. The book design was provided and implemented in LaTeX by Wiley.

We also would like to thank vectos Corporation for its support and patience during the time-consuming preparation of the manuscript.

It is the publisher that does whatever is required to deliver the final product to the reader. Pak-Hang Wan, Sarah Hinton and Robert Hambrook gave us much support and encouraged us to finish the manuscript. Many other professionals at Wiley worked to finish the book.

Finally, we welcome emails from any reader with comments, suggestions or bug fixes.

Berlin, Germany
August 2000

Hartmut Brandt <brandt@fokus.gmd.de>
Christian Hapke <christian.hapke@vectos.de>

Abbreviations

AA Administrative Authority

AAL ATM Adaptation Layer

AAL5 ATM Adaptation Layer 5

AAL-CP ATM Adaptation Layer Common Part

ABR Available Bit Rate

AESA ATM End System Address

AFI Authority and Format Identifier

AINI ATM Inter-Network Interface

ANS ATM Name Service

ANSI American National Standards Institute

AoD Audio on Demand

API Application Programming Interface

ARP Address Resolution Protocol

ASP ATM Service Provider

ATM Asynchronous Transfer Mode

ATMARP ATM Address Resolution Protocol

ATMF ATM Forum

AvCR Available Cell Rate

AW Administrative Weight

BCD Binary Coded Decimal

BHLI Broadband Higher Layer Information

B-ICI B-ISDN Inter Carrier Interface

B-ISDN Broadband ISDN

B-ISUP Broadband ISDN User Part

BLLI Broadband Lower Layer Information

BUS Broadcast and Unknown Server

CAC Call Admission Control

CBR Constant Bit Rate

CDV Cell Delay Variation

CLIP Classical IP over ATM

CLIR Calling Line Identification Restriction

CLP Cell Loss Priority

CLR Cell Loss Ratio

COBI Connectionless Bearer-Independent

CP-AAL Common Part AAL

CPCS Common Part Convergence Sublayer

CR Cell Rate

CRC Cyclic Redundancy Check

CRM Cell Rate Margin

CTD Cell Transfer Delay

DCC Data Country Code

DFI Domain Specific Part Format Indicator

DFN Deutsches Forschungsnetzwerk

DNS Domain Name System

DS Database Summary

DSP Domain Specific Part

DTL Designated Transit List

ESI End System Identifier

ETSI European Telecommunications Standards Institute

Eurescom European Institute For Research And Strategic Studies in Telecommunications

FCS Frame Check Sequence

FSM Finite State Machine

GCAC Generic Call Admission Control

HO-DSP High-Order DSP

ICD International Code Designator

ID Identification

IDI Initial Domain Identifier

IDP Initial Domain Part

IE Information Element

IETF Internet Engineering Task Force

IG Information Group

ILMI Integrated Local Management Interface

IME Interface Management Entity

InATMARP Inverse ATM Address Resolution Protocol

IP Internet Protocol

IPv4 IP version 4

IPv6 IP version 6

ISDN Integrated Services Digital Network

ITU-T International Telecommunication Union

LAN Local Area Network

LANE LAN Emulation over ATM

LE LAN Emulation

LE Service LAN Emulation Service

LEC LAN Emulation Client

LECID LAN Emulation Client ID

LECS LAN Emulation Configuration Server

LES LAN Emulation Server

LGN Logical Group Node

LIS Logical IP Subnetwork

LLC Logical Link Control

LNNI LAN Emulation Network–Network Interface

LUNI LAN Emulation User–Network Interface

MAC Media Access Control

MAC Multiple Access Control

maxCR maximum Cell Rate

maxCTD maximum Cell Transfer Delay

MIB Management Information Base

MID Multiplexing Identifier

MPOA Multi Protocol Over ATM

MTP-3b Message Transfer Part 3 (broadband)

MTU Maximum Transport Unit

NDIS Network Driver Interface Specification

N-ISDN Narrowband ISDN

NNI Network–Node Interface

NSAP Network Service Access Point

ODI Open Data-link Interface

OUI Organisationally Unique Identifier

PDU Protocol Data Unit

PGL Peer Group Leader

PNNI Private Network Node Interface

POTS Plain Old Telephone System

prefix Network Address Prefix

PSA Proxy Signalling Agent

PTSE PNNI Topology State Element

PTSP PNNI Topology State Packet

PVC Permanent Virtual Connection

PVCC Permanent Virtual Channel Connection

PVPC Permanent Virtual Path Connection

QoS Quality of Service

RAIG Resource Availability Information Group

RCC PNNI Routing Control Channel

RCC Routing Control Channel

RFC Request for Comments

RPC Remote Procedure Call

SAAL Signalling ATM Adaptation Layer

SDL Structured Description Language

SEL Selector

SMS Selective Multicast Server

SNAP SubNetwork Attachment Point

SNMP Simple Network Management Protocol

SSCF Service Specific Coordination Function

SSCOP Service Specific Connection Oriented Protocol

SSCP Signalling Connection Control Part

SSCS AAL Service Specific Convergence Sublayer

stdin standard input

stdout standard output

SVC Switched Virtual Channel

SVCC Switched Virtual Channel Connection

SVP Switched Virtual Path

TCAP Transaction Capabilities

TLV Type/Length/Value

TNS Transit Network Selection

UBR Unspecified Bit Rate

ULIA Uplink Information Attribute

UNI User–Network Interface

UPC Usage Parameter Control

VBR Variable Bit Rate

VC Virtual Channel

VCC Virtual Channel Connection

VCI Virtual Channel Identifier

VoD Video on Demand

VP Virtual Path

VPC Virtual Path Connection

VPCI Virtual Path Connection Identifier

VPI Virtual Path Identifier

VR Variance Factor

1

Introduction

During the last few years the use of Asynchronous Transfer Mode (ATM) networks has increased dramatically. Both public and private network providers use ATM which integrates conventional line-switched networks, such as Narrowband ISDN (N-ISDN), and conventional data networks, such as Ethernet. ATM is based on the standards of the International Telecommunication Union (ITU) and the ATM Forum.

In ATM networks the information is transmitted between communicating entities using fixed-size packets, referred to as the ATM cells. An ATM cell has a length of 53 octets, consisting of a 5-octet header and a 48-octet payload field. Therefore, all data transmitted from a source must be segmented into cells of this size. Then these cells can be transported as a stream by the ATM network. If different streams need to be transported, the associated cells can be multiplexed inside the ATM network. This process increases the network utilisation and is essential to achieve an economically working network.

Connection-oriented networks like ATM substantially differ from connectionless networks like Internet Protocol (IP) in the need for control protocols. In connectionless networks all information that is needed to get the information from the source to the destination is contained in every datagram. In connection-oriented networks a connection must be established before data can be transmitted. This is done by means of control protocols. Naturally, in connection-oriented networks it is easier to support quality of service—much more information can be exchanged in the connection establishment phase between the network and the user (in pure connectionless networks one would need to carry this information in every datagram) and network resources can be verified and reserved in the establishment phase. The network can optimise the route and fit it to the needs of the user.

The philosophy and architecture of the ATM control plane stems from the telephone network. Many of the concepts and much of the terminology are adapted to a broadband environment (in ATM speak narrowband means the telephone network, broadband means Asynchronous Transfer Mode (ATM)). In contrast to IP, ATM was first standardised and then implementation begun. All this together has led to a quite complicated architecture.

The ATM protocol architecture is usually described as a cube consisting of planes (Figure 1.1). ATM networks distinguish between the data, control and management planes. The data plane defines the protocol layers that are used to process user data, whereas the control plane defines the stack of protocols that are used to establish, tear-down and modify user and control connections. The management planes are used to manage the layers of the user and control plane (layer management) and the planes as a whole (plane management).

In this book we focus on the control plane protocols used in private ATM networks. We expect the reader to be familiar with the general concepts of ATM: cells, virtual channels

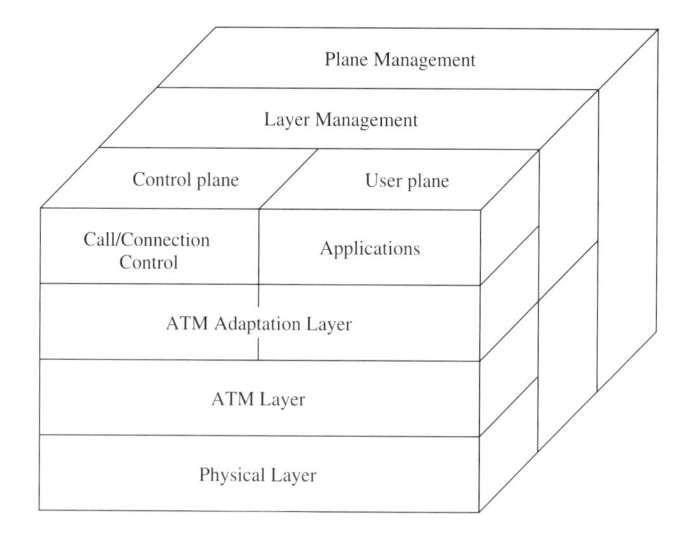

Figure 1.1: ATM protocol reference model

and paths, virtual channel and path connections, multiplexing and the ATM Adaptation Layer (AAL).

1.1 Organisation of the Book

This book describes the most important protocols of the ATM control plane that are used in private ATM networks. In Chapter 3 we start with the User–Network Interface (UNI) protocol, which is the central protocol to establish connections in an ATM network. Chapter 5 analyses the layer below the UNI—the Signalling ATM Adaptation Layer (SAAL). This layer provides transport services to the UNI. ATM is asymmetric in the sense that the protocols at the boundary of the network (the UNI) are different from the protocols inside the network. Chapter 6 concentrates on the routing and connection control protocols inside the network—the Private Network Node Interface (PNNI). UNI links need to be managed. This is the task of the Integrated Local Management Interface (ILMI) which is discussed in detail in Chapter 7. A couple of protocols use the services of the ATM control protocol suite to provide other network services. The most popular of these protocols, Classical IP over ATM (CLIP) and LAN Emulation over ATM (LANE), are described in Chapter 8.

Two appendices help the reader to find the way through the labyrinth of ITU-T standards (Appendix A) and to find source code for tracing tools and protocol software as well as to find standard documents (Appendix B).

1.2 Systems used for Experiments

In this book we show the results of several experiments with ATM systems. The experiments were performed in the ATM laboratory of GMD Fokus.

In our experiments we used up to four different ATM switches. The names of these switches are foreplay, forelle, forest and forever. All these switches are Fore ASX 200 switches running version 4 and 5 of the ForeThought software. The switches support the network side as well as the user side of the ATM protocols. Depending on the goal of the experiment, each switch was equipped with an appropriate number of OC3 and TAXI 100MBit/s ATM port modules. All ATM cables were optical fibres. Electrical twisted pair cables are unsuitable because it is difficult to trace the communication on such cables.

We used four different workstations as ATM end systems. The names of these workstations are kirk, spock, atmos and lovina. Kirk and spock are Sun Sparc Ultra 1 workstations with 128 MB RAM running Sun's operation System Solaris 2.5. Atmos is a Sun Sparc 10 workstation with 96 MB RAM running Solaris 2.5 and lovina is a Sun Sparc 20 workstation with 96 MB RAM running Solaris 2.5. Kirk, spock and atmos use one Fore SBA 100 ATM adapter (TAXI 100MBit/s). Lovina is equipped with a Fore SBA 200 ATM adapter (OC3). In addition, lovina contained two Tanya ATM interface cards [GMD] for tracing. Different software was used for the different experiments. The software of the Fore ATM interface cards was always running and supports the protocols AAL5 and ILMI. For the SAAL and UNI experiments the software described in Section 1.4 was employed.

The workstations and switches were used exclusively for our experiments. Side effects with other communications were not possible.

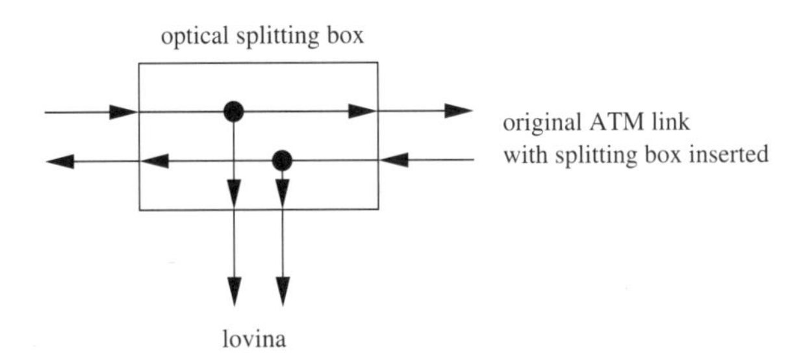

Figure 1.2: ATM link with optical splitting box inserted

Lovina plays a special role in our experiments in addition to its role as ATM end system. In all experiments lovina was used to capture ATM traces of other systems communicating via one or more optical ATM links. How does this capturing work? In each ATM link to be traced an optical splitting box was inserted. Such a splitting box splits the optical signals of a link (one link = two fibres, one for each direction) from two input fibres to four output fibres. Figure 1.2 shows one ATM link with the optical splitting box inserted. The two additional output fibres were connected to the receivers of two additional ATM interface cards in lovina. The input on these two ATM interface cards was captured and decoded with the software described in the next section. The communication of the original ATM link is not affected. All traces presented in this book were generated using this principle.

1.3 Protocol Tracing Tools

1.3.1 Introduction

The Tina framework was written by the Begemot Computer Associates team. Tina is a set of tools and library routines designed for ATM and network testing. The Tina tool framework was designed to support network providers and administrators on maintaining complex networks. Although Tina was originally used in the ATM context, it can be also used in non-ATM environments.

The toolkit currently runs on UNIX workstations and employs the approach of combining tools by pipelining to achieve a certain functionality. The kit uses two different binary interfaces named cell stream and packet stream.

Tina stems from the paradigm that there needs to be a single inexpensive tester for all kinds of ATM and network testing. Testing means monitoring, interoperability testing, performance measurements and simulations, as well as measurements of a living ATM network (troubleshooting, monitoring, accounting, etc.).

The framework of interfaces and tools runs on a variety of platforms ranging from specialised ATM testing hardware to standard workstations and personal computers. A workstation or PC equipped with a standard ATM adapter is sufficient for higher level protocol testing and fulfils the requirement of being inexpensive.

Typical tools of Tina follow the UNIX filter principle: they consume a certain stream of data at the standard input (stdin) and generate another at the standard output (stdout). The toolkit employs binary interfaces due to performance reasons. A wide variety of compile, decompile and dump tools exist for conversion between the binary data and their respective human readable forms.

How can one start using Tina? Download the Tina framework software as stated in Section B.2. Then unpack the package and follow the instructions in the "`README`" file. Then you can start using the software, e.g. you can start the tools of the framework. Or you can explore more details and background information from the delivered HTML documentation and man pages.

The Tina framework software runs on UNIX platforms such as Solaris, Linux or FreeBSD. We used the Solaris version for the experiments in this book. Each tool comes with its own manual page that can be displayed with

```
man name_of_the_tool.
```

With

```
man TINA
```

you can get an introduction to the Tina manual pages.

The Tina framework is based on cell and packet streams. A cell stream is a binary interface between software or hardware modules describing a uni-directional sequence of ATM cells. An ATM cell consist of all the fields found in standard ATM cells plus some flags and a 64-bit timestamp. The packet stream is a binary interface that describes a sequence of variable-length messages. It uses the same 64-bit timestamp as a cell stream. These binary streams are byte-order dependent, i.e. the streams must be converted if they are to be exchanged between little-endian and big-endian systems.

The timestamps are essential for the measurements. The timestamp used by Tina is a 64-bit signed integer value. Because absolute times are never smaller than 0, actually 63 bits of the value can be used. The granularity of a timestamp is 1 nanosecond. This gives a maximum duration of a cell or packet stream of more than 290 years. That was long enough for our experiments, especially because the start moment is context dependent and can be moved.

The tools of the Tina toolkit can be connected by pipes to exchange data. Sometimes it is also important to transport control information which has a special semantics inside a running system. One application of the control information is a flushing mechanism, and another is the contribution of clock information. Both control cells and control packets are used in Tina. They are almost transparent to the user. To use the Tina toolkit it is not essential to understand the control concept. The user only needs to know that there can be control cells and packets and that they can safely be ignored.

The following is a short description of the tools we used for our experiments. Many more tools are shipped with the Tina framework. The Tina documentation provides more details.

1.3.2 Sun Batman API to Packet Stream Stub `baps`

The `baps` tool is used to capture AAL5 packet stream data from the ATM network using Sun's network interface cards with the second preliminary Application Programming Interface. The `-d` option is used to specify the device of the ATM interface card, the default is "`/dev/ba0`". The packet stream is written to stdout. One or more VPIs/VCIs may be specified for attachment by the option `-c`. The packets from different VPIs/VCIs get fed into the outgoing stream in the order they appear at the API. Timestamping is done in the user process software.

1.3.3 Fore API to Packet Stream Stub `faps`

The `faps` tools is used to capture AAL4 and AAL5 packet stream data out of the ATM network using Fore network interface cards with the preliminary (SPANS-based) Application Programming Interface. By default AAL5 packets are assumed. One or more VPIs/VCIs may be specified for attachment by the `-c` option. The packets from different VPIs/VPCs get fed into the outgoing stream (stdout) in the order they appear at the API. Timestamping is done in the user process. The `-d` option is used to specify the device of the Fore card; the default is "`/dev/fa0`".

1.3.4 Reading ATM Cells with HP 75000 BSTS and `hpcs`

The Broadband Series Test System 75000 (BSTS) from Hewlett Packard is a high end ATM tester. It is specialised for testing at ATM layers and AAL layers. A special feature of the BSTS is the capability of capturing incoming ATM cells. The BSTS can write up to 131072 cells directly from the link to the capture buffer. The BSTS adds a high precision absolute timestamp (resolution: 100 ns) to every received cell. This is much better than the software-generated timestamps of `baps` and `faps`.

The program `hpcs` controls the capture process. It lets the BSTS fill the capture buffer for a limited time or until the buffer is full. Then `hpcs` reads the capture buffer and outputs the cells as a cell stream. This cell stream can be stored to a disc or directly used as input for other modules (e.g. `a5r`).

A problem with the BSTS is that it cannot receive new cells while the program reads cells from the capture buffer. This makes it impossible to capture cells over a long timeframe without interruption. Therefore, we did not use this tool and its high resolution timestamps for the experiments in this book.

1.3.5 Cell Stream Dump with `csdump`

This tool dumps all cells in the input cell stream from stdin in a human-readable format, which can be specified by the user, to stdout. Header fields, the timestamp and the payload can be dumped.

1.3.6 AAL5 Reassembly Tool `a5r`

`a5r` reassembles a cell stream consisting of AAL5 SAR cells into AAL5 frames. The cell stream could be a received stream from an ATM interface card. `a5r` also outputs a packet stream containing periodic samples of error statistics.

Normally `a5r` is used to reassemble AAL5 frames based on the cells received from stdin and write them to the packet output stream on stdout. Each packet receives the timestamp from the last cell of this packet. At the same time `a5r` emits a statistics stream on the file descriptor 3. A special option `-c` can be used to specify a specific channel (VPI/VCI). Cells on other channels will be ignored.

1.3.7 Generic Packet Stream Dump `psd`

`psd` dumps the packet stream from stdin, which is assumed to contain binary data, in hexadecimal onto stdout. The output is human readable.

1.3.8 Trace SSCOP Protocol tool `sscopdump`

`sscopdump` dumps a packet stream from stdin, which is supposed to contain a trace of SSCOP messages, onto stdout. There are different modes of operation: trace the protocol in a human-readable format, dump only user data (can by used as packet stream to analyse the higher layer information, e.g. UNI) and try to report problems like retransmissions and error states. See also Chapter 5.

1.3.9 Generic Decoder for UNI 3.1, UNI 4.0, Q.2931 and PNNI 1.0 Signalling `sigdump`

`sigdump` dumps the contents of the packet stream on stdin, which is assumed to contain signalling protocol messages, onto stdout. The output is human readable by default. The user of the tool can specify the protocol type and its version. Details on the protocols can be found in Chapters 3 and 6.

`sigdump` might be used in two ways. In principle, VPI/VCI 0/5 should be captured and piped trough `sscopdump` first to get only the data stream at the top of the SAAL layer. `sigdump` supports the `-i` option that strips SSCOP trailers and alleviates the need for `sscopdump`.

`sigdump` extensively checks the correctness of message coding. Errors are flagged with a bang (!). Decoding proceeds until the end of the message, if possible. Length checking is done to the extension made possible by specifications. Trailing (unused) bytes are printed if present.

At verbose level 1 (option `-v`) a HINT section is appended to each message giving advice about potential problems with information elements resulting from inter-message dependencies, which cannot be checked by `sigdump`. `sigdump` does not implement the signalling state machine and therefore cannot do a full check of message contents. The `maybe` statement in the output indicates that `sigdump` has detected a portion of the message which suggests that this message may belong to a different signalling version or protocol.

Records include invalid `i` (but present) information elements, mandatory `m` (but missing) elements, potentially illegal `pi` or potentially mandatory `pm` elements, depending on message type and signalling context.

1.3.10 ATM Forum PNNI 1.0 Routing Decoder `pnnidump`

`pnnidump` dumps the packet stream from stdin, which is assumed to contain PNNI 1.0 routing protocol messages, onto stdout. The output is human readable by default. See Chapter 6 for more details about PNNI routing protocol.

PNNI signalling messages are not dumped by this tool. Use `sigdump` instead.

1.3.11 ILMI Dump tool `ilmidump`

`ilmidump` dumps the content of the standard packet stream from stdin, which is assumed to contain ILMI protocol messages, onto stdout. The output is human readable. See Chapter 7 for more details about ILMI.

1.4 ATM Protocol Software

The ATM control protocols are quite complex. Therefore publicly available implementations are quite rare and new features that the standardisation organisations continue to define are introduced slowly. Three implementations of UNI software are:

- The Begemot FreeUNI implementation. This software implements UNI 3.1 and some of the ITU-T supplementary services, as well as parts of the Generic Functional Protocol [Q.2932.1]. It is available via the Internet (see Appendix B).
- The HARP Host ATM Research Platform from Network Computer Services (`www.msci.magic.net/harp`).
- The LinuxATM software (`lrcwww.epfl.ch/linux-atm`).

For our experiments we have used the Begemot FreeUNI implementation. This package runs on Sun Solaris, FreeBSD and Linux, and is easily portable to other operating systems.

1.5 Standardisation Process

The original ATM standardisation efforts were driven by the ITU-T in an attempt to provide a more scalable bandwidth and more services than ISDN more than 15 years ago. This process was mainly dictated by political discussions and not by technical arguments. One of the main goals of ATM (or B-ISDN, as it was called later) was compatibility or interworking with existing telecommunication systems including Plain Old Telephone System (POTS) and narrowband ISDN. All this has led to a very complicated system of standards with a gap of several years between standardisation and implementation. First implementations of ATM protocols usually appear on the market several years after the standard was finalised. One of the possible readings of ATM is **A**fter **T**he **M**illennium. But the millennium has arrived and the gap has not closed.

Some years ago a couple of major ATM equipment manufacturers came together to found the ATM-Forum—an organisation to fill the missing gaps in the ITU-T recommendations and to streamline the standards into a form in which they could be implemented. At the beginning of this process the ATM-Forum was quite successful—standards came out that were small and (if compared with the ITU-T recommendations) easy to implement. Since than the organisation has become bigger and slower. Often, new ATM-Forum standards are ITU-T documents with small changes.

A number of other organisations also take part in the work on ATM: ETSI—the European Telecommunications Standards Institute (`www.etsi.org`), Eurescom—the European Institute For Research And Strategic Studies in Telecommunications(`www.eurescom.de`), the Internet Engineering Task Force (IETF) and others.

Anyway, we have to live with these standards because some of them (most notably UNI, ILMI, PNNI, LANE) are now widely adopted. Experience in the implementation of complex protocols has grown, and more and more, even of the esoteric features, get implemented. Major telecom operators use ATM now in their backbone networks and offer ATM services to their customers. ATM has proved to be an excellent technology to build large, scalable and robust private networks.

2

Overview of ATM Signalling

2.1 Signalling Interfaces and Protocols

The ATM control plane consists of many protocols that run between the different entities of the network. ATM is asymmetric in the sense that the protocols, and even the cell format, are different on the interfaces between end user systems and network nodes, and between the network nodes. In IP, on the other hand, datagrams are the same, whether they are exchanged between a computer and a router, or between two routers (although routing protocols have diverged now).

The situation is even more complex because in ATM usually two network types are taken into account: public networks and private networks. In the past public networks used to be built and operated by government organisations—the public telcos. Each country had at most one public telecom operator. Nowadays most of these organisations in Europe have become private companies and there is not just one of them in a given country but many. So the difference between public and private networks has become somewhat fuzzy. One might say that a public network is a network that is managed and operated by dedicated companies (soon even this may be not true because these companies are trying to move over into electricity, television and other markets). A private network is owned and operated by a company for which this network is simply a part of the company's infrastructure.

Because of the large number of available protocols and interface variants it is not easy to show all the different ATM interfaces in one figure. Figure 2.1 tries to show at last those that are analysed in this book.

ATM end systems (usually computers or multimedia workstations, often referred to as "the user") can be connected either to a private or to a public network. For the establishment of user data connections the User–Network Interface (UNI) protocol is used between the end system and the first switch in the network (the access switch). This protocol runs on a dedicated, pre-configured channel (although, theoretically it is possible to dynamically establish this channel via Meta-Signalling [Q.2120]). The discussion about this protocol is the central goal of this book.

If an end system is attached to a private network usually a second protocol is used for interface configuration, namely the Integrated Local Management Interface (ILMI). This protocol is based on the Simple Network Management Protocol (SNMP) and provides automatic configuration facilities for end systems and end system switch ports. This protocol is not used in public ATM networks (although it could be).

The biggest difference between public and private networks from the protocol point of view is in the protocols used between the network nodes. Private systems use Private Network Node

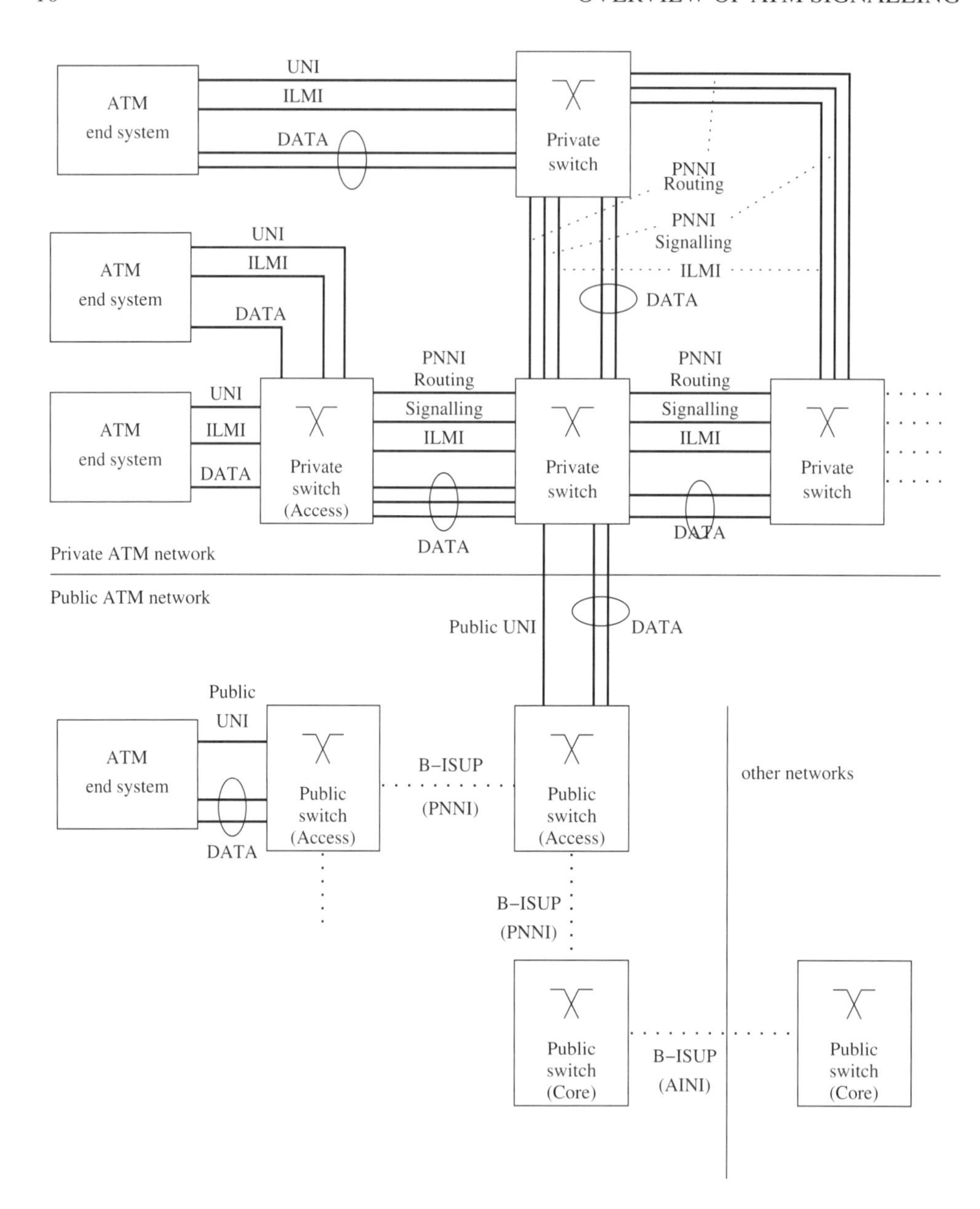

Figure 2.1: Interfaces in an ATM network

Interface (PNNI), whereas most public systems use Broadband ISDN User Part (B-ISUP). It is also possible to use PNNI in public networks—PNNI scales quite well. PNNI consists of two protocols: PNNI signalling, which is based on UNI signalling, and the PNNI routing protocol. The signalling protocol's task is to establish user connections between switches. The routing protocol is use for the dynamic topology management of ATM networks.

Between two ATM networks several protocols may be used. Two public networks usually are connected via B-ISUP. Private networks are attached to public, B-ISUP based networks via UNI. Two private or two public networks can also be connected by B-ISDN Inter Carrier Interface (B-ICI) (see [BICI2.0] and [BICI2.1]), which is based on B-ISUP, or ATM Inter-Network Interface (AINI) [AINI1.0], which is based on PNNI.

In this book we focus on the interfaces and protocols in private networks. The complex public network protocols would fill another book. The interested reader should refer to the standard documents, especially the Q.7XX and Q.22XX, Q.26XX and Q.27XX series of ITU-T recommendations (see Appendix B for availability).

2.2 Example ATM Connection

To establish a connection through a private ATM network a number of protocols have to play together. The most basic protocols are UNI, PNNI and ILMI. The following is a schematic example of a call setup in a private network with an overview explanation of these protocols.

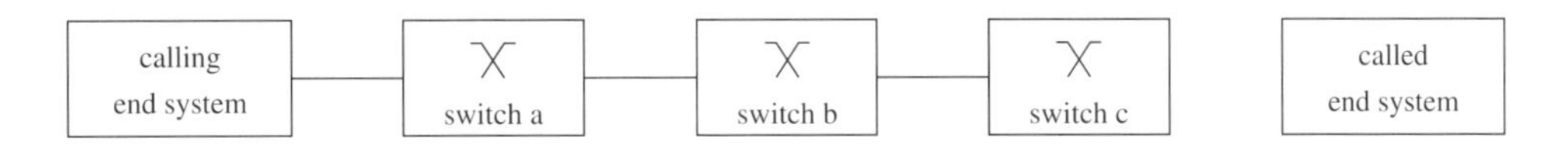

Figure 2.2: ATM call example: configuration

Figure 2.2 shows our example configuration. The ATM network consists of three switches and two end systems. Let us assume for the moment that the second end system is not yet connected to the network. Let us also assume that PNNI routing is up and running. In this case the network is in a stable state—each node knows each other node (provided that they are all in the same PNNI hierarchy level), and each node knows how to reach the single end system.

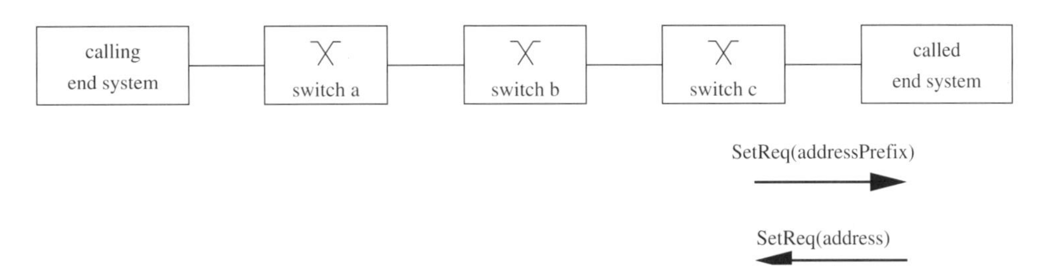

Figure 2.3: ATM call example: ILMI address registration

When the second end system is plugged in, the ILMI protocol instances in both the switch and the end system connect to each other (Figure 2.3). They exchange information about the physical and ATM layers and establish the end system address. This is done by the switch providing the end system with its switch prefix (this is one part of the entire address), and the switch responding with the entire address (the switch prefix plus an end system specific part). Both of these messages are acknowledged (not shown here). At the end of this procedure the end system and the switch know the address of the end system. Switches can set more than one address prefix in the end system and end systems can register more than one address in the switch. Basing addresses on switch provided prefixes ensures that all, or at least many, of the addresses at a given switch have the same prefix and can be advertised to the other switches using only the common prefix. The downside is that moving an end system from one switch to another changes its address. This problem can be overcome by using either fixed addresses or by using symbolic names. If fixed addresses are used (without common and, between different switches, unique prefixes), reachability information for each single address has to be distributed to the other switches. Symbolic names can be used in analogy to IP—the ATM-Forum has defined an ATM Name Service (ANS) which is based on, and almost similar to, the Domain Name System (DNS) for IP networks.

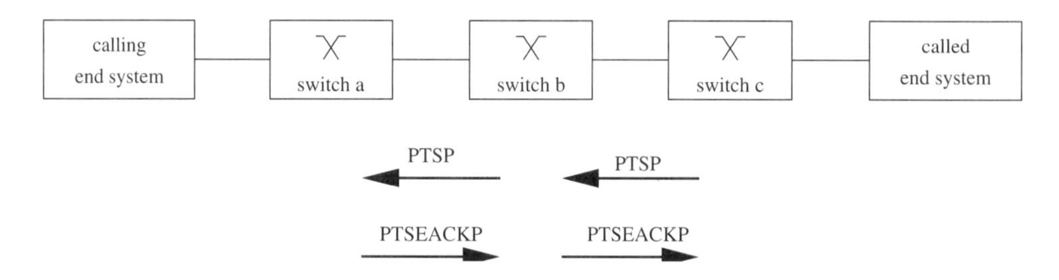

Figure 2.4: ATM call example: PNNI reachability information flooding

Now the switch has to distribute the information that an end system with the new address is reachable to all switches in the network (Figure 2.4). This is done by sending PNNI Topology State Packet (PTSP) messages to all adjacent switches. These in turn will send the information to their neighbour. This process is called flooding. Each of the messages is acknowledged. Besides flooding of reachability information the PNNI routing protocol also provides the facility for semi-automatic network configuration, periodic link status checks and the exchange of topology information. These operations are not shown in this example, because they are to complex and will be analysed in Chapter 6.

Once this process ends, the network is again in a stable state. Now the left end system needs to establish a connection to the right one. This is done by sending a SETUP message over the UNI channel to the switch to which the system is connected (see Figure 2.5). After checking the SETUP message for correctness and after checking for availability of resources the switch may send a CALL PROCEEDING message back to the end system to inform it that the call is going to be forwarded. Based on the information it got during reachability information flooding the switch computes a Designated Transit List (DTL), which is a list of all network nodes through which the called end system can be reached. It includes this list into the SETUP packet and sends this to the next hop switch on the PNNI signalling channel. The next switch answers with a CALL PROCEEDING (after checking the message itself and the availability of resources) and forwards the SETUP to the next hop switch. In this way the SETUP travels through the network until it reaches the last switch just before the called system. This switch

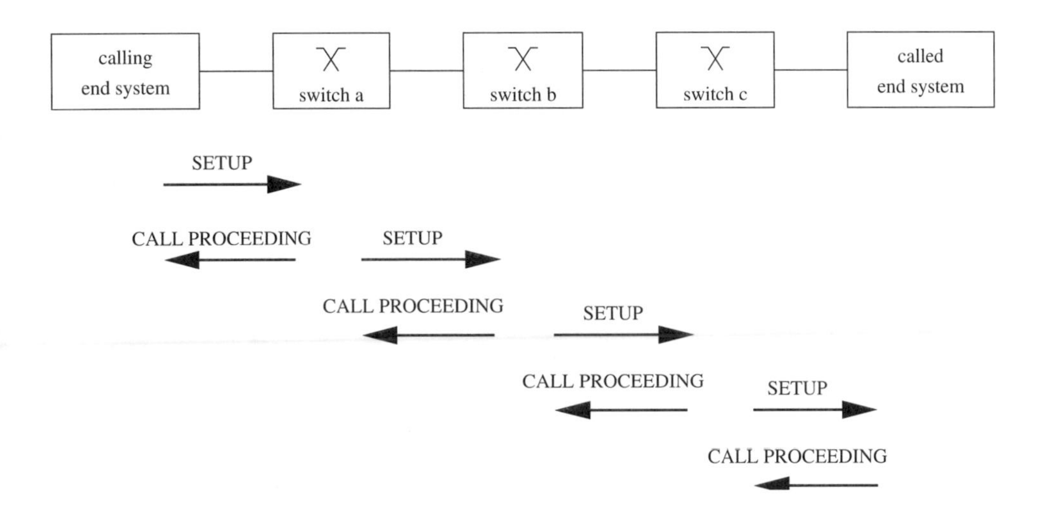

Figure 2.5: ATM call example: sending the SETUP

removes all PNNI specific information elements, builds a UNI SETUP message and sends this message to the called user. This user in turn may answer with a CALL PROCEEDING to indicate that he is going to process the call.

The SETUP message contains all the information that is needed to allocate resources in the network and may provide the called end system with additional information about the call (which user level protocol will be used, and which AAL; application dependent identifiers and parameters). Information for the network includes: the kind of call (point-to-point or point-to-multipoint), the kind of connection (CBR, VBR, UBR, ABR), timing requirements of the call (real time, non-real time, end-to-end transit delay), cell rates (peak, mean), quality of service requirements, and others. The switch includes an application on top of the signalling stacks, namely the Call Admission Control (CAC). This application checks the SETUP parameters for consistency and ensures that sufficient resources are available on the given switch and its links for that call. Some of the SETUP parameters are also needed for Usage Parameter Control (UPC), which controls the traffic sent by the user to ensure that it matches the SETUP parameters. Traffic parameters may also be needed for the Multiple Access Control (MAC) or the physical layer if some links are wireless links.

Now the called end system can "alert" the user. For some types of connections this does not make much sense (for example, connections that are going to carry IP packets). For others, like video telephony calls, it may be useful to inform the calling system that the user is alerted. This is done by sending an ALERTING message backwards through the network (Figure 2.6).

Alerting obviously stems from narrowband ISDN on which ATM signalling is based. Because one of the main applications of N-ISDN is the provision of plain telephone services, the alerting feature is needed to signal the calling user that the phone at the other end is ringing. In the context of ATM, where much more information must be provided (given that the main application area of ATM will be multimedia), alerting will rather be provided at the application level. Nevertheless alerting via signalling is available for ATM.

When the called end system decides that the call has to be completed it responds to the SETUP request with a CONNECT message (Figure 2.7). This message travels backwards

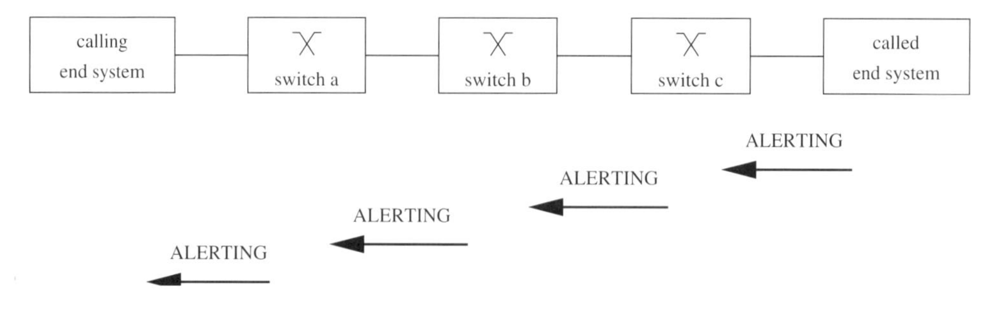

Figure 2.6: ATM call example: alerting

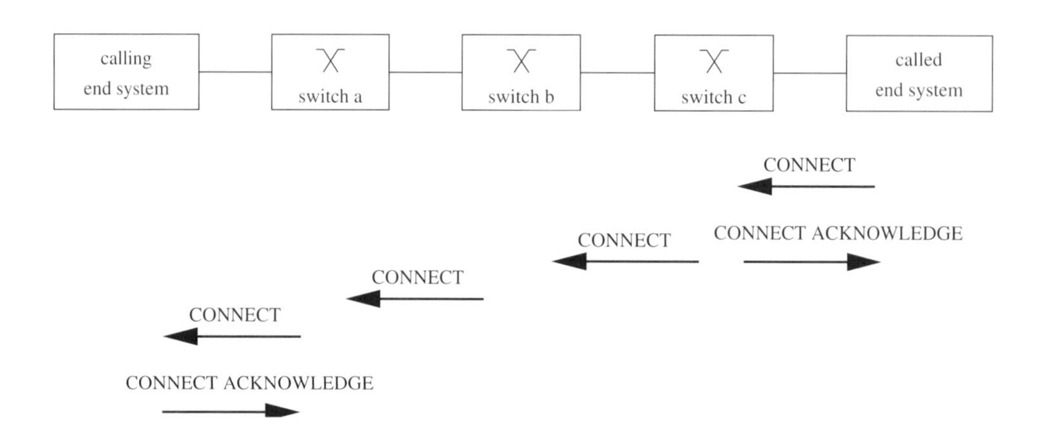

Figure 2.7: ATM call example: connecting

through the network. At the UNI interfaces the CONNECT is answered with a CONNECT ACKNOWLEDGE. At this point the connection is established and data can be sent.

The CONNECT message may be used to carry parameters from the called user back to the calling user. This is especially useful if certain parameters of the connection are negotiated during the setup. Negotiation is possible for traffic parameters, like cell rates, as well as for end-to-end parameters, like AAL maximum message size and lower layer user protocols and their parameters. Negotiation is done by including alternative parameters or parameter ranges in the SETUP message. These parameters may be adjusted by the network and the called user, who puts these adjusted parameters into the CONNECT message. CONNECT carries these parameters back to the calling user who may adapt the traffic to the negotiated parameters.

When one of the end systems decides that the call should be released, it sends a RELEASE message to its switch (Figure 2.8). The switch releases all resources that where allocated to this connection, answers with a RELEASE COMPLETE and forwards the RELEASE to the next hop. In this way the clearing process moves through the network until the last hop is released.

The user who initiates the release may include a cause for that release into the RELEASE message. This code is carried to the other side of the connection, where it may be used

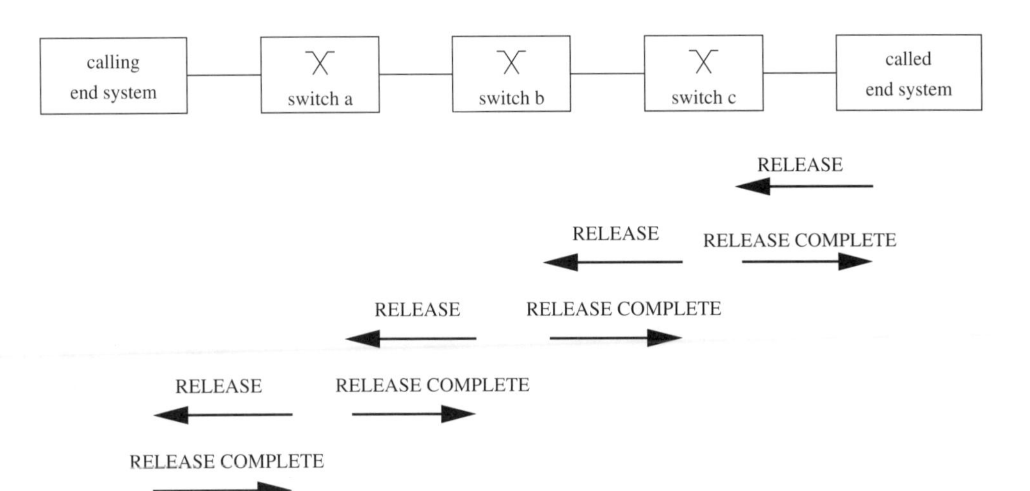

Figure 2.8: ATM call example: clearing the call

by the application process. If no cause is specified by the user, a default value of "normal, unspecified" is provided by the network.

3

UNI: User–Network Interface

3.1 Overview

Figure 3.1 shows the layers of the control plane. The control plane shares the lower two layers (physical and ATM) with the user plane. The ATM Adaptation Layer (AAL) consists of several sublayers. The lower sublayers again share with the user plane (usually this is AAL5), whereas the higher layers are specific to the control plane (note that it is possible to use the Service Specific Connection Oriented Protocol (SSCOP) as user transport protocol)—it provides a reliable message transfer. The UNI layer contains the state machines necessary to manage the user connections. This layer is slightly different in switches and in end systems. On top of the UNI either the user applications (in end systems) or switch functions like routing, call admission control and management are located.

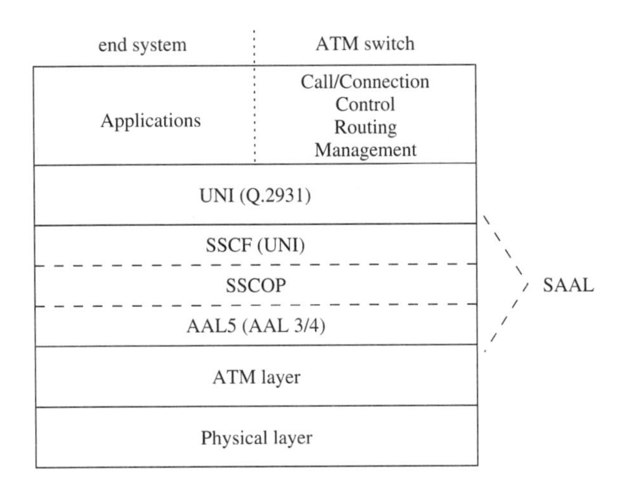

Figure 3.1: ATM control plane

Figure 3.2 shows the control plane for a user connection, which involves several ATM switches and crossconnects and two end systems. One point that often causes confusion is that the UNI is a protocol that runs between the end system and the access switch (the first switch of the ATM network): it is not an end-to-end protocol. The signalling stack of the source end system does not talk to the UNI stack of the destination system, but to the stack in the switch. The figure also shows that this is not always that easy: there may be crossconnects between the end system and the switch or between switches. These crossconnects can even be normal ATM switches that have configured a Virtual Path Connection (VPC) for the given signalling association.

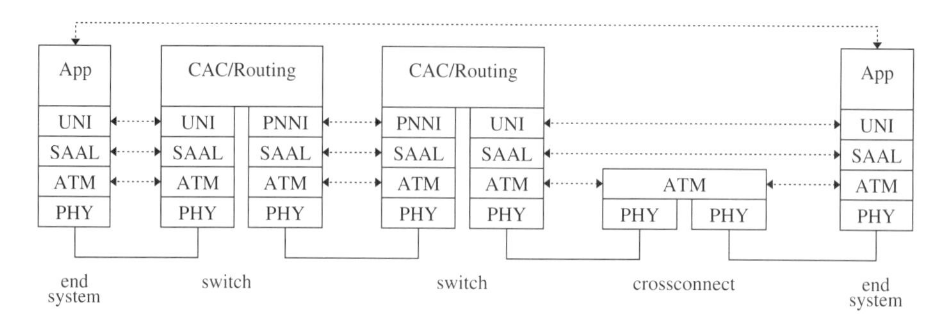

Figure 3.2: Example of the control plane for a user connection

In this chapter we describe the UNI part of the control plane, i.e. the state machine and messages used between an ATM end system and an ATM switch.

ATM protocols are standardised by two major standardisation bodies: the ITU-T and the ATM-Forum. Because of this, there are actually several standards, which differ slightly. The ITU-T standards for UNI signalling are [Q.2931] and [Q.2971] with numerous additions in the Q.29XX series of standards. The ATM-Forum standards are [UNI3.1] and [UNI4.0]. Fortunately, with UNI4.0 these standards have been converged—UNI4.0 is a quite short document which summarises many of the Q.29XX standards and defines some minor changes and additions. Therefore, this chapter will be based on UNI4.0.

ATM connections come in two different flavours: point-to-point connections (Figure 3.3) and point-to-multipoint connections (Figure 3.4). Although there was some discussion about multipoint-to-multipoint connections, the standardisation bodies have not yet come up with standards in this area and the semantics of such connections are not very clear.

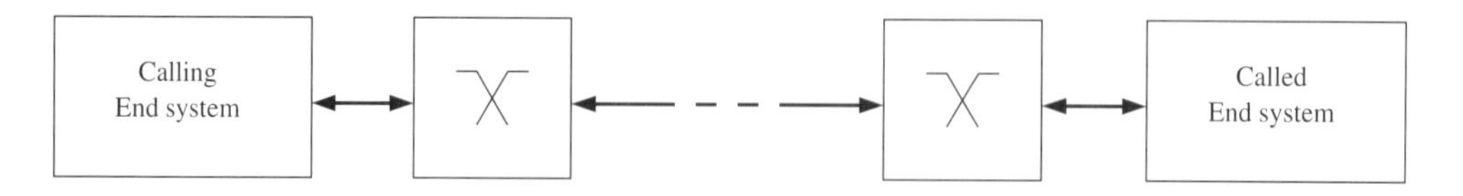

Figure 3.3: Point-to-point connection

ATM connections are always bi-directional in the sense that the VPI/VCI values along the connection are reserved in both directions. However, they can be, and in the case of point-to-multipoint connections must be, used uni-directionally—the traffic in one of the directions may be zero.

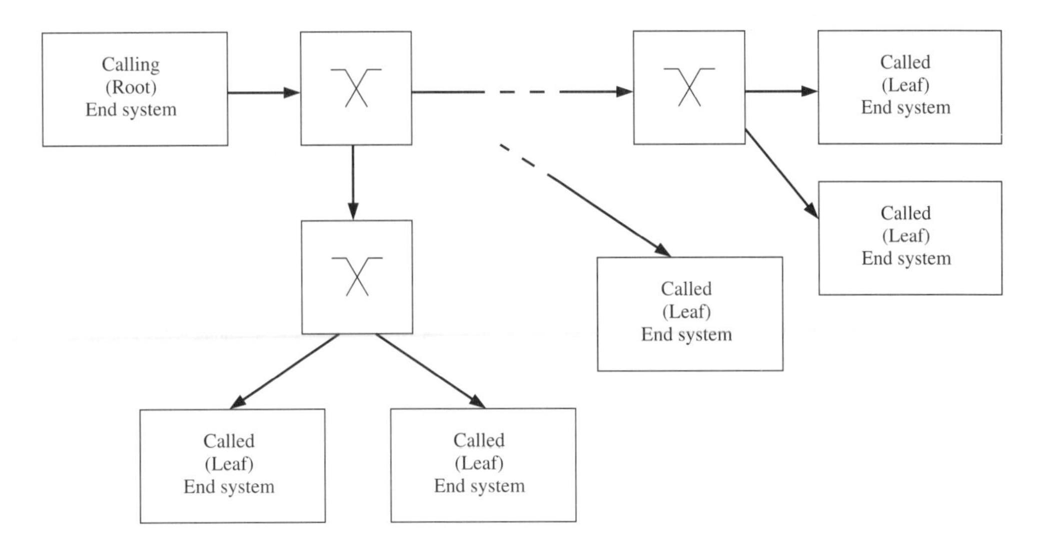

Figure 3.4: Point-to-multipoint connection

For most of the procedures of the UNI we will show traces that were taken with the method described in Section 1.2. The traces always show the communication over *one* UNI with timing information. Timing is relative. This means that the part of the entire UNI communication that is of interest was cut out from the entire communication protocol and the begin was set to zero to allow better tracking of the messages by the reader. An arrow between the two characters S (for Switch) and E (for End system) shows the direction of the message. The timestamp and the direction of each message is shown only in the first line that belongs to that message. Lines without the arrow and the timestamp are continuation lines.

3.2 Configuration

Put simply UNI is a protocol running between an ATM end system and the first ATM switch (access switch) in the network. This protocol is used to manage connections between ATM end systems, i.e. establish, modify and release them. In practice things are not so easy. A common configuration of ATM end systems and switches is shown in Figure 3.5.

In ATM usually two kinds of networks are distinguished: public ATM networks and private ATM networks. The difference between them is somewhat fuzzy. Public networks are usually associated with the big telecom operators; private networks are intranets that are more or less under control of the user. The private ATM network is interfaced to the public ATM network via a UNI, which in this case is called a public UNI. Of course, end systems can be interfaced to the public network directly (see ATM end system A) or via VPI crossconnects (systems B1 and B2). A VPI crossconnect is an ATM switch that switches only VPIs, not VCIs, without signalling support. Connections through a VPI crossconnect are established by administrative means.

On the private network side end systems can be either directly attached to the network (system C), with the help of a proxy signalling agent as defined in [UNI4.0] (system D), or via a VP multiplexer like systems E1 and E2. Proxy signalling is used for ATM systems

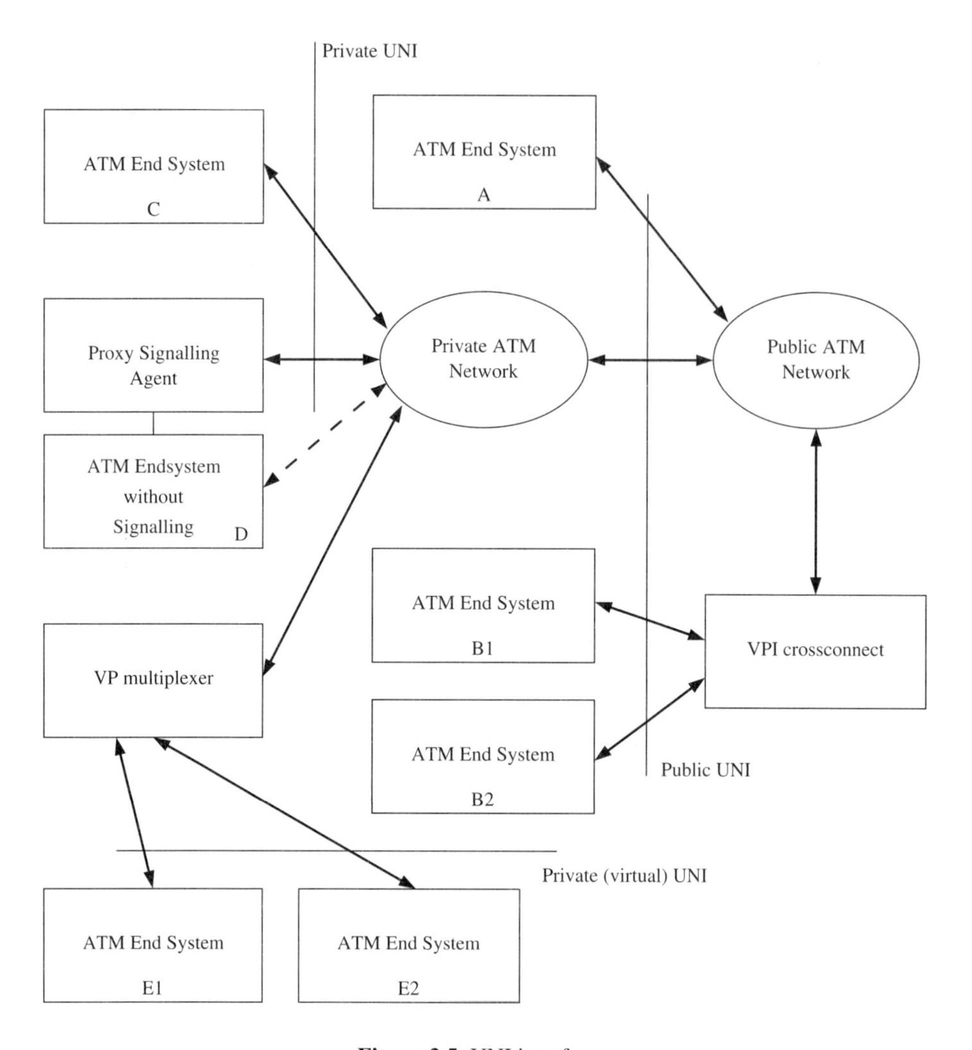

Figure 3.5: UNI interfaces

that do not support signalling by themselves (for example, small video devices) or for high performance end systems. Virtual UNIs via a VP multiplexer are used mainly for wireless access to ATM.

The UNI by which end systems are connected to the private ATM network is called a private UNI and differs slightly from the public UNI (mostly in the supported addressing schemes).

3.2.1 Signalling Channels and Modes

The UNI signalling protocol uses one VC for communication between the end system and the ATM switch. The default value for the VCI is 5, although there exists a special metasignalling protocol that can be used to dynamically create signalling channels with other VCI values [Q.2120] and most switches support the assignment of the signalling channels to other VCIs.

Generally two modes of signalling exist: non-associated signalling and associated signalling. They differ in the use of the VPs:

- *Associated signalling.* In this case a signalling VC controls only the VP in which it is allocated. This means that on a given physical interface each VP can have its own signalling VC—in the end system and in the switch there will be one UNI protocol instance for each of these VCIs. Switched channels are allocated in this mode only with the same VPI value as the signalling VC.
- *Non-associated signalling.* In this case a signalling VC controls the VP in which it is allocated and may control other VPs as well.

Whereas the ITU-T standards support both modes, ATM-Forum UNI4.0 supports only non-associated signalling. Note that each VP is controlled by a maximum of one signalling VC and that each VP can carry no more than one signalling VC (with the exception of proxy signalling; see Section 3.2.2).

The usual configuration for end system UNIs is one signalling channel with $VCI = 5$ and $VPI = 0$ which controls the entire VPI/VCI space (note that this space may be limited by management (Chapter 7) or administrative means).

Besides the VPI and VCI values signalling uses a Virtual Path Connection Identifier (VPCI) to identify an ATM connection. This identifier is carried in UNI messages between the switch and the end system instead of the VPI. The connection identifier information element contains two subfields: a VCI value and a VPCI value. Usually this information element is sent from the switch to the user when a new connection is established (although it could also be sent by the user). Whereas the VCI value in this information element corresponds directly to the VCI field in the ATM header of the data cells, the VPCI value needs to be mapped to a VPI value. On the user side there is usually a $1:1$ relationship between the VPCI and the VPI: $VPI = VPCI$. On the switch side, however, this may be different if VP crossconnects are used between the switch and the user, so switches may need to have a mapping table between VPCI values and VPI values and port numbers. VP crossconnects can be used for different tasks, namely to connect one end system to more than one switch (or network), to connect more that one end system to one physical switch port, or to connect one end system with more than one interface to one switch port.

Figure 3.6 shows an example where one ATM end system (or the border switch of a private ATM network) is connected via a VP crossconnect to two switches (or public networks). This scenario could be used to enhance reliability or to select the cheaper network on a per connection basis depending on connection parameters. In this case the end system is connected by one physical link to the crossconnect. This link carries a number of VPIs and on the end system side VPI and VPCI values are numerically equal. The crossconnect maps these VPs to two different ports and to other VPI values, so the two switches need mapping tables that associate VPCI values with VPI values.

In Figure 3.7 an end system with two ATM interfaces is connected via a crossconnect to one port of an ATM switch. This scenario could be used for high availability or high performance applications (in the case when the link between the crossconnect and the switch has a higher bandwidth than the links between the end system and the crossconnect). In this example the end system keeps a mapping table that relates VPCI values to VPI/interface pairs. The switch on the other side needs a table that maps VPCI values to VPI values.

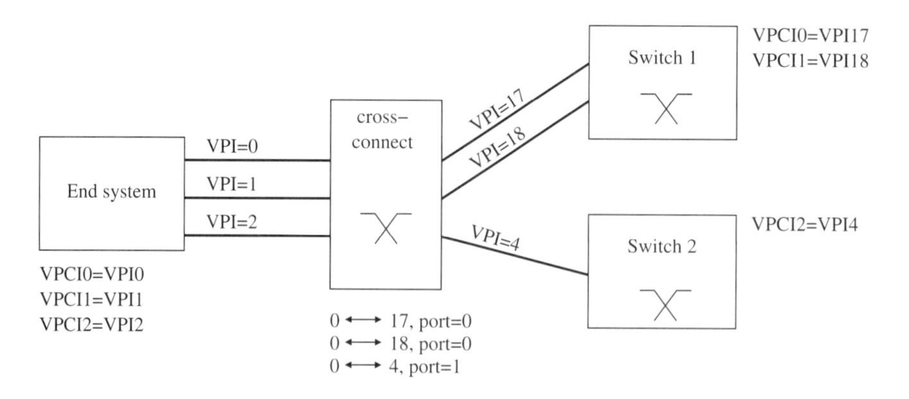

Figure 3.6: VPCI/VPI relationship for connection to multiple switches

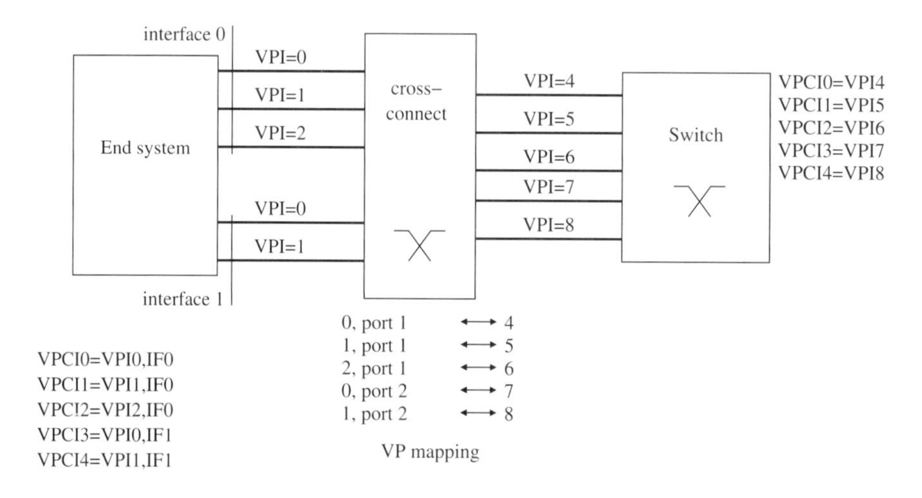

Figure 3.7: VPCI/VPI relationship for the connection of interfaces

A third example in Figure 3.8 shows how two end systems can be connected to one switch port via a VP crossconnect. In this case again there is no need for a mapping at the end system; VPCI and VPI values are numerically equal, but the switch needs a mapping between VPCI/UNI pairs and VPI values.

3.2.2 Proxy Signalling

The proxy signalling capability is an optional feature that was introduced by the ATM-Forum in UNI4.0. Proxy signalling allows one user, called the Proxy Signalling Agent (PSA), to perform signalling on behalf of other end systems that lack signalling capabilities. As shown in Figure 3.9 the PSA has one or more signalling VCIs, each of which controls a set of one or more VPs, which can be across different UNIs. Applications of this feature are the support of small ATM devices that lack the computing power for full signalling support and the support

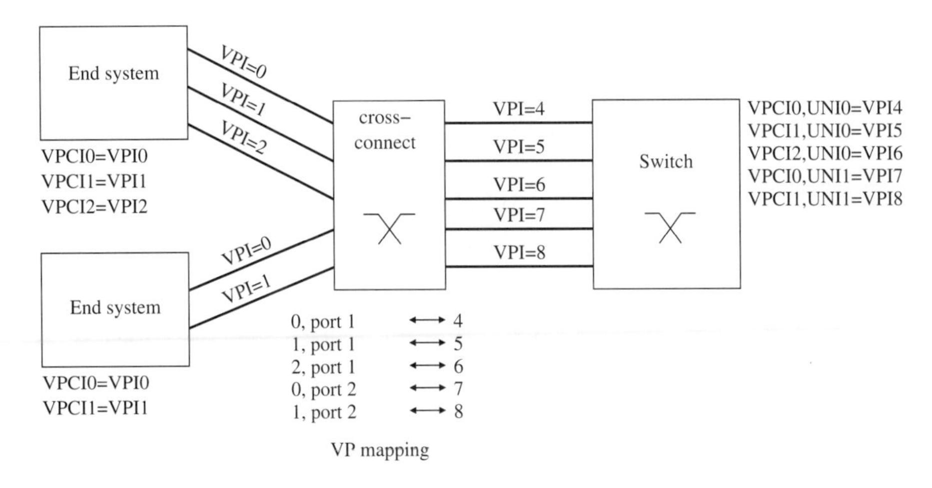

Figure 3.8: VPCI/VPI relationship for the connection of multiple end systems

of high performance equipment to use multiple physical interfaces that share the same ATM address.

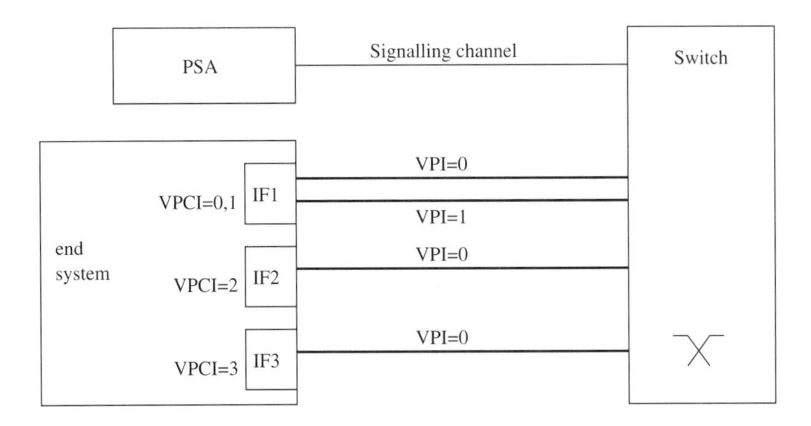

Figure 3.9: Proxy signalling

In Figure 3.9 a high performance end system is connected to an ATM switch via three physical links. Each of these links has one or more VP allocated. A PSA is used to control all three links. This PSA may reside on the end system (in this case the signalling channel should be carried in a VP on one of the three links), or may be implemented in another end system, which can even be remote connected to the ATM switch. It would also be possible for the PSA to control more than one of these high performance systems. In this case there would be one signalling channel for each of the systems between the switch and the PSA (and, consequently, more than one signalling channel in the same VP).

Both the PSA and the switch need to know the mapping between VPCI values, interface numbers and VPI values, as shown in Table 3.1. This table has to be configured by management means.

Table 3.1: VPCI mapping

VPCI	Interface	VPI
0	1	0
1	1	1
2	2	0
3	3	0

3.2.3 Virtual UNIs

Virtual UNIs are an optional feature introduced by the ATM-Forum in UNI4.0. Virtual UNIs allow several end systems to share one physical port on a switch by using a VP crossconnect. This feature is intended to implement wireless ATM access as shown in Figure 3.10 (see [WATM]).

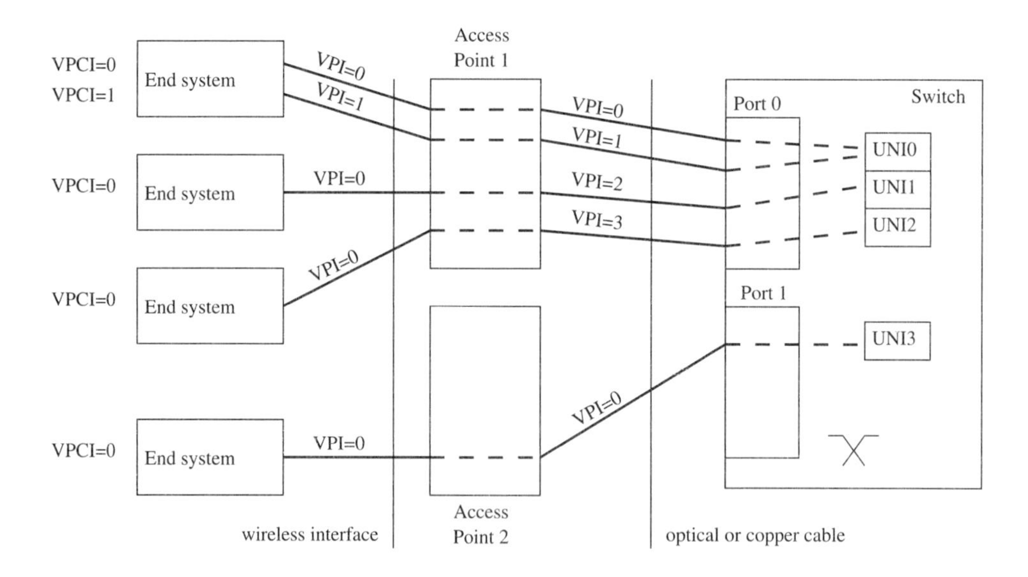

Figure 3.10: Wireless ATM access with virtual UNIs

In this figure two access points are connected to an ATM switch. For wireless networks these access points contain the radio equipment. From the ATM point of view their main task is VP switching—they act as a VP crossconnect. Each of the end systems sees at least the standard VP with VPI = 0, which also contains the signalling channel VCI = 5. Additionally, end systems may also have other VPs which may be allocated via UNI. On the switch side the different end users are distinguished by their VPI value—the switch maintains a mapping table between users, VPI values and VPCI values. The accumulated bandwidth of all users, which are switched by the access point onto one switch port, may not exceed the capacity of that port.

3.3 UNI Messages

The UNI protocol uses variable size messages that are transported between the protocol entities by a lower layer transport protocol. Each message consists of a fixed size header and a variable number of variable sized information elements (see Figure 3.11).

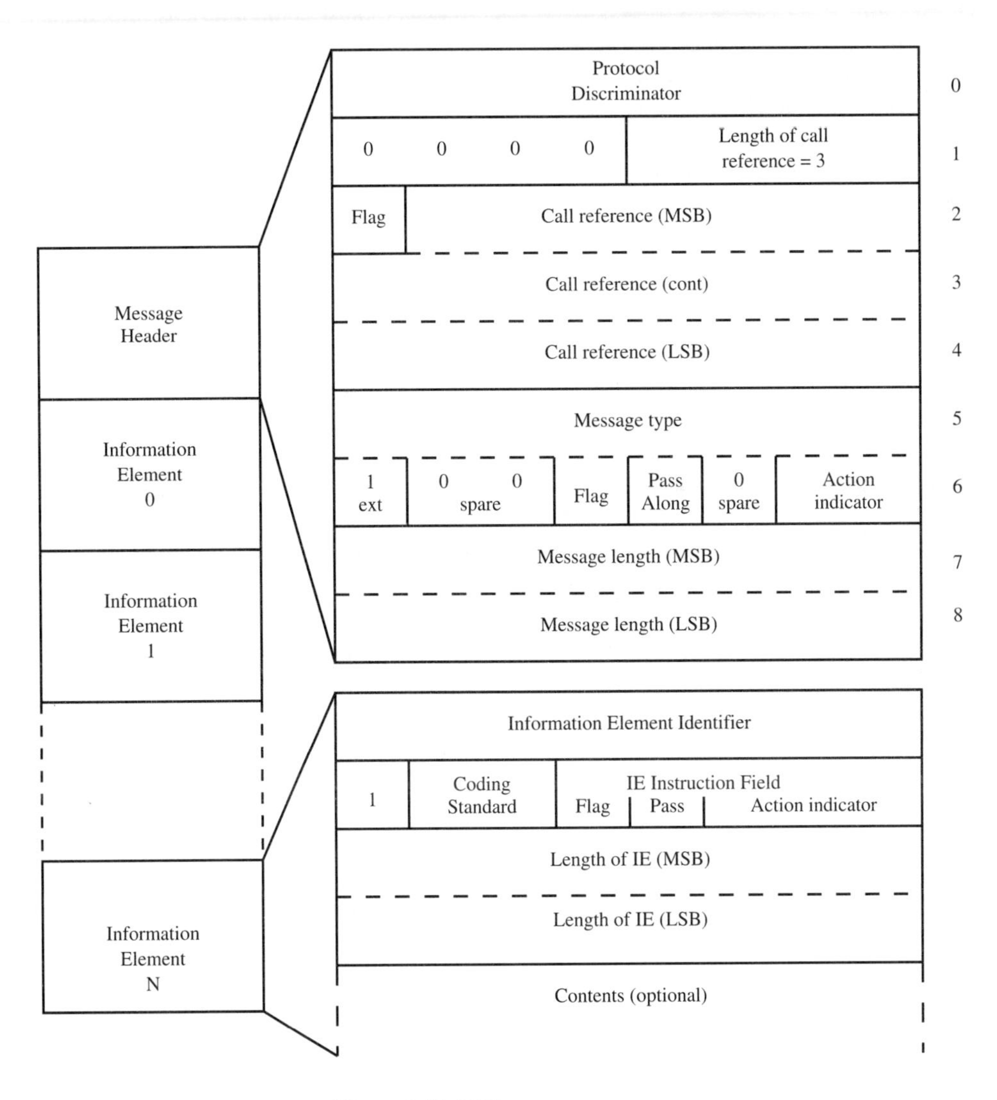

Figure 3.11: UNI message structure

The message header contains the overall length of the message in bytes; the individual information elements contain their own length. This enables some kind of consistency check. A message can contain any number of information elements, even zero; the number of these elements is limited by the size of the message length field (16 bit) which would enable a message size of 64kbyte + 9byte, and by the maximum PDU size of the underlying SAAL which is 4096 bytes.

3.3.1 Message Header

The size of the message header is 9 byte. Each message must contain at least this header; shorter messages trigger exception handling of the protocol.

The first byte of the message header is the protocol discriminator. It takes the fixed value of 9 and can be used to route UNI messages to the UNI protocol when there are other protocols running on the same SAAL connection besides UNI.

The next four bytes contain the call reference. The call reference is an identifier that identifies an ATM call on the given UNI interface. This call reference is always unique on that interface. For a given call the call reference along the connection through the network is different at each hop—the call reference has only a local meaning at a given UNI interface.

The first byte of the call reference contains four reserved bits and the length of the call reference (in bytes) as a four-bit number. This length is always three. The remaining three bytes contain the actual call reference value as a 23-bit big-endian binary number and, in the highest bit, the call reference flag. This flag is used to avoid collisions in the allocation of call reference values. The problem is that each of the two UNI protocol stacks on the two sides of a link allocate the call reference values for their outgoing calls independently. This makes it possible that they allocate the same value for two different calls. To prevent confusion in this situation, each message contains the call reference flag that is set to 0, if the message is sent by the UNI stack that also allocated the call reference, and 1 if the message is sent to that side. So in fact all 24 bits are needed to identify the call to which the given message belongs.

Two call reference values are used for special purposes: the global call reference and the dummy call reference. The first of these has a value of 0 and in the second one all bits are set to 1. The global call reference is used for the reset procedure and the dummy call reference for connectionless services (see [Q.2932.1]).

The next two bytes of the message header describe the message type. The first byte is the actual eight-bit message type, and the second byte contains flags to handle exceptions. Table 3.2 gives a list of messages defined for basic signalling.

The message action indicator in the second byte of the message type field gives a UNI stack the opportunity to change the exception handling of the peer entity. Normally the flag bit in this byte is set to zero, which means: follow the default exception handling procedures. Setting this flag to 1 enables the use of the action indicator which can take the following values:

0 Clear (release) the call in the case of an exception.

1 Discard this message and ignore it. This can be used if it is not known, whether the peer UNI stack supports a given optional UNI feature or not.

2 Discard this message and report status. In this case a STATUS message is sent back to the sender of the original message, containing a indication of the problem.

This also can be used to exploit optional features and to detect whether they are supported or not.

3 This value is reserved and handled like value 0.

Table 3.2: UNI messages

Code	Message name
Basic point-to-point messages (Q.2931)	
0x00	escape code (not supported by UNI4.0)
0xFF	extension code
Call establishment:	
0x01	ALERTING
0x02	CALL PROCEEDING
0x03	PROGRESS (narrowband interworking)
0x05	SETUP
0x07	CONNECT
0x0D	SETUP ACKNOWLEDGE (narrowband interworking)
0x0F	CONNECT ACKNOWLEDGE
Call clearing:	
0x46	RESTART
0x4D	RELEASE
0x4E	RESTART ACKNOWLEDGE
0x5A	RELEASE COMPLETE
Miscellaneous:	
0x7B	INFORMATION (narrowband interworking)
0x6E	NOTIFY
0x75	STATUS ENQUIRY
0x7D	STATUS
Basic point-to-multipoint messages (Q.2971)	
0x80	ADD PARTY
0x81	ADD PARTY ACKNOWLEDGE
0x82	ADD PARTY REJECT
0x83	DROP PARTY
0x84	DROP PARTY ACKNOWLEDGE
0x85	PARTY ALERTING
Messages for leaf-initiated joins (UNI4.0)	
0x90	LEAF SETUP FAILURE
0x91	LEAF SETUP REQUEST

Table 3.2: UNI messages (continued)

Code	Message name
Messages for generic functional protocol (Q.2932.2)	
`0x15`	CO-BI SETUP
`0x62`	FACILITY
Messages for bandwidth modification (Q.2963)	
`0x88`	MODIFY REQUEST
`0x89`	MODIFY ACKNOWLEDGE
`0x8A`	MODIFY REJECT
`0x8B`	CONNECTION AVAILABLE

The Pass Along bit is defined only for PNNI (see Chapter 6). It indicates that the message, if not recognised by the receiving instance, should be forwarded, provided that the next interface is a PNNI. No error checking occurs. The same holds for the Pass Along bit in the information element header—if the IE is not recognised it should be forwarded without checking. For UNI both bits are reserved and should be set to zero.

The last element in the message header is the message length field. This field contains the total length of the message minus the header as a 16-bit big-endian number.

3.3.2 Information Elements

Each Information Element (IE) starts with a header of four bytes (see Figure 3.11). The first byte of this header contains the eight-bit information element identifier. Table 3.3 lists the most common information element identifiers. This identifier is followed by the flag byte and the 16-bit information element length field.

Other information elements are defined in the Q.29XX series of standards and in new and upcoming ATM-Forum documents.

The flag byte contains two fields: an instruction field like that in the message header, and the coding standard identifier. The coding standard identifier describes by which standard the actual information element contents are covered (see Table 3.4). Although there are four values defined only two of them are actually used. The use of these values is somewhat confusing: all the ITU-T defined information elements contain the indication 0, meaning "ITU-T standardised coding". The ATM-Forum, however, changed some of these information elements and added new elements. For some of these a coding of 3 (network specific or ATM-Forum specific) is used and for some of them not. For some of the ATM-Forum no coding is specified, but it is suggested that 3 should be used.

The instruction field contains a bit which, if 0, selects default error handling and, if 1, the error handling defined by the three-bit action indicator. The action indicator tells the receiver of a message what to do in the case of errors in this information element (see Table 3.5). It can be set to values that report STATUS to detect whether the peer UNI handles certain optional features and extensions.

For the PNNI (see Chapter 6) bit 4 is defined as the Pass Along Request bit. If a PNNI protocol instance receives a message which contains an information element that it cannot

Table 3.3: UNI information elements

Code	Information Element Name	Code	Information Element Name
0x04	narrowband bearer capability	0x62	broadband sending complete
0x08	cause	0x63	broadband repeat indicator
0x14	call state	0x6C	calling party number
0x1C	Q.2932 facility	0x6D	calling party subaddress
0x1E	progress indicator	0x70	called party number
0x27	notification indicator	0x71	called party subaddress
0x42	end-to-end transit delay	0x78	transit network selection
0x4C	connected number	0x79	restart indicator
0x4D	connected subaddress	0x7E	user-to-user info
0x54	endpoint reference	0x7F	generic identifier transport
0x55	endpoint state	0x81	minimum ATM traffic descriptor
0x58	AAL parameters	0x82	alternative traffic descriptor
0x59	ATM traffic descriptor	0x84	ABR setup parameters[a]
0x5A	connection identifier	0x89	broadband report type
0x5C	quality of service parameter	0xE4	ABR additional parameters[a]
0x5D	broadband higher layer info	0xE8	LIF call identifier[a]
0x5E	broadband bearer capability	0xE9	LIF parameters[a]
0x5F	broadband lower layer info	0xEA	LIF sequence number[a]
0x60	broadband locking shift	0xEB	connection scope selection[a]
0x61	broadband non-locking shift	0xEC	extended QoS parameters[a]

[a]Only in UNI4.0.

understand, it looks at the Pass Along Request bit. If this bit is set and the next hop of the connection is also a PNNI hop, the information element is transferred without any error checking to the message to be sent at the next hop. In this way PNNI is able to support information elements which it does not yet know. For UNI this bit should be set to zero (it is a reserved bit).

Table 3.4: Information elements codings

Code	Coding Standard
0	ITU-T standardised coding
1	ISO/IEC standard
2	national standard
3	network specific; ATM-Forum standard

Table 3.5: IE action indicator

Code	Action
0	clear call
1	discard and ignore IE
2	discard and ignore IE, send STATUS
5	discard and ignore message
6	discard and ignore message, send STATUS

3.3.3 Information Element Coding

UNI messages contain many parameters which are coded in different ways. A number of rules can be extracted from the standards:

- Integer values of different sizes are supported. The size of these integers is not always a power of two—there are also odd sized integers (three-bit and five-bit, for example). Integer are mostly unsigned values.
- If an integer value is longer than one byte and has a fixed size, it is coded as a big-endian binary values. 24-bit values, for example, are coded in three bytes.
- Sometimes values are coded with an extension mechanism: The most significant bit of a byte is set to 1 if this is the last byte of the value and to 0 if other bytes follow.
- There are spare bits and reserved bits. Reserved bits must be zero, spare bits should be zero, but are ignored.
- Some information elements contain subfields, which are identified by a 1-byte subfield identifier. These subfields can come in any order, repetitions are ignored.
- The BLLI information element has two-bit subfield identifiers. Q.2931 defines no order for the subfields; UNI4.0 specifies an order.

Note, that all these different coding variants can be mixed in one information element.

Some information elements can be repeated in a message. There are two different mechanisms to do this: explicit indication of repetition and implicit indication of repetition. The first kind of repetition uses a special information element, i.e. the Broadband repeat indicator. This indicator can specify how the repeated information elements are to be treated. At the moment only one interpretation is defined: as a prioritised list with descending priority where one information element has to be selected. The main use of this feature is protocol negotiation by means of the broadband lower layer information IE. Implicit repetition is done by simply including several information elements of the same kind into a message. In this case usually all these IEs are used for processing at the receiver. In both cases all information elements of the same kind must follow each other and must, for explicit repetition, directly follow the broadband repeat indicator.

3.3.4 Coding Examples

Figure 3.12 shows one of the more complex information elements, namely the Broadband Lower Layer information element (BLLI). This IE is used to carry information about lower protocol layers to the called user. Up to three of these IEs can be included in a call so that the receiver can select the one he supports. The BLLI element is an example of an information element with subfield identifiers but a fixed order of the subfields. Its structure is further complicated by the fact that ITU-T and the ATM Forum continue to produce standards whith changes to this IE.

Byte	Coding	Meaning
0x5F	01011111	information element identifier
0x80	1-------	last byte in this group
	-00-----	ITU-T standardised coding
	---00000	default error handling
0x00	00000000	
0x05	00000101	information element length
0x50	0-------	information continued
	-10-----	layer-2 subfield
	---10000	user-specified protocol
0x83	1-------	end of information subfield
	-0000011	user protocol identifier
0x66	0-------	information continued
	-11-----	layer-3 subfield
	---00110	X.25 packet layer
0x20	0-------	information continued
	-01-----	normal packet sequence numbering
	---00000	spare
0x8C	1-------	end of information subfield
	-000----	spare
	----1100	default packet size 4096

Figure 3.12: Example of a BLLI element

Figure 3.13 shows the ATM traffic descriptor IE. This is used to communicate cell rates, burst sizes and other characteristics of the expected traffic on a given connection to the network. Resource reservation and allocation, as well as usage parameter control, is done based on this information element.

The traffic descriptor is an example of an IE with subfields. Each subfield of the information element starts with a subfield identifier. This identifier defines the length and format of the subfield and its meaning. Most of the subfields are optional but there is a table in the appendix of UNI4.0 that lists all the legal combinations of subfields. Most subfields in the traffic descriptor contain a three-byte integer. There is also a one-byte subfield used to

Byte	Coding	Meaning
0x5F	01011001	information element identifier
0x80	1-------	last byte in this group
	-00-----	ITU-T standardised coding
	---00000	default error handling
0x00	00000000	
0x08	00001000	information element length
0x82	10000010	subfield identifier: forward PCR
0x00	00000000	
0x00	00000000	
0x80	10000000	128 cells/second
0x83	10000010	subfield identifier: backward PCR
0x00	00000000	
0x00	00000000	
0x80	00000000	0 cells/second

Figure 3.13: Example of a traffic descriptor IE

specify tagging and frame discard options and the best-effort indicator that consists only of its identifier.

Figure 3.14 shows an (almost) minimal SETUP message with two BLLI information elements. In this example a point-to-point connection to the national telephone number 112 (this is the fire department in Germany) is requested. The connection should be a unidirectional (the backward peak cell rate is specified as zero), non-realtime VBR with a forward peak cell rate of 128. The requested QoS class is 0 (unspecified QoS). The called user can select among two user-specified layer 2 protocols: protocol 1 and protocol 2. The two BLLI elements are preceeded by a broadband repeat indicator.

Byte	Coding	Meaning
0x09	00001001	protocol discriminator
0x03	0000----	reserved
	----0011	length of call reference
0x00	0-------	messages sent by call reference originator
	-0000000	
0x00	00000000	
0x01	00000001	call reference 1
0x05	00000101	SETUP
0x80	1-------	end of this information subfield
	-00-00--	spare
	---0--00	default error handling
0x00	00000000	
0x32	00110010	message length 50
0x5F	01011110	*broadband bearer capabilites*
0x80	1-------	last byte in this group
	-00-----	ITU-T standardised coding
	---00000	default error handling
0x00	00000000	
0x03	00000011	information element length
0x81	0-------	information continued
	-00-----	spare
	---10000	BCOB-X
0x8A	1-------	last byte in this group
	-0001010	non-realtime VBR
0x80	1-------	last byte in this group
	-00-----	no clipping
	---000--	spare
	------00	point-to-point connection
0x5F	01100011	*broadband repeat indicator*
0x80	1-------	last byte in this group
	-00-----	ITU-T standardised coding
	---00000	default error handling
0x00	00000000	
0x01	00000001	information element length
0x82	1-------	last byte in this group
	-000----	spare
	----0010	prioritised list

Figure 3.14: Example of a SETUP message

Byte	Coding	Meaning
0x5F	01011111	*broadband lower layer information*
0x80	1-------	last byte in this group
	-00-----	ITU-T standardised coding
	---00000	default error handling
0x00	00000000	
0x02	00000010	information element length
0x50	0-------	information continued
	-10-----	layer-2 protocol
	---10000	user-specific protocol
0x81	1-------	last byte in this group
	-0000001	protocol 1
0x5F	01011111	*broadband lower layer information*
0x80	1-------	last byte in this group
	-00-----	ITU-T standardised coding
	---00000	default error handling
0x00	00000000	
0x02	00000010	information element length
0x50	0-------	information continued
	-10-----	layer-2 protocol
	---10000	user-specific protocol
0x82	1-------	last byte in this group
	-0000010	protocol 2
0x70	01110000	*called party number*
0x80	1-------	last byte in this group
	-00-----	ITU-T standardised coding
	---00000	default error handling
0x00	00000000	
0x04	00000100	information element length
0xA1	1-------	last byte in this group
	-010----	national number
	----0001	ISDN (E.164) number
0x31	00110001	
0x31	00110001	
0x32	00110010	number = "112"

Figure 3.14: Example of a SETUP message (continued)

Byte	Coding	Meaning
0x5F	01011001	*ATM traffic descriptor*
0x80	1-------	last byte in this group
	-00-----	ITU-T standardised coding
	---00000	default error handling
0x00	00000000	
0x08	00001000	information element length
0x82	10000010	subfield identifier: forward PCR
0x00	00000000	
0x00	00000000	
0x80	10000000	128 cells/second
0x83	10000010	subfield identifier: backward PCR
0x00	00000000	
0x00	00000000	
0x80	00000000	0 cells/second
0x5C	01011100	*QoS parameters*
0x80	1-------	last byte in this group
	-00-----	ITU-T standardised coding
	---00000	default error handling
0x00	00000000	
0x04	00000010	information element length
0x00	00000000	forward QoS class 0
0x00	00000000	backward QoS class 0

Figure 3.14: Example of a SETUP message (continued)

3.3.5 Interdependence of Information Elements

A problem that often occurs is that a connection cannot be established with the network returning error codes like "unsupported combination of traffic parameters". In general information elements cannot be treated as independent entities. This is especially true for the information elements that specify the parameters of the connection to be established. These information elements are:

- Broadband bearer capabilities
- ATM traffic descriptor
- Quality of Service parameters
- End-to-end transit delay
- Alternative ATM traffic descriptor
- Minimum traffic descriptor
- Extended QoS parameters

For a standard connection setup the first three of these are specified; the alternative and minimum traffic descriptors are used if traffic parameter negotiation is required, and the last one if there is a need for fine-grained QoS specification (for example cell loss ratios). In the case of ABR connections there are also:

- ABR setup parameters
- ABR additional parameters

As it turns out, the parameters specified in all these information elements depend on each other to some extent—it is not possible to specify arbitrary combinations. [UNI4.0] lists the allowable combinations of traffic parameters in Annex 9. However, specifying legal parameter combinations does not necessarily mean that the connection request will succeed—some network nodes still may not support these concrete traffic parameters. If a SETUP is rejected with error codes like "unsupported combination of traffic parameter" or "bearer capability not implemented" one should first check whether the traffic parameters are specified correctly.

3.4 Connection States

ATM connections generally have three phases: the establishment phase, the active phase and the release phase. During the establishment phase signalling messages are exchanged between the end systems and the network to negotiate connection characteristics and parameters. In this phase data cannot yet be sent, although at the called side there is no means to detect whether the connection has been fully established. In the active phase data actually can be sent and, this is an optional feature of the network, connection parameters can be modified. For point-to-multipoint connections new leaf nodes can be added to the connection tree or removed from it. In the release phase signalling messages are exchanged again to tear-down the connection.

These different connection phases are handled on the protocol side by different states of the protocol instance, namely the call states (see Table 3.6).

Table 3.6: UNI call states

Code	State	Remark
U0/N0	Null	
U1/N1	Call initiated	
U2/N2	Overlap sending	not in UNI4.0
U3/N3	Outgoing call proceeding	
U4/N4	Call delivered	
U6/N6	Call present	
U7/N7	Call received	
U8/N8	Connect request	
U9/U9	Incoming call proceeding	
U10/N10	Active	
U11/N11	Release request	
U12/N12	Release indication	
U13/N13	Modify requested	in Q.2963
U14/N14	Modify received	in Q.2963
U25/N25	Overlap receiving	not in UNI4.0

Each state has two short names: the UX are for the user side UNI instance, and the NX for the network site UNI instance. Calls in the active phase are in state U10/N10, U13/N13 or U14/N14; calls in the release phase are in state U11/N11 or U12/N12. U0/N0 means that the call does not exist. All other states belong to the call establishment phase.

Besides these call states, there is also a state associated with the global call reference. This call reference can be used to reset the entire UNI stack (and thereby release all connections) or a single ATM connection (that is, a VPI/VCI pair). The process handling this reset can be in one of three states as shown in Table 3.7 (see also the section on the structure of a UNI stack (Section 3.10) and the restart procedure (Section 3.7)).

Table 3.7: Global call states

Code	State
REST0	Null
REST1	Restart request
REST2	Restart

3.5 Point-to-Point Calls

The result of a point-to-point call is a bi-directional connection between two ATM end systems. Bi-directional does not necessarily mean that traffic will be sent in both directions. It only means that the VCI/VPI values are reserved for both directions while the traffic descriptor may specify a cell rate of zero. It is usual to differentiate outgoing and incoming calls—the meaning of outgoing and incoming is taken from the end system's point of view. The end system that initiates the call is said to make an outgoing call; the end system receiving the call is said to get an incoming call.

During the establishment phase the call goes through the following steps:

1. The calling end system places an outgoing call by sending a SETUP message on the signalling channel to the network. The network may acknowledge the receipt of the SETUP and the start of processing it by responding with a CALL PROCEEDING (Figure 3.15).

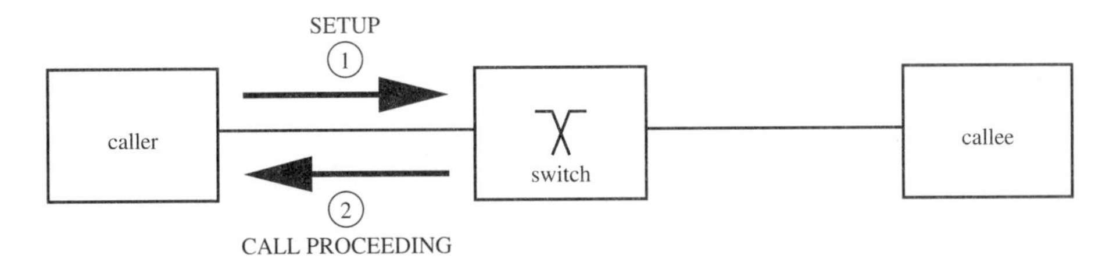

Figure 3.15: Setting up a point-to-point connection, part 1

2. The call is forwarded across the network, resources may be reserved along the path, and the last switch in the network creates an incoming call on the called end system by sending a SETUP message on the signalling channel to that end system (our example shows only

one switch). The called system in turn can acknowledge the receipt of the SETUP and the start of processing it with a CALL PROCEEDING (Figure 3.16).

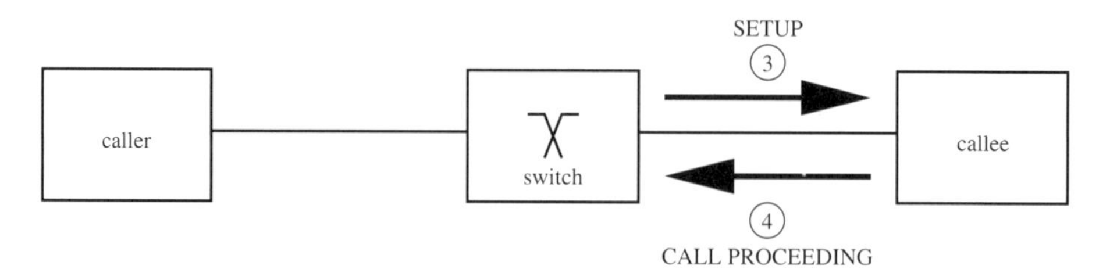

Figure 3.16: Setting up a point-to-point connection, part 2

3. The called system may send an ALERTING message back to the calling system to indicate that it is calling the user (or starting the application or whatever meaning one puts on the telephone term "alerting"). This message is forwarded to the calling system (Figure 3.17).

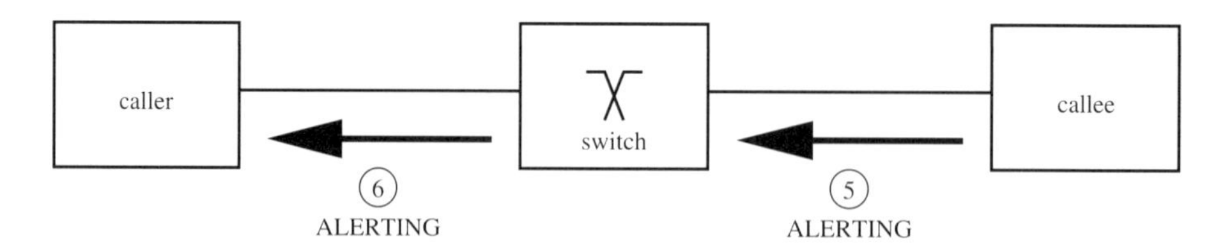

Figure 3.17: Setting up a point-to-point connection, part 3

4. When the called system has finished everything that has to be done to use the connection (starting applications, allocating resources, etc.) it sends a CONNECT message to the network. This message is acknowledged by a CONNECT ACKNOWLEDGE and forwarded to the calling end system. At this point the connection is in the active phase for the called system and the user can start sending ATM cells. However these cells are not guaranteed to arrive until the CONNECT is received by the calling system—there is no guarantee, that the CONNECT returning to the calling user travels at a higher or even the same speed as the user plane ATM cells.

   The calling end system receives the CONNECT message and returns a CONNECT ACKNOWLEDGE. At this point the connection is fully established (Figure 3.18).

Releasing a connection is much simpler:

1. One of the end systems (this may be either the calling one or the called one) decides to release the connection and sends a RELEASE message to the network (Figure 3.19). This message may include a cause information element that indicates the reason for releasing the connection. There is no way to reject a connection release (except if the message has an error). Note that it is also possible for one of the switches to release the connection in the case of an error.

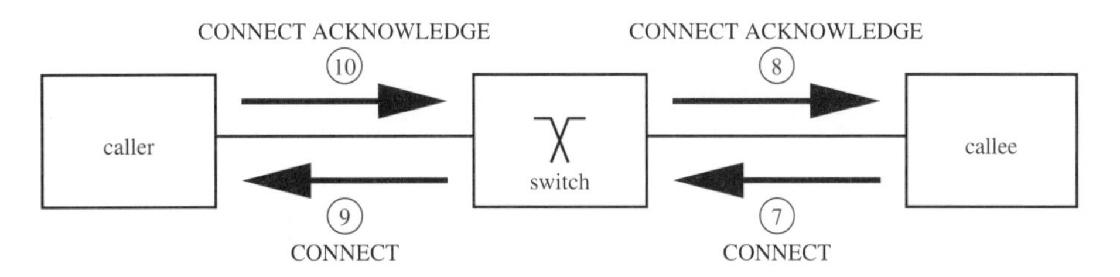

Figure 3.18: Setting up a point-to-point connection, part 4

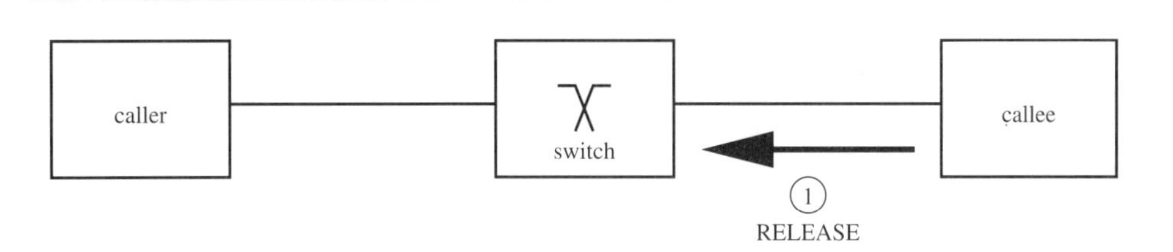

Figure 3.19: Releasing a point-to-point connection, part 1

2. This message is acknowledged with a RELEASE COMPLETE after releasing all the resources for that connection. At the same time the RELEASE is forwarded in the direction of the other end system and resources are freed as the message travels along the path (Figure 3.20).

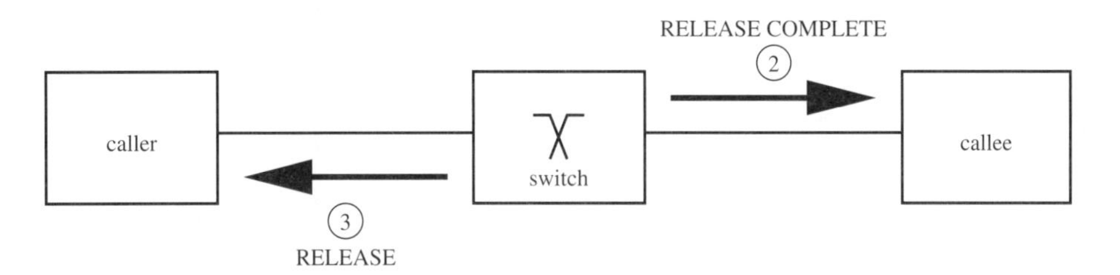

Figure 3.20: Releasing a point-to-point connection, part 2

3. Finally the second end system receives the RELEASE message, releases all the resources for this connection and acknowledges the message with a RELEASE COMPLETE. At this point all resources of this connection have been released in the network (Figure 3.21).

In the next section the protocol operation on both sides of the call will be discussed in detail.

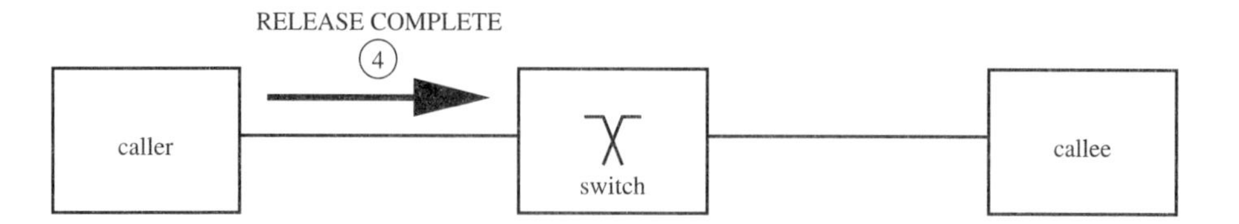

Figure 3.21: Releasing a point-to-point connection, part 3

3.5.1 Outgoing Calls

The flow of messages between the initiating end system and the network for an outgoing call is shown in Figure 3.22.

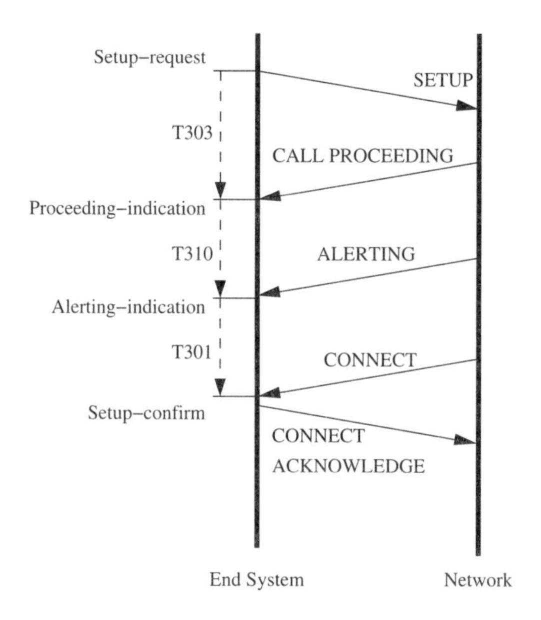

Figure 3.22: Message flow for an outgoing call

The corresponding state changes for the user side are shown in Figure 3.23 and for the network side in Figure 3.24.

The end system initiates the call by sending a SETUP message to the network with a fresh call reference and starting timer T303. This timer is usually 4 seconds long and protects against loss of the SETUP message. The following trace shows a typical SETUP message sent to the network:

```
1  E ⇒ S   0.000 uni cref={you,12} mtype=setup mlen=51
2                traffic={fpcr01=0,bpcr01=0,be}
3                bearer={class=bcob-x,traffic=noind,timing=noind,clip=not,user=p2p}
4                called={type=unknown,plan=aesa,addr=spock}
```

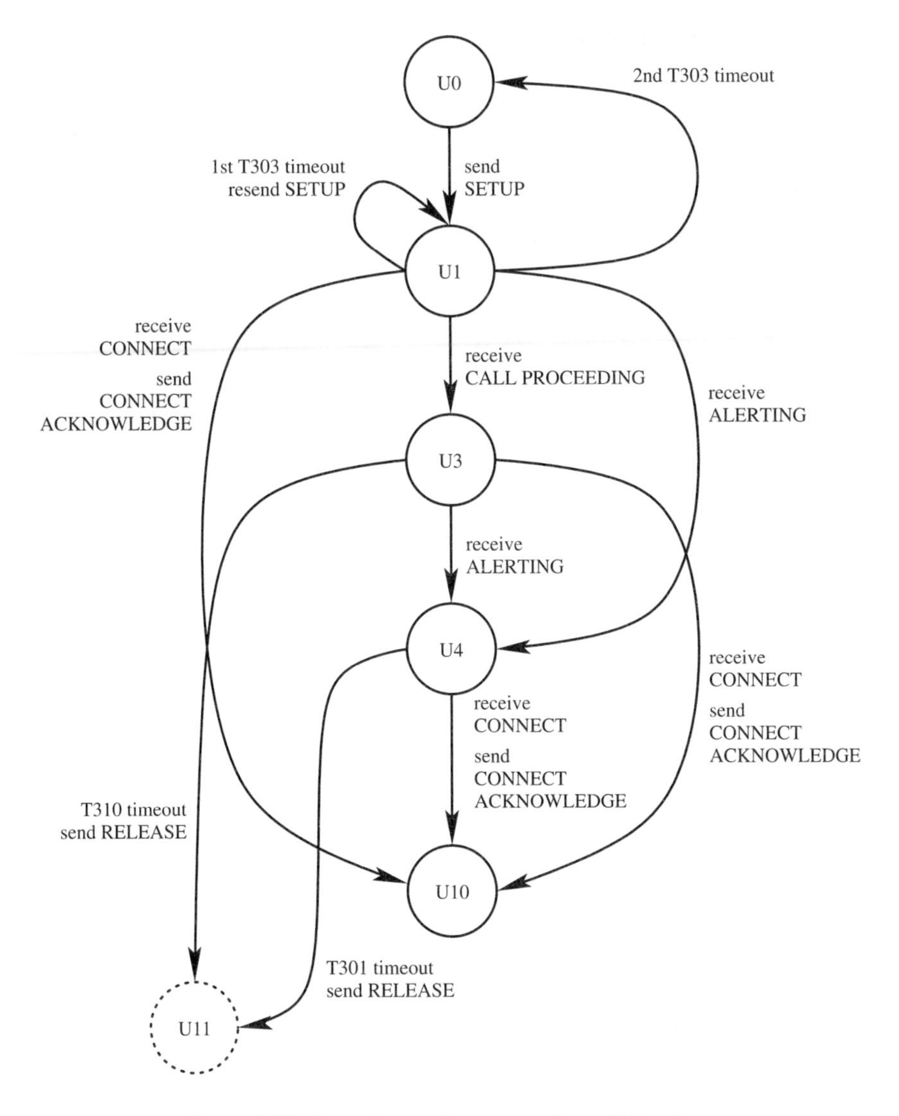

Figure 3.23: State changes for an outgoing call (user)

```
5                    calling={type=unknown,plan=aesa,addr=kirk}
6                    qos={forw=class0/unspecified,back=class0/unspecified}
7   E ⇐ S   0.170  uni cref={me,12} mtype=call_proc mlen=9
8                    connid={vpass=explicit,pex=exclusive_vpci_vci,vpci=0,vci=56}
9   E ⇐ S   0.204  uni cref={me,12} mtype=alerting mlen=0
10  E ⇐ S   0.234  uni cref={me,12} mtype=connect mlen=0
11  E ⇒ S   0.250  uni cref={you,12} mtype=conn_ack mlen=0
```

In this example the end system “kirk” establishes a UBR connection to the end system “spock”. In lines 1 to 6 the SETUP message is sent to the switch. At this point timer T303 is

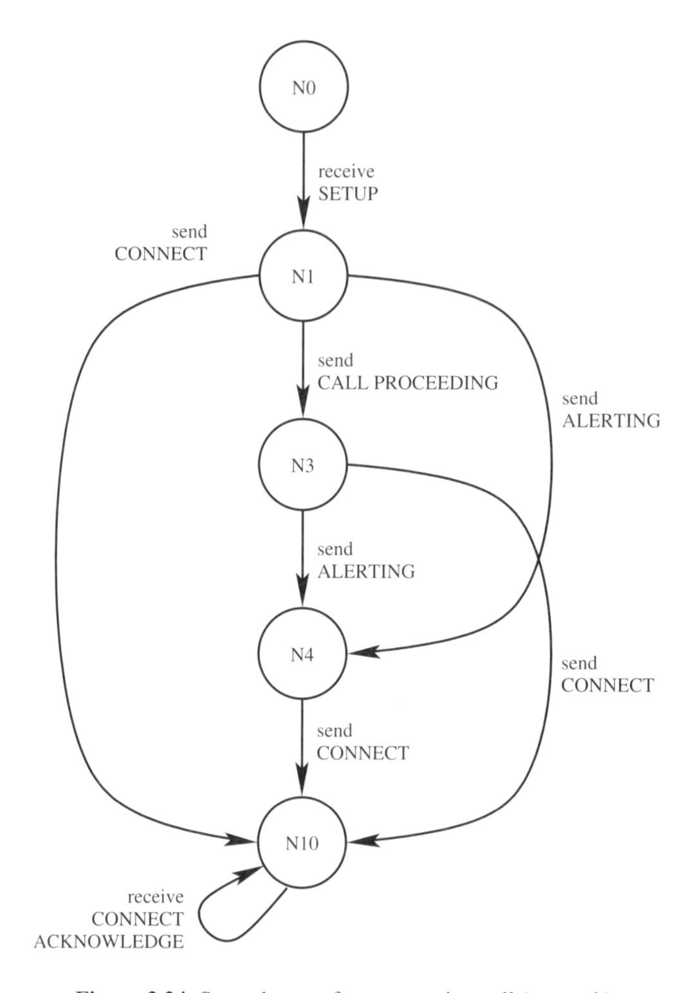

Figure 3.24: State changes for an outgoing call (network)

running and the call is in state U1 (call initiated) and N1 on the switch side. After 170 milliseconds the switch sends a CALL PROCEEDING message to the end system, indicating that the message seems OK and that it is processing the SETUP. Now the call progresses to state U3 (outgoing call proceeding) and N3 on the network side. Timer T303 is stopped and T310 started instead. T310 is considerably longer than T303 (30–120 seconds instead of 4). It must be noted that the CALL PROCEEDING message has only local relevance; receiving it means two things: the SETUP message was received and is syntactically OK and the access switch is going to forward the SETUP message across the network.

The CALL PROCEEDING contains a connection identifier IE. This tells the end system the VPCI and VCI that will be used for that connection. Generally these values are negotiable (see Section 3.5.2).

When the called end system receives the SETUP message it can respond with an ALERTING message. This ALERTING message obviously has come from narrowband ISDN, where the user is alerted and the calling user should receive a tone indication that the remote phone is ringing. In the context of data networks and ATM this seems archaic. Nevertheless the ATM-Forum has decided to take the entire ITU-T Q.2931 standard including such features into its UNI4.0 standard.

When an ALERTING message is sent by the called system, it is forwarded backwards through the network to the calling system. When the caller receives this message, it stops timer T310, optionally starts T301 and goes into state U4 (call delivered) (see Figure 3.23). The switch goes to state N4 when sending the ALERTING. Starting timer T301 is specified in Annex H of Q.2931. Originally the network side and the user side of the UNI protocol were specified somewhat asymmetrically. Annex H specifies optional extensions to the protocol for symmetric operation. In most UNI implementations this Annex is implemented. So the alerting phase is protected by a timer on the caller's side. This timer is even longer than T303 and T310—its minimum value is 3 minutes.

Our example end systems are faster, so the CONNECT (which was originally sent by the called system to the network) arrives quite fast in line 10 (now the switch is in state N10). Kirk answers with a CONNECT ACKNOWLEDGE in line 11 after stopping T301. At this point the call is active in state U10 and data can be received and sent. Like the CALL PROCEEDING, the CONNECT ACKNOWLEDGE has local meaning only. In fact, the switch silently ignores it, because it moved to state N10 when it sent the CONNECT message (this is one of the remaining asymmetries of the protocol).

One thing must be noted: although we have used the (almost) simplest SETUP message, from the protocol point of view we have seen all, even the optional, messages. One of the problems with this type of connection creation is that it takes a quite a long time to come to a state where data can be sent. In the next trace we will go the shortest way to get a working connection.

To establish a connection both the CALL PROCEEDING and the ALERTING messages can be omitted. Whether to send a CALL PROCEEDING in response to a SETUP is usually a configuration feature of the access switch; to send ALERTING or not is a feature of the end system software. In the following trace we have configured the switch to not send CALL PROCEEDING and the end system to not send ALERTING (see also Figure 3.25 for the message flow).

```
1  E ⇒ S   0.000 uni cref={you,12} mtype=setup mlen=76
2                traffic={fpcr01=0,bpcr01=0,be}
3                bearer={class=bcob-x,traffic=noind,timing=noind,clip=not,user=p2p}
4                called={type=unknown,plan=aesa,addr=plan=aesa,addr=spock}
5                calling={type=unknown,plan=aesa,addr=kirk}
6                qos={forw=class0/unspecified,back=class0/unspecified}
7  E ⇐ S   0.347 uni cref={me,14} mtype=connect mlen=9
8                connid={vpass=explicit,pex=exclusive_vpci_vci,vpci=0,vci=57}
9  E ⇒ S   0.420 uni cref={you,12} mtype=conn_ack mlen=0
```

In this scenario the SETUP is directly answered with a CONNECT. When the CONNECT is received, T303 is stopped and the call goes from state U1 (call initiated) directly to state U10 (active).

Whether or not to configure a switch to send a CALL PROCEEDING depends on the expected total round trip time of a call setup. If no CALL PROCEEDING is sent by the access switch, the first answer from the called end system must come back to the calling

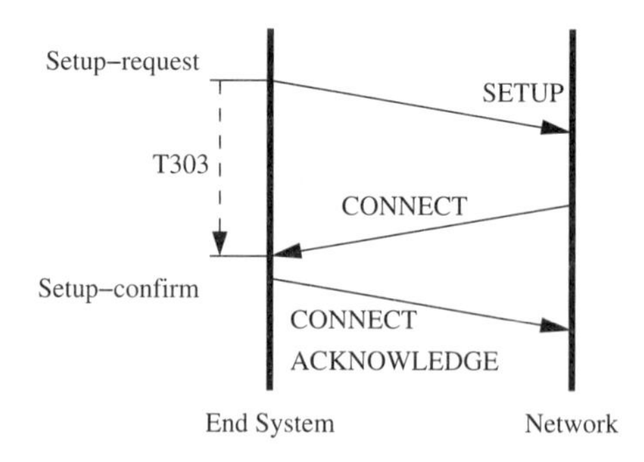

Figure 3.25: Short message flow for outgoing call

system within the runtime of T303 (3 seconds). If a CALL PROCEEDING is sent, this is relaxed to T310 (30–120 seconds). Three seconds may be too short if a complex multihop topology is involved (especially, if geostationary satellite hops are needed to reach the called system).

3.5.2 Connection Identifier Selection

Generally the VPCI and VCI to be used for a connection can be negotiated over the UNI. The originator of a connection can include a connection identifier (Figure 3.26) in its SETUP message. The peer UNI can then decide whether this connection identifier is acceptable or not and, if not, can reject the call. In most UNI implementations, however, connection identifiers are selected by the switch: an outgoing SETUP (i.e. one sent by the end system) does not include a connection identifier; an incoming SETUP (sent by the switch) always includes a connection identifier. If the SETUP contains no connection identifier, the first message sent in response to this SETUP is required to contain one. This message may be either a CALL PROCEEDING, an ALERTING or a CONNECT. Later messages can repeat the connection identifier IE.

For negotiation the sender of the SETUP message includes a connection identifier information element with one of the following three codings:

1. the IE indicates "exclusive VPCI; any VCI" (1);
2. the IE indicates "exclusive VPCI; exclusive VCI" (0);
3. the IE indicates "exclusive VPCI; no VCI" (4).

In all three cases the receiver of the SETUP checks whether the indicated VPCI value is acceptable. If it is, then for the first case the receiver selects an appropriate VCI value. For the second case the receiver checks whether the indicated VCI value is also acceptable. The third case is allowed only when the bearer capabilities IE indicates that a transparent VP is to be established. In this case the VCI field is ignored. In any case, if the indicated value is not acceptable or an appropriate value cannot be allocated, the SETUP is rejected. If the checks are acceptable, the resulting connection identifier is reported back in the first message that answers the SETUP. The sender of the SETUP is required to check that the returned values actually match what was sent.

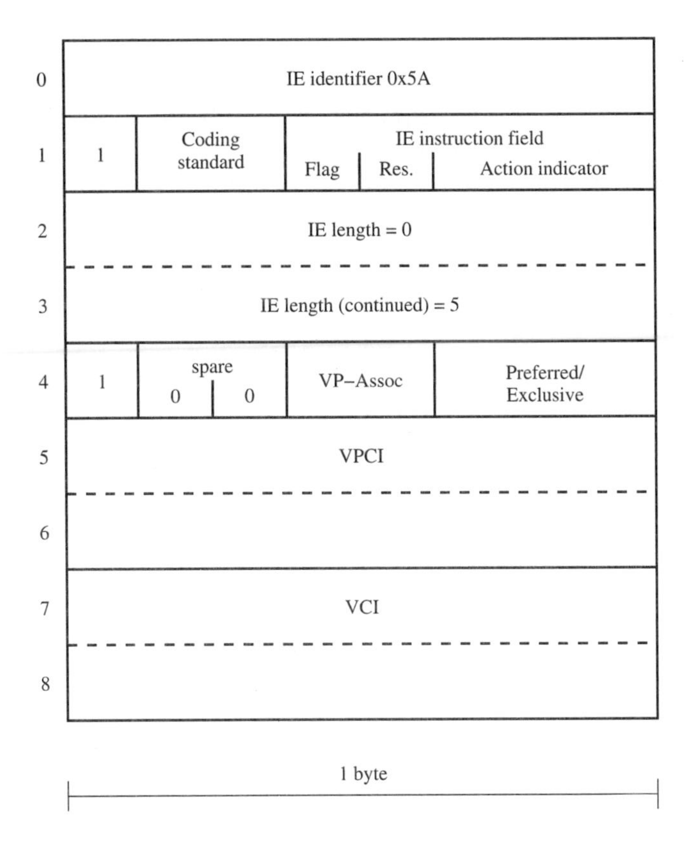

Figure 3.26: Connection identifier IE

For associated signalling (not supported by UNI4.0 but by PNNI) a connection identifier IE must always be included in the SETUP. Here only the first two codings are allowed and the VPCI field in the information element is ignored (because the VCI is always allocated in the VPC carrying the signalling channel). The indication whether signalling is associated (coded as 0) or non-associated (coded as 1) is contained in the connection identifier.

3.5.3 Negotiation of Connection Characteristics

Several parameters of the connection can be negotiated between the calling system and the called system. These parameters fall into two groups: parameters that have meaning for resource allocation in the network and for the end systems, and parameters that are meaningful only for the end systems. Table 3.8 shows some of these parameters.

For each of these four groups a different mechanism for negotiation is used:

- *AAL parameters.* The calling system includes an AAL Parameters IE in the SETUP message. On receipt of this message the called system checks whether the negotiable parameters are acceptable. If they are too low to be usable, the called system should reject the SETUP; if they are too high, they are adjusted accordingly and the changed information element is included in the CONNECT message that is sent back to the caller. Not all AAL parameters are negotiable. For AAL3/4 these include: forward and backward maximum CPCS-PDU size (can only be reduced), MID (can only be reduced) and SSCS type. For AAL5: forward and backward maximum CPCS-PDU size and SSCS type. The recent Amendment 4 to Q.2931 also allows negotiation of the AAL type. For this type of negotiation two AAL parameter information elements can be included in the SETUP message.

Table 3.8: Parameters that can be negotiated

Parameter	Network resource
AAL parameters	
AAL3/4 Forward maximum CPCS size	no
AAL3/4 Backward maximum CPCS size	no
AAL3/4 MID range	no
AAL5 Forward maximum CPCS size	no
AAL5 Backward maximum CPCS size	no
Traffic parameters	
Forward and backward PCR	yes
Forward and backward SCR	yes
Forward and backward MBS	yes
Tagging	yes
Frame discard	yes
Best effort indication	yes
Broadband lower layer information	
Layer 2 protocol	no
Layer 2 parameters	no
Layer 3 protocol	no
Layer 3 parameters	no
End-to-end transit delay	
Cumulative delay	yes

- *Traffic parameters.* These are negotiated by including a minimum ATM traffic descriptor IE or an alternative ATM traffic descriptor IE in the SETUP message in addition to the normal ATM traffic descriptor.

  In the case of the minimum traffic descriptor the called system can adjust the original value of a parameter to a value between the original one and the minimal one. Suppose, for example, that the traffic descriptor contained a forward PCR of 1024 cells/second and the minimum traffic descriptor 256 cells/second. In this case the called system can choose any value between 256 and 1024 to be returned in the traffic descriptor of the CONNECT message.

  If an alternate traffic descriptor is included in the SETUP, the called system has the option to choose that one instead of the normal traffic descriptor if it cannot support the values in the normal traffic descriptor.

- *Lower layer information.* This information is negotiated by including up to three Broadband Lower Layer Information (BLLI) information elements in the SETUP, preceeded by a repeat indicator. The called system must choose one of these to return it in the CONNECT message. The BLLIs are ordered with decreasing precedence.
- *End-to-end transit delay.* This information element once included in the SETUP, is updated at each node as it travels along the network. Each switch and the called end system add their delay into the cumulative transit delay. The information element can specify a maximum allowable value. If the cumulative delay turns out to be higher than the maximum, the call is rejected.

It must be noted that some of the parameters that are marked with "no" in Table 3.8 may influence network resources if the call is an interworking call with narrowband ISDN.

3.5.4 Incoming Calls

The flow of messages between the network and the end user for an incoming call is shown in Figure 3.27. The state changes for this case can be seen in Figure 3.28 and in Figure 3.29 for the network side.

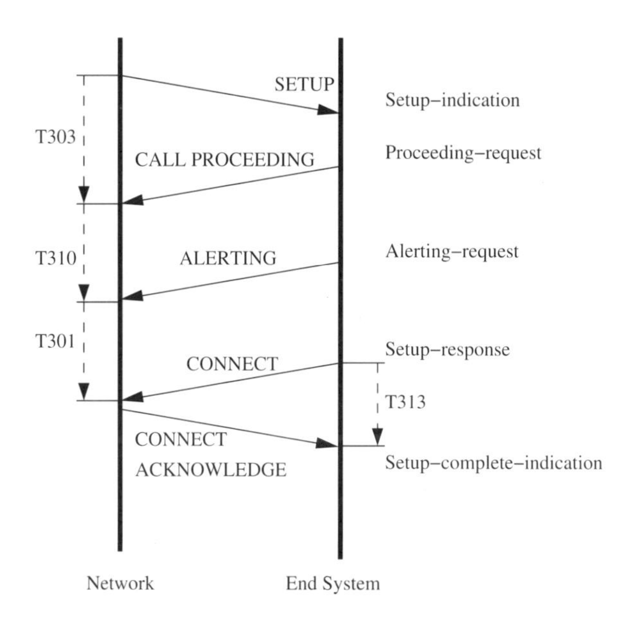

Figure 3.27: Message flow for an incoming call

The new call instance is created when the network sends the SETUP message to the called system (lines 1 to 7 in the following trace). The switch instance is now in state N6 (call present) and timer T303 with a value of 4 seconds is started to guard against loss of the SETUP (or a hanging end system). Upon receipt of the SETUP the user side UNI stack

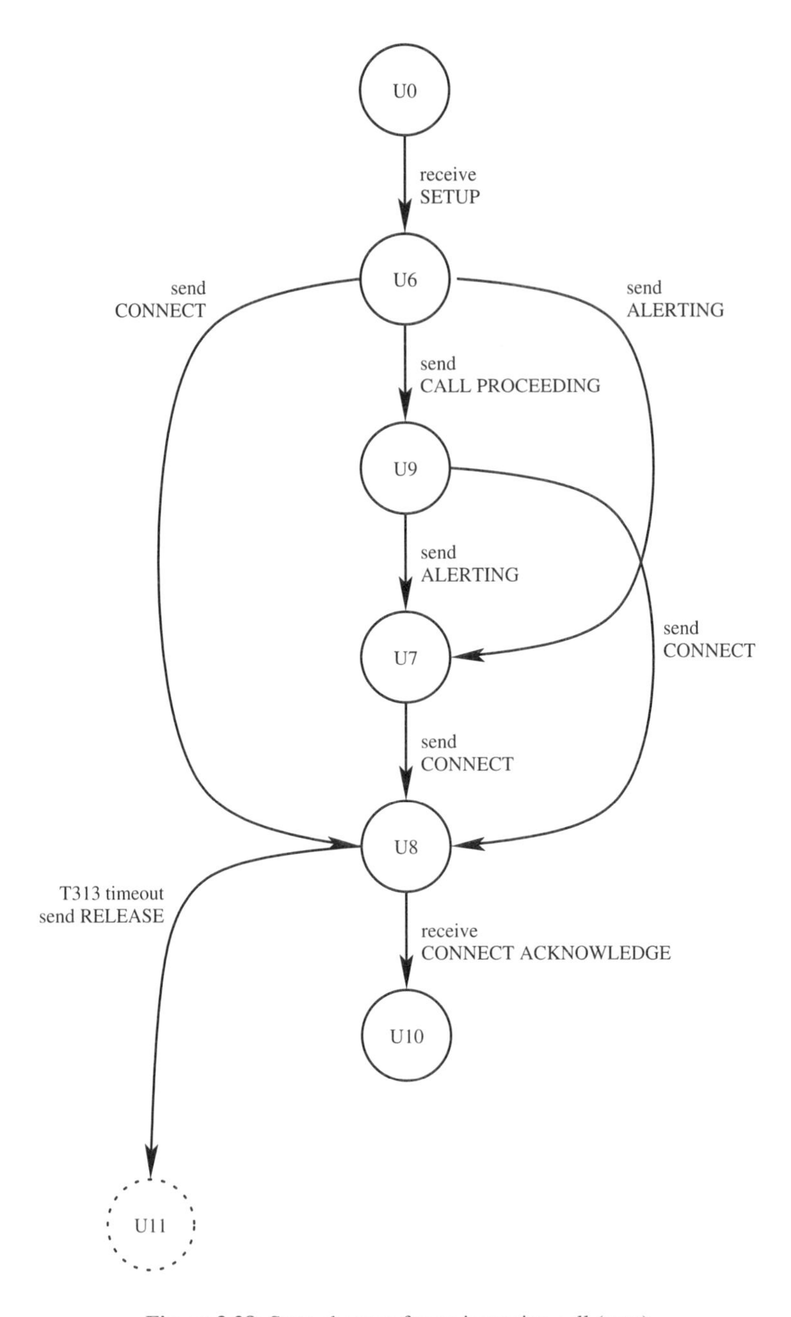

Figure 3.28: State changes for an incoming call (user)

creates a call instance and moves it to state U6. Depending on the time it takes to process the incoming SETUP, the end system can optionally send a CALL PROCEEDING (line 8). If this is done, the user side goes into state U9 (incoming call proceeding) and the network side starts timer T310 (10 seconds) and goes to state N9. Now the user system can send

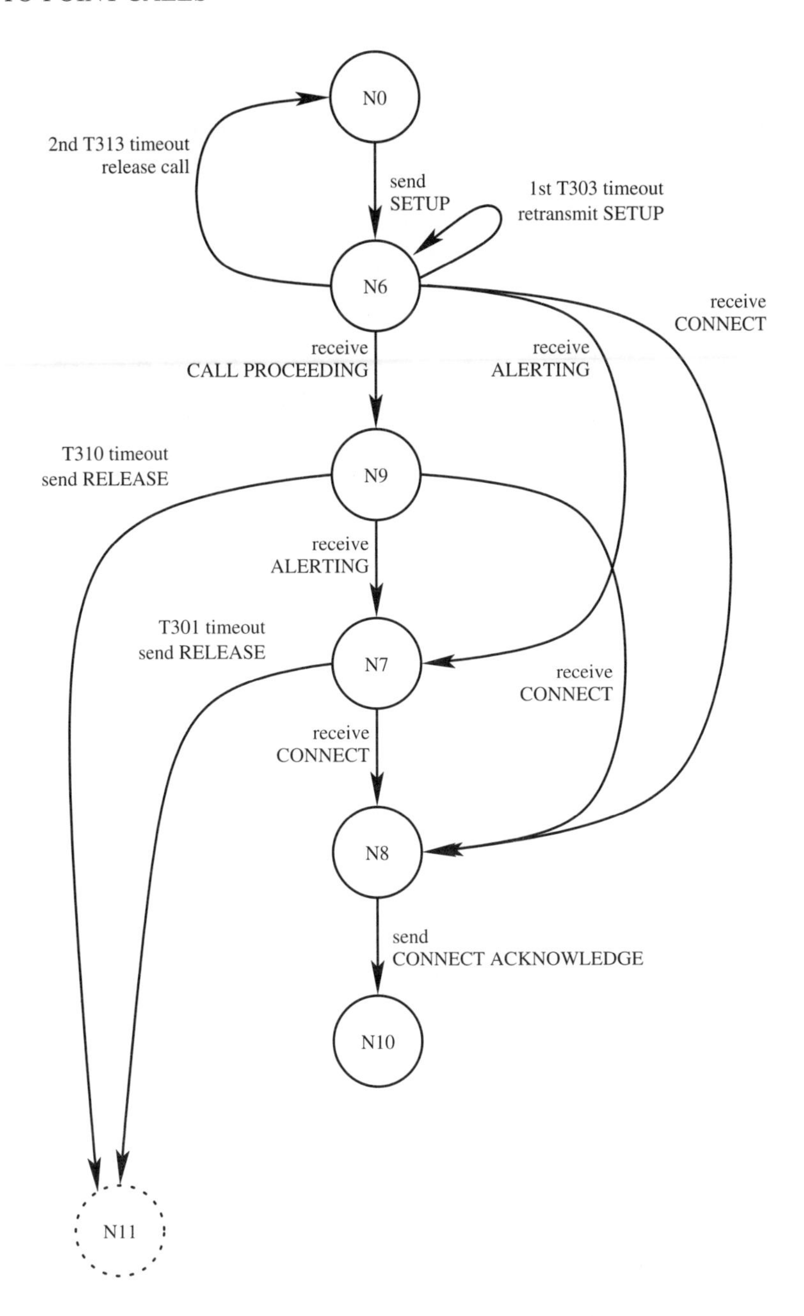

Figure 3.29: State changes for an incoming call (network)

an ALERTING message to inform the calling user that it is alerting the user (or starting applications). This message (line 9) moves the user side of the UNI to state U7 (call received) and the network side to N7. The network stops whatever timer is running (T303 or T310) and starts T301 with a minimum timeout of 3 minutes. In the next step the calling end system sends a CONNECT message across the UNI, indicating that it is willing to accept the connection. It starts timer T313 (4 seconds) goes into state U8 and waits for the acknowledgement of its message. The network, upon receipt of the CONNECT, forwards it to the calling user and answers the CONNECT with a CONNECT ACKNOWLEDGE (line 11). At this point the connection is in the active state (U10 and N10).

```
1   E ⇐ S   0.000 uni cref={you,81445} mtype=setup mlen=85
2                 traffic={fpcr01=0,bpcr01=0,be}
3                 bearer={class=bcob-x,traffic=noind,timing=noind,clip=not,user=p2p}
4                 called={type=unknown,plan=aesa,addr=plan=aesa,addr=spock}
5                 calling={type=unknown,plan=aesa,addr=kirk}
6                 connid={vpass=explicit,pex=exclusive_vpci_vci,vpci=0,vci=255}
7                 qos={forw=class0/unspecified,back=class0/unspecified}
8   E ⇒ S   0.127 uni cref={me,81445} mtype=call_proc mlen=0
9   E ⇒ S   0.201 uni cref={me,81445} mtype=alerting mlen=0
10  E ⇒ S   0.277 uni cref={me,81445} mtype=connect mlen=0
11  E ⇐ S   0.340 uni cref={you,81445} mtype=conn_ack mlen=0
```

Note that in this example the switch is offering the connection identifier to the end system.

As in the case of an outgoing call, all the optional messages can be omitted leading to faster establishment of the call (see also Figure 3.30):

```
1   E ⇐ S   0.000 uni cref={you,81447} mtype=setup mlen=85
2                 traffic={fpcr01=0,bpcr01=0,be}
3                 bearer={class=bcob-x,traffic=noind,timing=noind,clip=not,user=p2p}
4                 called={type=unknown,plan=aesa,addr=plan=aesa,addr=spock}
5                 calling={type=unknown,plan=aesa,addr=kirk}
6                 connid={vpass=explicit,pex=exclusive_vpci_vci,vpci=0,vci=254}
7                 qos={forw=class0/unspecified,back=class0/unspecified}
8   E ⇒ S   0.151 uni cref={me,81447} mtype=connect mlen=0
9   E ⇐ S   0.272 uni cref={you,81447} mtype=conn_ack mlen=0
```

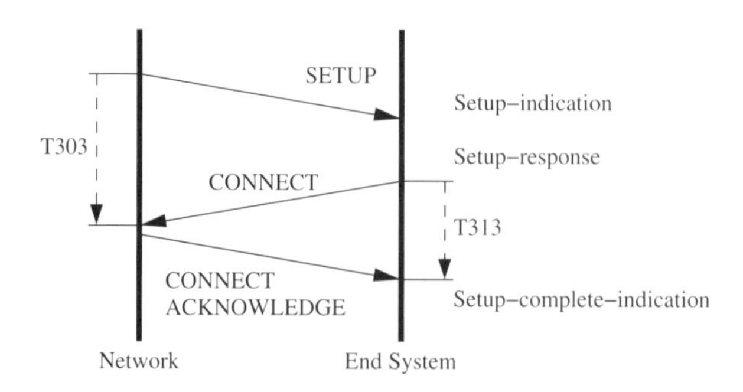

Figure 3.30: Short message flow for an incoming call

3.5.5 Unsuccessful Calls

Calls can be rejected for a number of reasons in various stages. They can be rejected by network nodes, by the called end system and by the called user.

3.5.5.1 Timer Expiry

As we have seen, several timers are used to safeguard the different states of a connection establishment at the outgoing and incoming interface. What happens when a timer expires depends on which timer has expired and whether it was on the network or user side. The following paragraphs show some of the most common situations.

T303 Timer T303 is the timer that is started when a SETUP is sent over the UNI. Its value is normally 4 seconds. On the first expiry of this timer the SETUP message is usually resent (this is optional in UNI4.0). On the second expiry the call is cleared locally (which means any resources in the end system or the switch that were reserved for this call are released). If this happens in a switch when creating an incoming call, a release towards the calling user is also initiated.

One may ask why do we need a retransmission at the UNI layer when there is a reliable transport protocol (SSCOP) beneath the UNI? There are two reasons for this: first, in the configuration of the UNI the SSCOP may actually lose messages (see Section 5.2) and second, the called end system, although able to process the SSCOP, may be too busy and drop the SETUP. Both of these reasons are of course quite weak and, obviously because of this weakness, the retransmission feature was made optional.

T310 Timer T310 is the timer that is started when a CALL PROCEEDING was received (and a SETUP is acknowledged) and the UNI instance is waiting for an ALERTING or CONNECT. This timer is different at the user and the network sites: 30–120 seconds vs. 10 seconds. If this timer expires, normal call clearing is initiated by sending a RELEASE and entering state U11 or N11. If this happens at a switch, call clearing towards the calling user is also initiated.

T301 Timer T301 is the timer that safeguards the alerting phase. Because it is supposed to involve user action (although this seems peculiar in the ATM context) it is considerably longer than the other timers—its minimum value is 3 minutes. It must be noted that often networks implement alerting timers at higher layers. These timers must be aligned appropriately. If the timer expires, call clearing is initiated by sending a RELEASE and entering state U11 or N11. A switch will also start call clearing towards the calling user.

T313 This timer is started at the user site when a CONNECT is sent to the network. It has a value of 4 seconds and if the timer expires, call clearing is initiated by sending a RELEASE to the switch and entering state U11. This timer is not used at the switch side

The following trace shows an example of an incoming call with a SETUP retransmission (T303 expiry) and a T313 expiry.

```
 1  E ⇐ S   0.000 uni cref={you,37} mtype=setup mlen=85
 2                traffic={fpcr01=0,bpcr01=0,be}
 3                bearer={class=bcob-x,traffic=noind,timing=noind,clip=not,user=p2p}
 4                called={type=unknown,plan=aesa,addr=plan=aesa,addr=spock}
 5                calling={type=unknown,plan=aesa,addr=kirk}
 6                connid={vpass=explicit,pex=exclusive_vpci_vci,vpci=0,vci=60}
 7                qos={forw=class0/unspecified,back=class0/unspecified}
 8  E ⇐ S   4.102 uni cref={you,37} mtype=setup mlen=85
 9                traffic={fpcr01=0,bpcr01=0,be}
10                bearer={class=bcob-x,traffic=noind,timing=noind,clip=not,user=p2p}
11                called={type=unknown,plan=aesa,addr=plan=aesa,addr=spock}
12                calling={type=unknown,plan=aesa,addr=kirk}
13                connid={vpass=explicit,pex=exclusive_vpci_vci,vpci=0,vci=60}
14                qos={forw=class0/unspecified,back=class0/unspecified}
15  E ⇒ S   4.197 uni cref={me,37} mtype=connect mlen=0
16  E ⇒ S   8.511 uni cref={me,37} mtype=release mlen=9
17                cause={loc=user,cvalue=recovery_on_timer_expiry,
18                  class=protocol_error,timer=T313}
19  E ⇐ S   8.617 uni cref={you,37} mtype=release_comp mlen=6
20                cause={loc=user,cvalue=normal_call_clearing,class=normal_event}
```

The SETUP sent in line 1 obviously got lost and is resent after T303 timeout in line 8. This time the SETUP is answered by the user with a CONNECT in line 15, but the CONNECT ACKNOWLEDGE from the switch is lost, so the user times out and sends a RELEASE in line 16. This RELEASE is acknowledged in line 19 and the call is cleared.

3.5.5.2 Call Rejection by the Called User

A call may be rejected by the called user for many reasons:

- *Bad called address or subaddress.* According to [Q.2931] the called user is required to check whether the called addresses or subaddresses in an incoming SETUP really match the addresses of the endpoint. If they do not match, the call should be rejected. This check is optional in [UNI4.0].
- *Bad connection characteristics.* The called user should check the connection characteristics that are specified in the call (cell rates, bearer capabilities, QoS parameters, etc.). If they are not usable, the user should reject the call. This is optional in [UNI4.0].
- *Bad AAL parameters.* The called user should check the AAL parameters IE and reject the call if he cannot support the specified parameters (or they cannot be adjusted as described in Section 3.5.3).
- *Bad lower layer information.* The called user should reject the call if he cannot support the specified protocols or parameters (see also Section 3.5.3).
- *Bad higher layer information.* The call should be rejected by the called user if he cannot support the protocol or protocol parameters specified in this information element.
- *CLIR.* The user may reject calls with Calling Line Identification Restriction (CLIR).
- *Missing optional IEs.* The called user may reject calls where certain optional information elements are missing, for example calling party number or calling party subaddress.
- *User busy.* The call may be rejected because the user is busy. This seems to be an anachronism from narrowband ISDN.

The call can be rejected in two ways. Normally a RELEASE is sent to the network including an appropriate cause information element (for example, "User rejects all calls with call line identification restriction (CLIR)") and, if applicable, diagnostics. The switch releases resources, initiates call clearing towards the calling user and sends a RELEASE COMPLETE. If immediately after receiving the SETUP the end system can determine that the call has to be rejected, it can also send a RELEASE COMPLETE with an appropriate cause IE and without allocating any resources to the call. In this case the network will release its resources and clear the call towards the calling user. The following trace shows a rejection because of bad AAL parameters (the user does not support AAL3/4) after sending a CALL PROCEEDING.

```
1   E ⇐ S   0.000 uni cref={you,55723} mtype=setup mlen=103
2                 aal={aaltype3/4={fmaxcpcs=4096,bmaxcpcs=4096,midrange={0,63},
3                    sscstype=data_sscs_assured}
4                 traffic={fpcr01=0,bpcr01=0,be}
5                 bearer={class=bcob-x,traffic=noind,timing=noind,clip=not,user=p2p}
6                 called={type=unknown,plan=aesa,addr=plan=aesa,addr=spock}
7                 calling={type=unknown,plan=aesa,addr=kirk}
8                 connid={vpass=explicit,pex=exclusive_vpci_vci,vpci=0,vci=212}
9                 qos={forw=class0/unspecified,back=class0/unspecified}
10  E ⇒ S   0.291 uni cref={me,55723} mtype=call_proc mlen=0
11  E ⇒ S   0.324 uni cref={me,55723} mtype=release mlen=6
12                cause={loc=user,cvalue=aal_parameters_can_not_be_supported,
13                   class=service_or_option_not_implemented}
14  E ⇐ S   0.401 uni cref={you,55723} mtype=rel_compl mlen=6
15                cause={loc=user,cvalue=normal_call_clearing,class=normal_event}
```

In this trace the call is rejected because the called party number is wrong. This usually means a misconfiguration of the switch or a failure of PNNI or ILMI:

```
1   E ⇐ S   0.000 uni cref={you,55797} mtype=setup mlen=85
2                 traffic={fpcr01=0,bpcr01=0,be}
3                 bearer={class=bcob-x,traffic=noind,timing=noind,clip=not,user=p2p}
4                 called={type=unknown,plan=aesa,addr=plan=aesa,addr=lovina}
5                 calling={type=unknown,plan=aesa,addr=kirk}
6                 connid={vpass=explicit,pex=exclusive_vpci_vci,vpci=0,vci=201}
7                 qos={forw=class0/unspecified,back=class0/unspecified}
8   E ⇒ S   0.397 uni cref={me,55797} mtype=rel_compl mlen=7
9                 cause={loc=user,cvalue=incompatible_destination,
10                   class=invalid_message,ie=called}
```

3.5.5.3 Call Rejection by the Network

The call can also be rejected by the network. There are four major groups of reasons for call rejection:

- *Message errors.* The contents of the SETUP message are wrong. The message or information elements may be wrongly coded, length information may be inconsistent, mandatory information elements may be missing or the call reference may be in use. Not all coding errors lead to a rejection of the call. Errors in optional information elements are usually reported via a STATUS message and the information elements are discarded. The call is progressed in that case. If mandatory information elements are bad or missing, the call is rejected with an appropriate diagnostics field in the cause information element.

- *Unsupported feature.* The calling end system may try to use optional features or features that are supported in [UNI4.0] but not in the ITU-T standards. Examples are: requesting a switched virtual path, ATM anycast or a second called party subaddress IE. The cause information element must be examined in this case. Note that not all unimplemented optional features lead to call rejection. In the case of a second subaddress IE, this IE maybe discarded by the switch and the call forwarded without that IE.
- *Unsupported connection parameters.* The user may try to request a connection that cannot be established because of its parameters. Examples are: unimplemented bearer classes or QoS parameters, cell rates are unavailable or maximum end-to-end transit delay exceeded. One major problem are the cause codes 65 ("Bearer capability not implemented") and 73 ("Unsupported combination of traffic parameters"). These usually mean that the user has requested a combination of bearer capabilities, traffic parameters or QoS classes which are illegal. The allowed combinations are listed in Annex 9 of [UNI4.0].
- *Destination problems.* If the user specifies an unknown called party number in the SETUP or the destination system is down, the call is rejected by the network with cause codes 27 ("Destination out of order"), 41 ("Temporary failure"), 1 ("Unallocated number") or 3 ("No route to destination"). Other cause codes are also possible (the standards are somewhat fuzzy about the usage of specific cause codes).

Depending on where and when the problem is detected, the call may be rejected with a RELEASE or a RELEASE COMPLETE. The latter happens only if the first switch is able to find the problem before sending a CALL PROCEEDING.

In the following trace the user requests a CBR connection and specifies sustainable cell rates. According to Annex 9 of [UNI4.0] and Appendix F of [UNI3.1] this is illegal, so the call is rejected by the access switch.

```
1   E ⇒ S   0.000 uni cref={you,27} mtype=setup mlen=91
2                 traffic={fpcr01=334,bpcr01=334,fscr01=334,bscr01=334,fmbs01=4,bmbs01=4}
3                 bearer={class=bcob-x,traffic=cbr,timing=noind,clip=not,user=p2p}
4                 called={type=unknown,plan=aesa,addr=plan=aesa,addr=spock}
5                 calling={type=unknown,plan=aesa,addr=kirk}
6                 qos={forw=class0/unspecified,back=class0/unspecified}
7   E ⇒ S   0.157 uni cref={me,27} mtype=rel_compl mlen=6
8                 cause={loc=private_network_serving_local_user,
9                    cvalue=unsupported_combination_of_traffic_parameters,
10                   class=service_or_option_not_implemented}
```

3.5.6 Clearing a Call

Call clearing (or connection release) may be initiated by either side of the connection as well as by the network. Normally a call is cleared on a UNI by sending a RELEASE message. The only exception is call rejection via RELEASE COMPLETE (see Sections 3.5.5.2 and 3.5.5.3).

The sender of a RELEASE message starts timer T308 (30 seconds) and enters state U11 or N11 (release request). Upon receipt of the RELEASE the peer should release all resources for this connection. If it is a switch initiate call clearing towards the other end of the connection

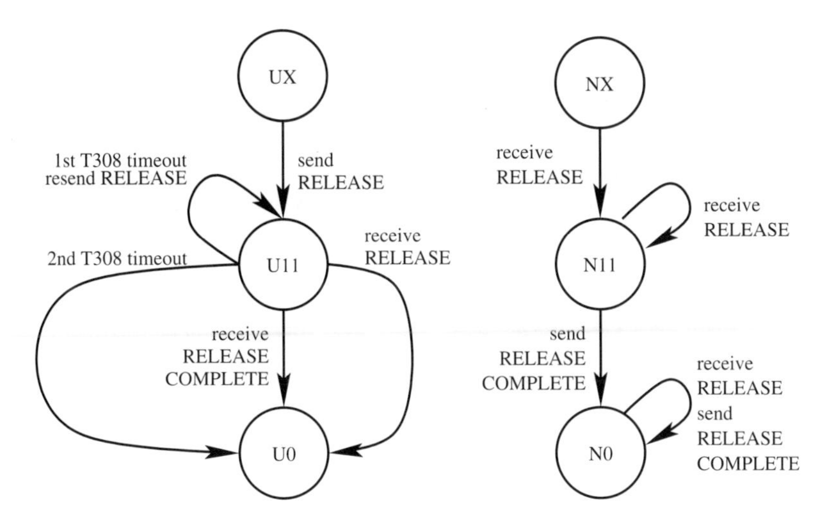

Figure 3.31: Call clearing initiated by a local user

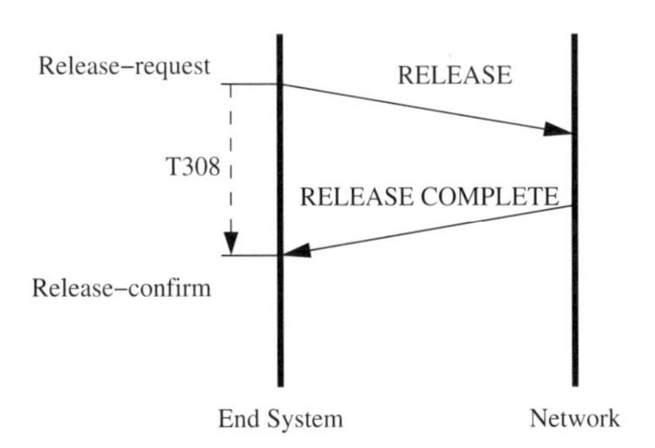

Figure 3.32: Message flow for call clearing by a local user

and answer with a RELEASE COMPLETE. The sender of the RELEASE should then release its resources and stop the timer. At this point the connection no longer exists on this connection segment. It should be noted that RELEASE COMPLETE has only local significance—it does not indicate, that the entire connection has been released.

The following trace shows a release initiated by the user. The corresponding state changes are shown in Figure 3.31 and the message flow in Figure 3.32.

```
1   E ⇒ S   0.000 uni cref={you,33} mtype=release mlen=6
2                 cause={loc=user,cvalue=normal_call_clearing,class=normal_event}
3   E ⇐ S   0.127 uni cref={me,33} mtype=rel_compl mlen=6
4                 cause={loc=private_network_serving_local_user,
5                    cvalue=normal_call_clearing,class=normal_event}
```

The user sends a RELEASE in lines 1 and 2 containing a cause information element that informs the remote user about the reason for the connection tear-down. In lines 3 to 5 the switch answers with the appropriate RELEASE COMPLETE.

What happens if timer T308 expires? This can happen when either the RELEASE or the RELEASE COMPLETE message are lost (the switch could also be too busy to answer in time, although this seems unlikely with a timeout of 30 seconds). On the first expiry of T308 the original RELEASE message is retransmitted, optionally including a second cause information element with cause code 102 ("Recovery on timer expiry") as can be seen in the following trace.

```
1  E ⇒ S   0.000  uni cref={you,33} mtype=release mlen=6
2                 cause={loc=user,cvalue=normal_call_clearing,class=normal_event}
3  E ⇒ S   30.472 uni cref={you,33} mtype=release mlen=15
4                 cause={loc=user,cvalue=normal_call_clearing,class=normal_event}
5                 cause={loc=user,cvalue=recovery_on_timer_expiry,
6                    class=protocol_error,timer=T308}
7  E ⇐ S   30.601 uni cref={me,33} mtype=rel_compl mlen=6
8                 cause={loc=private_network_serving_local_user,
9                    cvalue=normal_call_clearing,class=normal_event}
```

This RELEASE finds the call at the switch either in state N10 (or another state N1...N10) if the RELEASE was lost, or in state N0 (i.e. it does not find the call altogether), if the RELEASE COMPLETE was lost. In any case a new RELEASE COMPLETE is transmitted. On the second expiry of T308 all local resources for the call are released and, as the standard states, "the virtual channel placed in a maintenance condition". Additionally the restart procedure can be invoked (see Section 3.7). Generally the VPI/VCI values that were associated with the call should be marked as in use, because there is no evidence to the UNI stack that the network has really released the channel.

If a call is released by the remote peer, the operation of the protocol is symmetric to the described one (see Figures 3.33 and 3.34).

Of course, it may happen that both the local user and the remote user release the call at the same time. This can lead to a so-called clear collision, as one can see from the state diagrams (Figures 3.31 and 3.33). This is handled by treating a RELEASE message in state U11 and N12 like a RELEASE COMPLETE. Figure 3.35 shows the message flow. The following is a trace that was obtained by delaying the release-response of the application in the end system.

```
1  E ⇐ S   0.000  uni cref={me,37} mtype=release mlen=6
2                 cause={loc=user,cvalue=normal_call_clearing,class=normal_event}
3  E ⇒ S   5.812  uni cref={you,37} mtype=release mlen=6
4                 cause={loc=user,cvalue=normal_call_clearing,class=normal_event}
```

There is also a faster way to release a call by using the exception handling procedures specified in the standard. If a RELEASE COMPLETE message is received in a state where it is not expected (and it is expected only in U0, U1, U11 on the user side and in N0, N6 and N11 or the switch side), all resources for the call are released, all timers stopped and, if this happens at a switch, the call released towards the remote user.

A call may also be released by the network in exceptional conditions. This may happen, for example, if the remote user disappears (i.e. its SAAL goes down and cannot be restarted) or

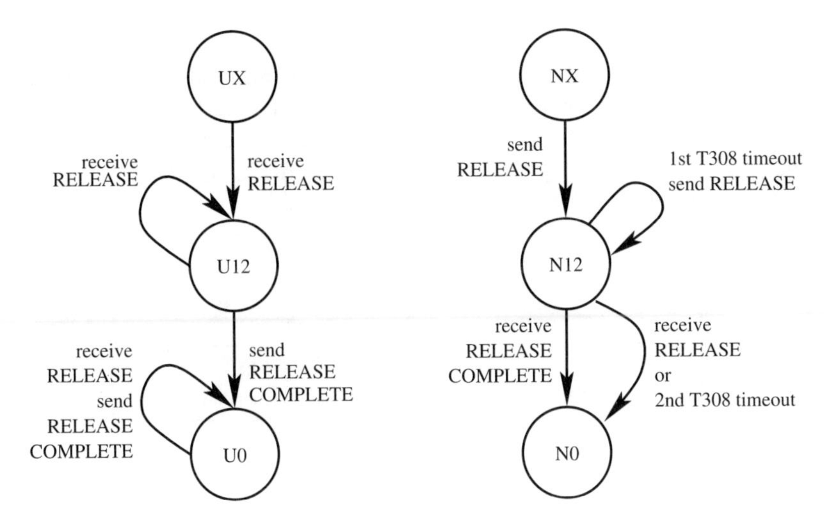

Figure 3.33: Call clearing initiated by a remote user

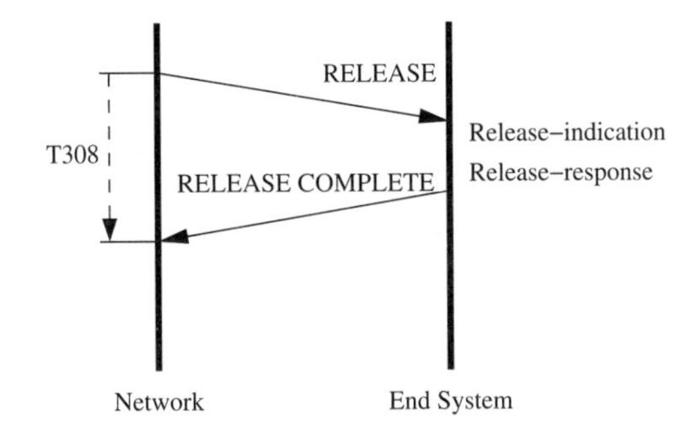

Figure 3.34: Message flow for call clearing by a remote user

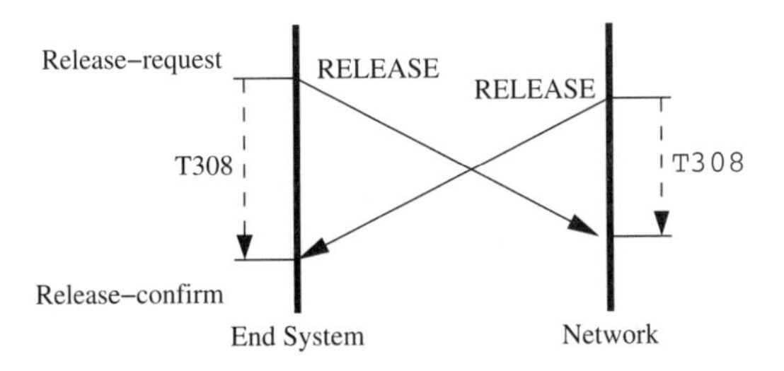

Figure 3.35: Call clearing collision

a link in the network is broken (excavators are good for doing this kind of experiment). The procedures used in this case are exactly the same as for normal call clearing, except that an appropriate cause information may be included in the RELEASE messages.

The cause information element is used to communicate all kinds of more or less useful information between network nodes, especially during call clearing. The structure of this information element is shown in Figure 3.36. The information element is specified in [Q.850] (cause values, location values), [Q.2610] (additions for B-ISDN), [UNI3.1] and [UNI4.0] (some additional cause values).

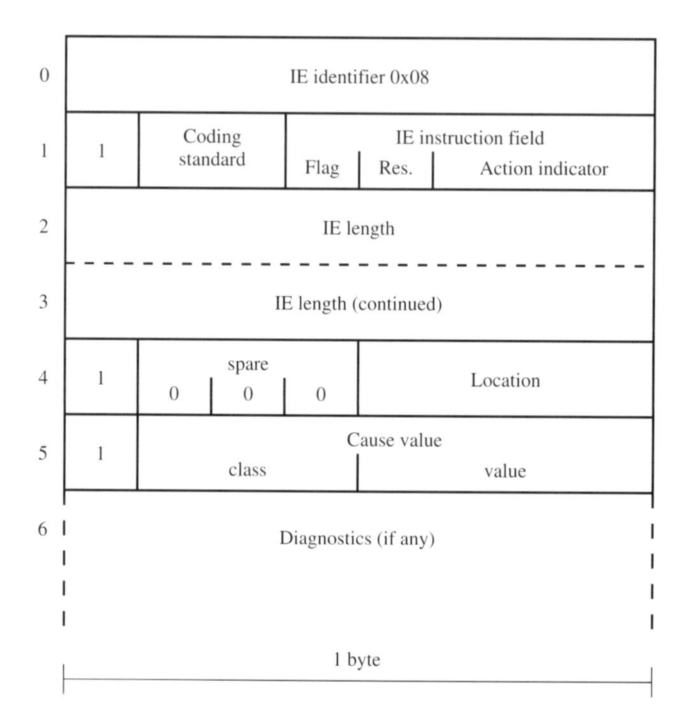

Figure 3.36: UNI cause information element

The location field of the information element describes where the reason for the release of the connection has been detected. The possible values are shown in Table 3.9. This enables a user to see who actually initiated the release and, in the case of unusual cause values, who detected the problem that led to the call clearing.

The cause value consists of two subfields: a class (see Table 3.10) and the cause value in this class. At the time of this writing approximately 80 cause values have been defined (see [Q.850] for most of these values).

Unfortunately the standards are in many cases quite reluctant to specify the exact cause value that should be used in a given situation. This has led to different UNI implementors returning different cause values for the same problem. In some cases however, the standard specifies exact values to use.

Table 3.9: Location values for the cause IE

Coding	Location
0x0	user
0x1	private network serving local user
0x2	public network serving local user
0x3	transit network
0x4	public network serving remote user
0x5	private network serving remote user
0x7	international network
0xa	network beyond interworking point

Table 3.10: Classes of cause values

Coding	Class
0	normal event
1	normal event
2	resource not available
3	service or option not available
4	service or option not implemented
5	invalid message
6	protocol error
7	interworking

The cause information element may include a diagnostic field. The contents and length of this field depend on the cause value: no explicit type information is included. The range of possible diagnostic information goes from a list of information element identifiers for cause value 96 ("Mandatory information element missing"), a connection identifier for value 82 ("Identified channel does not exist") to entire information elements for 2 ("No route to specified transit network") and 22 ("Number changed"). The diagnostic for cause 22 is especially interesting because it includes the new number where the end system can be reached and allows for a limited form of mobility.

The RELEASE message can include up to two cause information elements. This is useful to handle timeouts during call clearing—an additional information element describing the timer expiry can be included in the message.

3.5.7 Status Enquiry Procedure and STATUS Messages

In any state of a call (even in U0/N0) a UNI instance can ask its peer about the peer's vision of the call state. This is done by sending a STATUS ENQUIRY message and starting T322, which is normally 4 seconds. The peer should answer with a STATUS message with a cause information element that contains cause value 30 ("Response to STATUS ENQUIRY") and a call state information element. Upon receipt of this message the UNI stops T322 and does an implementation dependent consistency check. If no STATUS message is received when T322 expires, the STATUS ENQUIRY message may be retransmitted one or several times (again this is the implementor's decision). If after all retransmissions no answer is received, the call is cleared with a cause value of 41 ("Temporary failure"). The following is a trace of a status enquiry for a connection in the active state with one retransmission of the enquiry message.

```
1   E ⇒ S   0.000 uni cref={you,41} mtype=stat_enq mlen=0
2   E ⇒ S   4.233 uni cref={you,41} mtype=stat_enq mlen=0
3   E ⇐ S   4.301 uni cref={me,41} mtype=status mlen=11
4                 callstate={10-Active}
5                 cause={loc=private_network_serving_local_user,
6                    cvalue=response_to_status_enquiry,class=normal_event}
```

The "implementation dependent" procedure to check the compatibility of the peer's call state with the own vision of the call could, in the simplest case, either always report compatibility of the call states, or simply compare them. The problem with the second variant is that it fails if the status enquiry is invoked while a retransmission timer like T303 or T308 is running. In this case the call states of both peers are different, but this is expected. If, for example, the first SETUP was lost, the local call state will be U1 (call initiated), but the peer call state will still be N0. These are different, but obviously compatible. The standard, however, requires three things from the procedure:

1. If the STATUS message indicates a state other than U0/N0, but in the UNI protocol instance no such call exists (i.e. the call is in U0/N0), a RELEASE COMPLETE message should be sent. This forces the peer to release the call that is unknown to the receiving protocol instance.
2. A STATUS message which indicates states other than U0/N0 received in U11/N11 or U12/N12 (these are the releasing states) is effectively ignored.
3. If a STATUS message that indicates U0/N0 is received in any other state (this means that the receiver has a call that is unknown to the peer), the call is locally released.

STATUS messages can be sent not only upon enquiry, but also in exceptional situations (and, on user request, any time). This is especially useful to handle unimplemented optional features of UNI (see Section 3.9).

3.6 Point-to-Multipoint Calls

Point-to-multipoint calls are used to establish uni-directional connection trees from one source end system, called *root*, to a set of destination end systems called *leaves*. Uni-directional in this context means that although the VPI/VCI values are allocated for both directions of the connection and connection segments, data can flow only from the root to the leaves. The backward cell rates in the SETUP messages must be specified as zero. Signalling for the establishment of point-to-multipoint connections is defined in [Q.2971] which also contains numerous corrections of [Q.2931]. According to the ITU-T standards, point-to-multipoint connections are an optional part of ATM signalling for end systems as well as for switches; according to UNI4.0, which builds on [Q.2971], point-to-multipoint support is mandatory for switches.

For switches point-to-multipoint signalling support is not sufficient—they must also support cell duplication to split the traffic when the multipoint tree at this point involves multiple output ports. For switches that do not support point-to-multipoint (e.g. switches that conform to [Q.2931]) the signalling specification supports some kind of workaround to use these switches in point-to-point mode only. In the same manner end systems which are not capable of point-to-multipoint signalling nevertheless can be leaves in such a connection. An example of a point-to-multipoint connection is shown in Figure 3.37.

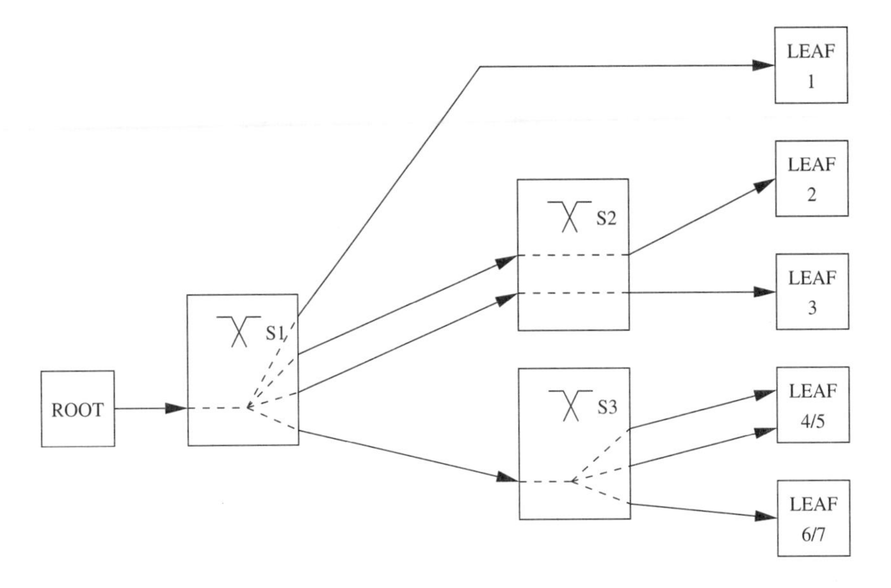

Figure 3.37: Point-to-multipoint example

In this example one root node is connected to seven leaves, where leaves 4 and 5 and leaves 6 and 7 live in the same end systems (from the UNI's point of view). This can happen, for example, if these end systems are interworking units or border switches of private ATM networks. There are two switches with point-to-multipoint support (S1 and S3) and one switch without (S2). Because S2 is unable to duplicate the traffic for leaves 2 and 3, this is done by S1. S2 does not know even, that the two connections belong to the same call. In the same way the end system hosting leaves 4 and 5 does not support point-to-multipoint signalling; the end system with leaves 6 and 7 does. In the first case the switch will duplicate the traffic and make two normal point-to-point SETUPs to the end system; in the second case the end system has to do the traffic duplication.

Once a point-to-multipoint call is created, additional leaves can be added to the call and existing leaves can be removed from it (dropped). Creating a call with two parties starts with the same steps as creating a point-to-point call. The only difference is in the contents of some of the exchanged messages (see Figures 3.15 to 3.18). Once the connection to the first leaf is established the second leaf will be added with the following steps:

1. The root node sends an ADD PARTY message containing the address of the new leaf to the network. This message is forwarded across the network to the called system. As long as the message travels along an existing branch of the call, it is sent as an ADD PARTY. As soon as the next node, which would either be not capable of point-to-multipoint signalling or a

node, previously not involved in this call (i.e. a new branch must be forked from the tree), the ADD PARTY is converted to a SETUP (Figure 3.38). Note that in the case of public signalling (B-ISUP) the actual message names differ from the names used here. To set up a connection to the first party at a given endpoint, a SETUP message is used. The called end system may acknowledge the receipt of the SETUP with a CALL PROCEEDING.

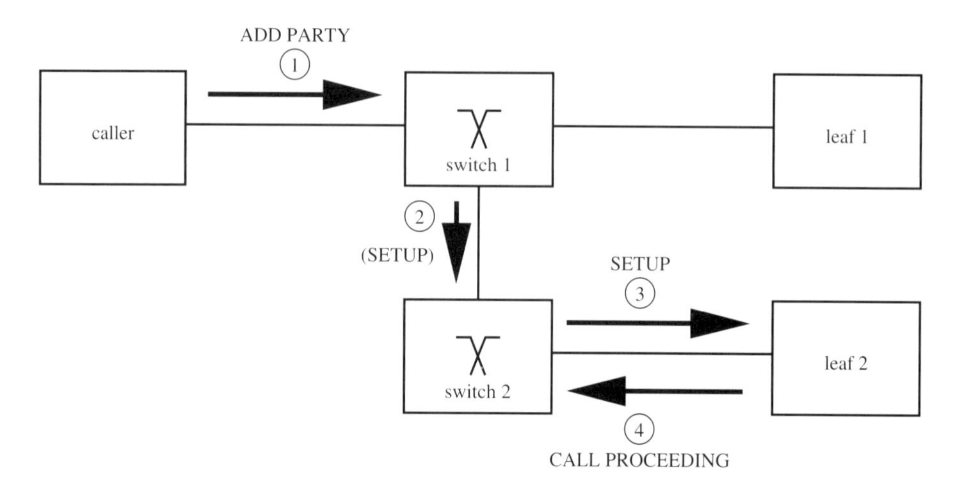

Figure 3.38: Adding a new party, step 1

2. The called leaf may send an ALERTING indication back to the root node to indicate that it is alerting the user or starting applications. This message is forwarded to the calling system and converted to a PARTY ALERTING message on the way (Figure 3.39).

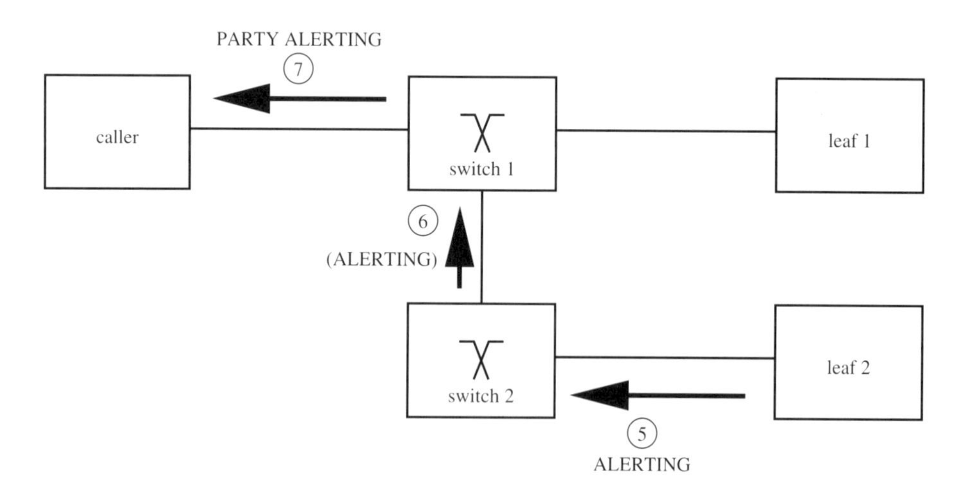

Figure 3.39: Adding a new party, step 2

3. When the called user has accepted the call, the end system sends a CONNECT message to the network. This message is sent back to the root node and converted to an ADD PARTY ACKNOWLEDGE on the way. The CONNECT message is acknowledged with a CONNECT ACKNOWLEDGE to the leaf. When the ADD PARTY ACKNOWLEDGE has been received by the root, the new branch is fully established (Figure 3.40).

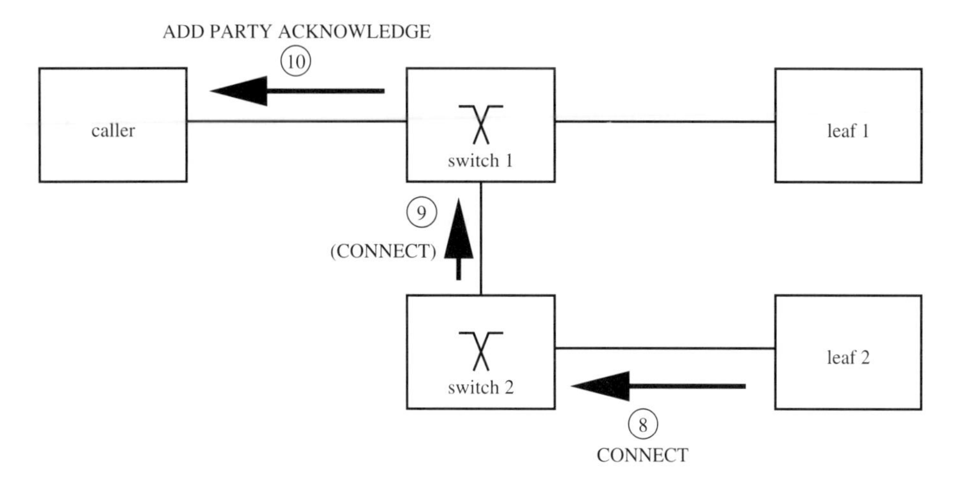

Figure 3.40: Adding a new party, step 3

Dropping a party from a point-to-multipoint tree can be initiated by both the root node and the leaf. The leaf does so by sending either a DROP PARTY message or a RELEASE message (if this is the only leaf at this endpoint). The root drops a party by sending a DROP PARTY or a RELEASE if the last leaf has to be dropped or the entire call has to be cleared. Root-initiated dropping of a party from a point-to-multipoint tree involves the following steps:

1. The root sends a DROP PARTY message to the switch (see Figure 3.41). This message is forwarded to the leaf and may be converted to a RELEASE along the path (Figure 3.42).

   The switch acknowledges the receipt of the DROP PARTY with a DROP PARTY ACKNOWLEDGE. This message has only local significance, which means that the resources for this leaf at this interface of the switch have been released. It does not mean that the remote party has been dropped already. With receipt of the acknowledge the party dropping procedure at the leaf is finished and all local resources are released.
2. The leaf acknowledges receipt of the RELEASE with a RELEASE COMPLETE after releasing all resources that were allocated to this connection (Figure 3.43).

In the next sections the operation of the protocol for the establishment of new parties and the dropping of parties will be described in detail.

3.6.1 End Point References and Party States

Because all messages for a given point-to-multipoint call at a given UNI use the same call reference, even if they belong to different parties, an additional information element needs to

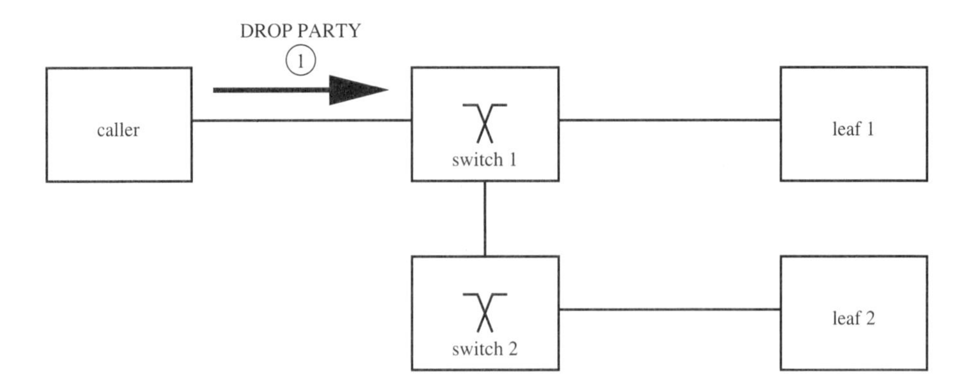

Figure 3.41: Dropping a party, step 1

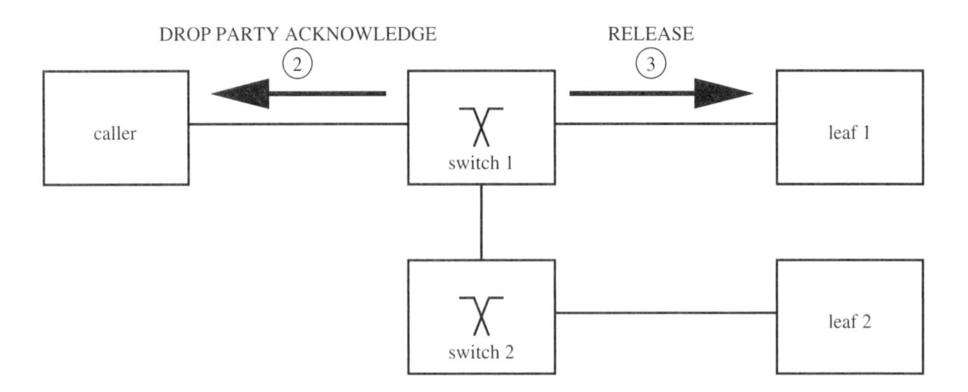

Figure 3.42: Dropping a party, step 2

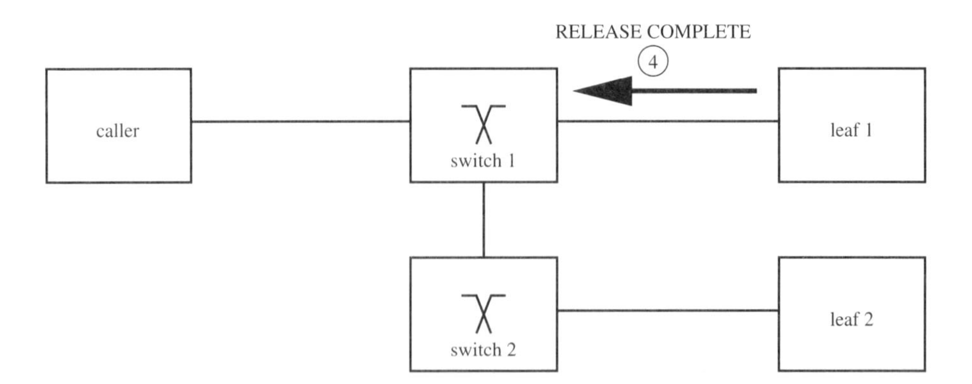

Figure 3.43: Dropping a party, step 3

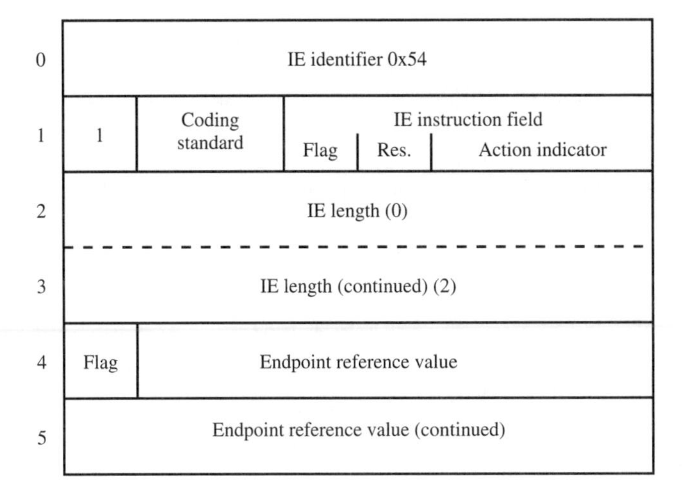

Figure 3.44: Endpoint reference IE

be used to distinguish messages for different parties. The endpoint reference IE contains a 15-bit integer value that uniquely identifies a leaf in a given point-to-multipoint call. As compared with the call reference, the endpoint reference has end-to-end meaning, which means that, it does not change when a message travels between the root and the leaf. Endpoint reference values are allocated by the root node. The value for the first leaf (which is connected when the call itself is established) should be zero. The format of the IE is shown in Figure 3.44.

The endpoint reference IE also includes a flag, which indicates, whether the given message is sent by the creator of the endpoint reference (i.e. the root node, flag 1) or has come from the leaf (flag 0).

To handle point-to-multipoint calls UNI stacks use an additional party object to describe the state of each party for a given call (see Section 3.10). For each party there is an associated party state (see Table 3.11).

Table 3.11: UNI party states

Code	Name	State
`0x00`	PU0/PN0	Null
`0x01`	PU1/PN1	Add party initiated
`0x04`	PU2/PN2	Party alerting delivered
`0x06`	PU3/PN3	Add party received
`0x07`	PU4/PN4	Party alerting received
`0x0b`	PU5/PN5	Drop party initiated
`0x0c`	PU6/PN6	Drop party received
`0x0a`	PU7/PN7	Active

As in the case of call states (see Section 3.4) each state has two names: PUX for the user side of the UNI and PNX for the network side. PU7/PN7 are the active states where actual data is delivered; PU1/PN1 to PU4/PN4 are the party establishment states and PU5/PN5 and PU6/PN6 are the drop states.

The state of a party may be communicated to the peer of the UNI via a STATUS message including an endpoint reference IE and an endpoint state IE. The endpoint state IE is shown in Figure 3.45.

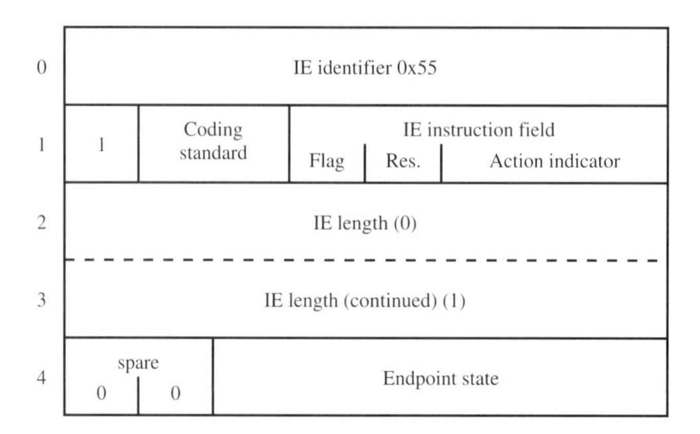

Figure 3.45: Endpoint state information element

3.6.2 Establishment of the First Party

For a point-to-multipoint call the call (and thus the connection to the initial party) is established in exactly the same way as a point-to-point call. The only difference is in the SETUP message:

- The bearer capability information element must request a configuration of point-to-multipoint instead of point-to-point.
- Backward cell rates in any of the traffic descriptors must be zero.
- The type of bearer indicated in the bearer capability information element may not be ABR. No ABR specific information elements may be included in the SETUP.

Additionally a party FSM is created for the first leaf, and instead of the optional timer T301 the (optional) timer T397 is used.

Refer to Section 3.5 for information, on how a point-to-point connection is established.

3.6.3 Adding a Leaf

New leaves can be added to a point-to-multipoint call only from the root node (see Section 3.6.7 for the leaf-initiated join feature). The messages for adding a new leaf use the same call reference as for the original call setup. Individual leaves are identified by the endpoint reference information element. These references are generated by the root node, usually starting at zero for the initial party.

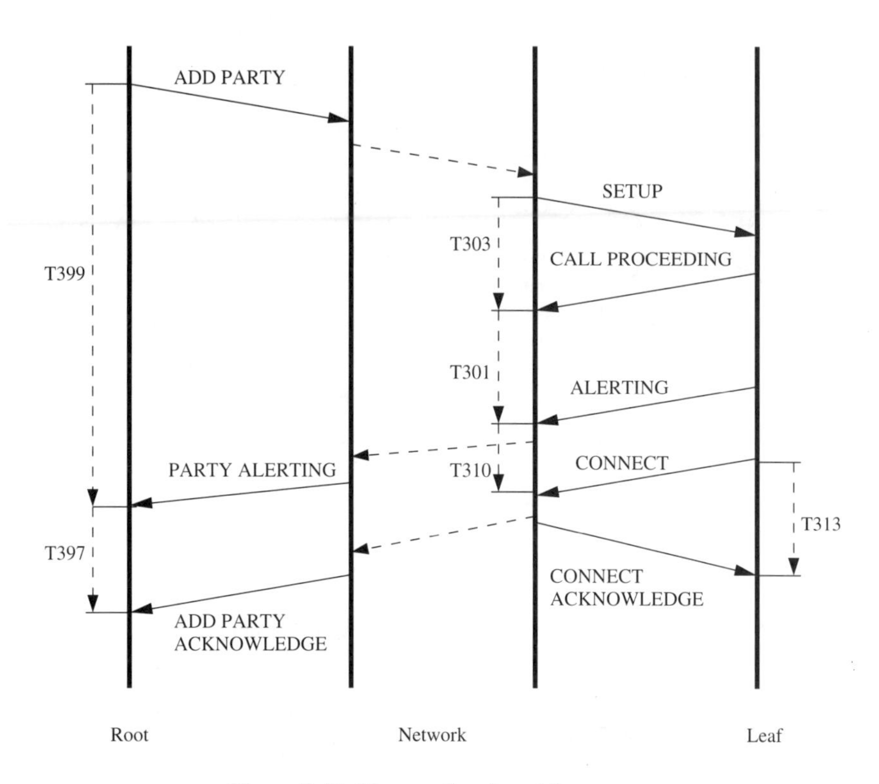

Figure 3.46: Message flow for adding a party

Adding a new leaf is initiated by sending an ADD PARTY message from the root node to the network and starting timer T399 (see Figure 3.46). The party FSM that is created to handle the party enters state PU1 (add party initiated). On the network side a new party object is created upon receipt of the ADD PARTY message and the state of the FSM is set to PN2 (add party received). The message is then forwarded in the direction of the leaf.

The ADD PARTY message contains a subset of the information elements that are allowed for a SETUP. No traffic-related IEs are allowed because the traffic characteristics for all leaves are the same and are established at the time when the call was originally created and the first leaf contacted. Negotiation of any parameters is also not allowed, so there is only on BLLI.

The ADD PARTY message may travel in the direction of the addressed leaf along already existing branches of the point-to-multipoint call. Usually at one point in the network a new branch has to be forked of from the point-to-multipoint tree. At this node the ADD PARTY is translated into an appropriate SETUP message. The main difference between an ADD PARTY and a SETUP from the resource point of view is that an ADD PARTY simply advises, that later in the network there will be another leaf getting the same traffic, while a SETUP will allocate a new virtual channel or path. This means that if an intermediate switch decides that the traffic now has to be duplicated, it must transform the ADD PARTY into a SETUP. This

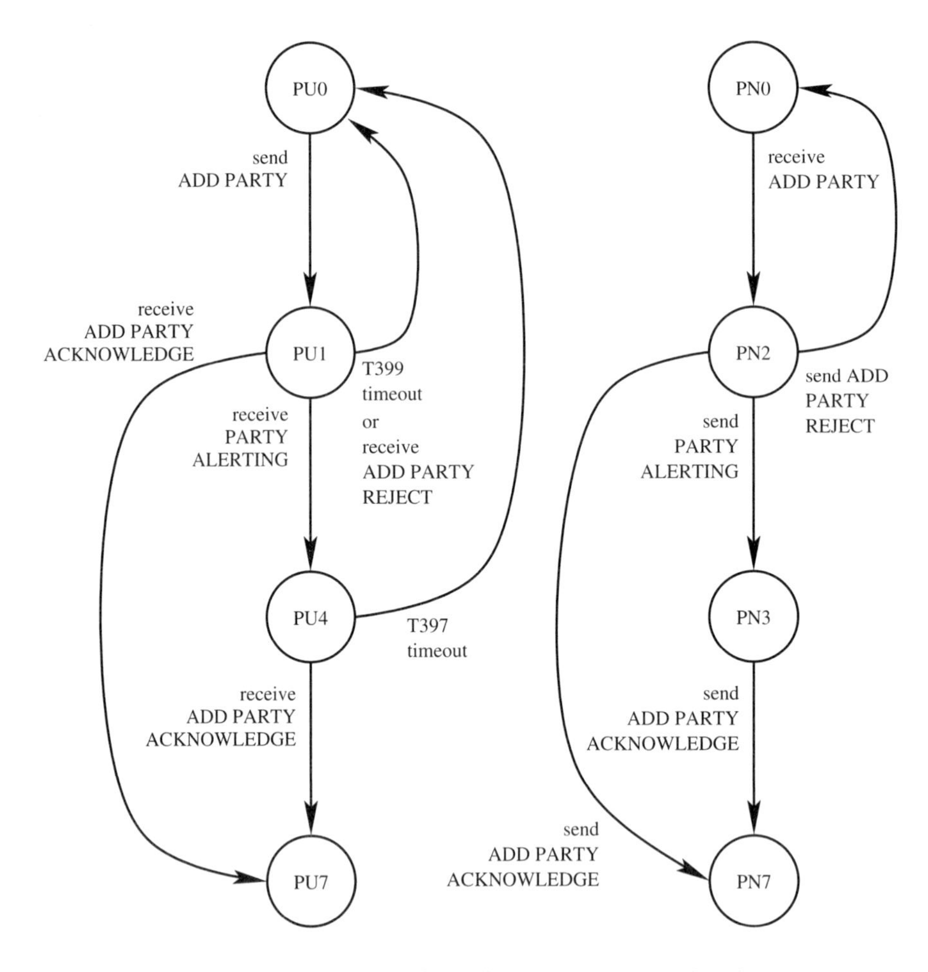

Figure 3.47: State diagram for adding a party at the root interface

also means that the information that was contained in the original SETUP, but is missing in the ADD PARTY, has to be cached at each intermediate node. However, not all information has to be cached—information elements that were used for negotiation, like repeated BLLIs, need not be inserted into the SETUP. Also information elements with end-to-end significance like Generic Identifier Transport are not cached at the network nodes but taken from the ADD PARTY. Of course, there are cases when an ADD PARTY never gets transformed before leaving the given network. If there are multiple leaves behind a public UNI in a private network (if we look from the viewpoint of the public network), the traffic duplication point will be in the private network and the ADD PARTY message will be sent over the public UNI to the private network (see later in this section).

When the leaf receives the SETUP the usual point-to-point establishment procedures apply, except that negotiation procedures do not apply (only for the initial leaf which is identified by an endpoint reference of zero negotiation is possible). The leaf may respond with a CALL PROCEEDING and an ALERTING. The ALERTING is sent back in the direction of the root node and gets transformed into a PARTY ALERTING. Upon receipt of the PARTY ALERTING the root stops timer T399 for this party and starts T397 instead. The network party is now in state PN3 (party alerting delivered) and the user party in state PN4 (party alerting received).

When the called leaf accepts the call it sends a CONNECT message to the network. This message is sent back to the root and transformed into an ADD PARTY ACKNOWLEDGE which finally reaches the root node. Upon receipt of this message the root node stops all party timers and enters PU7 (active). The network side party state is PN7. At this point the new leaf is fully connected to the call.

We now look at some traces. We assume that the call has been established already and we want to add another leaf.

```
1   E ⇒ S   31.247 uni cref={you,42} mtype=add_party mlen=57
2                  called={type=unknown,plan=aesa,addr=plan=aesa,addr=spock}
3                  calling={type=unknown,plan=aesa,addr=kirk}
4                  epref={type=local,flag=me,idval=1}
5   E ⇐ S   31.680 uni cref={me,42} mtype=party_alerting mlen=7
6                  epref={type=local,flag=notme,idval=1}
7   E ⇐ S   31.721 uni cref={me,42} mtype=add_party_ack mlen=7
8                  epref={type=local,flag=notme,idval=1}
9   E ⇒ S   33.112 uni cref={you,42} mtype=stat_enq mlen=7
10                 epref={type=local,flag=me,idval=1}
11  E ⇐ S   33.190 uni cref={me,42} mtype=status mlen=23
12                 callstate={10-Active}
13                 cause={loc=private_network_serving_local_user,
14                    cvalue=response_to_status_enquiry,class=normal_event}
15                 epref={type=local,flag=notme,idval=1}
16                 epstate={state=active}
```

In lines 1–3 the root node sends an ADD PARTY message the network. The leaf user obviously has decided to send an ALERTING message, so the root node receives a PARTY ALERTING in lines 5 and 6. The called user accepts the call. This shows up at the root node as an ADD PARTY ACKNOWLEDGE in lines 7 and 8. At the end we invoke the status enquiry procedure to verify the state of the call and the party—the call is in state U10/N10 (active) and the party is in state PU7/PN7 (active).

As we have mentioned previously, the traffic duplication point may actually be behind the destination UNI (as seen from the root node). One example is a private network with more than one leaf (see Figure 3.48). In this example a root node in the public network is connected to three leaves in a private network. For performance reasons the traffic should be duplicated in the private switch. That means that the second and the third leaves have to be added at the public UNI via ADD PARTY messages, not by transforming the ADD PARTYs into SETUPs. How this works is shown in Figure 3.49).

The ADD PARTY message is sent to the user side of the UNI and timer T399 is started on the switch. The user side may respond with a PARTY ALERTING in which case T399 is stopped and optionally T397 started. The PARTY ALERTING will be forwarded to the root node in this case. When the leaf has accepted the connection an ADD PARTY ACKNOWLEDGE is received from the user side of the UNI. The network stops all its timers and forwards the message to the root. The party is now in state PU7/PN7. Note that the state diagram is the same as for the root UNI.

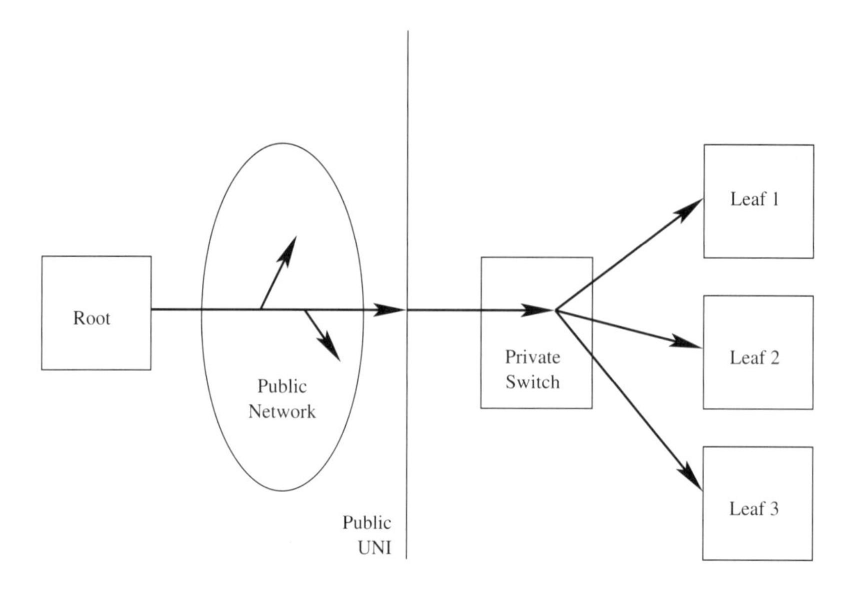

Figure 3.48: More than one leaf behind a destination UNI

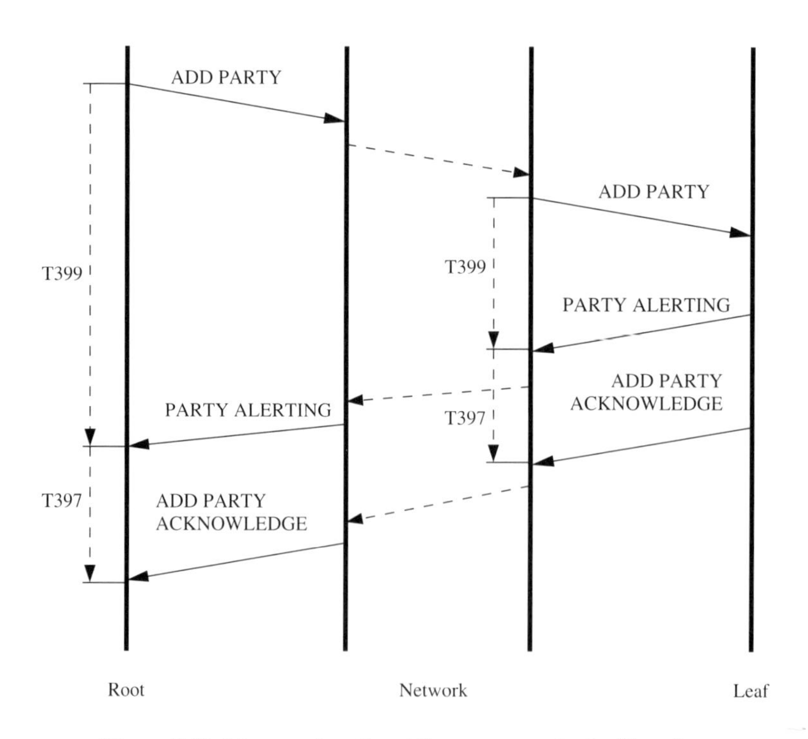

Figure 3.49: Message flow for adding a party at the leaf interface

The following is a trace taken on the destination system of the call in the previous example (spock). A second leaf is added to spock. The actions on the root side are almost identical to those in the previous trace, except that the endpoint reference is now 2. Note that the call reference is different, because this is the destination interface of the call.

```
1   E ⇐ S   67.612 uni cref={you,343} mtype=add_party mlen=57
2                  called={type=unknown,plan=aesa,addr=plan=aesa,addr=spock}
3                  calling={type=unknown,plan=aesa,addr=kirk}
4                  epref={type=local,flag=me,idval=2}
5   E ⇒ S   67.860 uni cref={me,343} mtype=party_alerting mlen=7
6                  epref={type=local,flag=notme,idval=2}
7   E ⇒ S   68.124 uni cref={me,343} mtype=add_party_ack mlen=7
8                  epref={type=local,flag=notme,idval=2}
9   E ⇒ S   70.301 uni cref={me,343} mtype=stat_enq mlen=7
10                 epref={type=local,flag=notme,idval=2}
11  E ⇐ S   70.312 uni cref={you,343} mtype=status mlen=23
12                 callstate={10-Active}
13                 cause={loc=private_network_serving_local_user,
14                    cvalue=response_to_status_enquiry,class=normal_event}
15                 epref={type=local,flag=me,idval=2}
16                 epstate={state=active}
```

In this trace the network sends an ADD PARTY in lines 1–4. The end system responds with a PARTY ALERTING in lines 2–6 and later with an ADD PARTY ACKNOWLEDGEMENT in lines 7 and 8. Then the end system verifies the party and call states by means of a STATUS ENQUIRY (lines 9 and 10), which is answered by the local switch in lines 11 to 16. The STATUS indicates, that the call and the party are both in their active states.

Of course, as for a point-to-point call, the PARTY ALERTING (or ALERTING at the destination UNI) messages are optional. In this case the establishment of a new leaf at the root node is shorter (same call as previous):

```
1   E ⇒ S   210.080uni cref={you,42} mtype=add_party mlen=57
2                  called={type=unknown,plan=aesa,addr=plan=aesa,addr=spock}
3                  calling={type=unknown,plan=aesa,addr=kirk}
4                  epref={type=local,flag=me,idval=6}
5   E ⇐ S   210.421uni cref={me,42} mtype=add_party_ack mlen=7
6                  epref={type=local,flag=notme,idval=6}
7   E ⇒ S   214.027uni cref={you,42} mtype=stat_enq mlen=7
8                  epref={type=local,flag=me,idval=6}
9   E ⇐ S   214.088uni cref={me,42} mtype=status mlen=23
10                 callstate={10-Active}
11                 cause={loc=private_network_serving_local_user,
12                    cvalue=response_to_status_enquiry,class=normal_event}
13                 epref={type=local,flag=notme,idval=6}
14                 epstate={state=active}
```

Here the ADD PARTY in lines 1–4 is directly answered with the ADD PARTY ACKNOWLEDGE (lines 5 and 6). The status enquiry procedure verifies that the call and the party are in the active state.

3.6.4 Rejecting an ADD PARTY Request

A request to add a party can be rejected either by the network or by the user in the same manner as the establishment of a point-to-multipoint call is rejected. The network may reject the new party, for example because of a wrong ADD PARTY message, insufficient resources or because the new leaf is non-existent. The leaf itself may reject the adding because the real leaf does not exist (in the case that the destination UNI is a public UNI where a private network is connected) or because the addressed application does not exist or resources are not available.

Let us look at what will happen if we address a non-existent leaf. The address `foobar` in the following example has been set to a non-existent address. The trace at the root interface is as follows:

```
1  E ⇒ S   29.701 uni cref={you,51} mtype=add_party mlen=57
2                 called={type=unknown,plan=aesa,addr=plan=aesa,addr=foobar}
3                 calling={type=unknown,plan=aesa,addr=kirk}
4                 epref={type=local,flag=me,idval=1}
5  E ⇐ S   29.789 uni cref={me,51} mtype=add_party_reject mlen=6
6                 cause={loc=private_network_serving_local_user,
7                    cvalue=no_route_to_destination,class=normal_event}
```

The root node sends an ADD PARTY with the unknown address to the network. Because the switch cannot find a route to the end system, it rejects the ADD PARTY with an ADD PARTY REJECT and an appropriate cause value.

The sequence of events is exactly the same when the remote user rejects the ADD PARTY request. He does so by sending a RELEASE COMPLETE message in response to the SETUP to the network. As the RELEASE COMPLETE travels back to the root node it gets converted to an ADD PARTY REJECT with the cause information element preserved.

3.6.5 Dropping a Leaf

A leaf can be dropped from a point-to-multipoint call either by the root node, the leaf itself or the network. The root node and the network can also drop all leaves at once by clearing the entire call. We will not look at network dropping because it is not different from the other variants except for the cause code and cause location value.

The network and user side party states for dropping a leaf by the root node are shown in Figure 3.50. The corresponding message flows at the root and leaf interface are shown in Figure 3.51.

The root node drops a leaf by sending a DROP PARTY message to the network indicating the endpoint reference of the leaf to be dropped. It then starts timer T398 and enters state PU5 (drop party initiated). When the network receives this message it initiates remote party dropping towards the leaf and responds to the user with a DROP PARTY ACKNOWLEDGE message. The root then node stops T398 and destroys the party object.

If the leaf to be dropped is the last leaf, which is in the active state (or one of the establishment states), the sequence of actions is different from the one described. There are two variants of what can happen: the root node can release the entire call by sending a RELEASE or the network will respond with a RELEASE instead of the DROP PARTY ACKNOWLEDGE. The second variant is shown in Figure 3.51. The root node sends a DROP

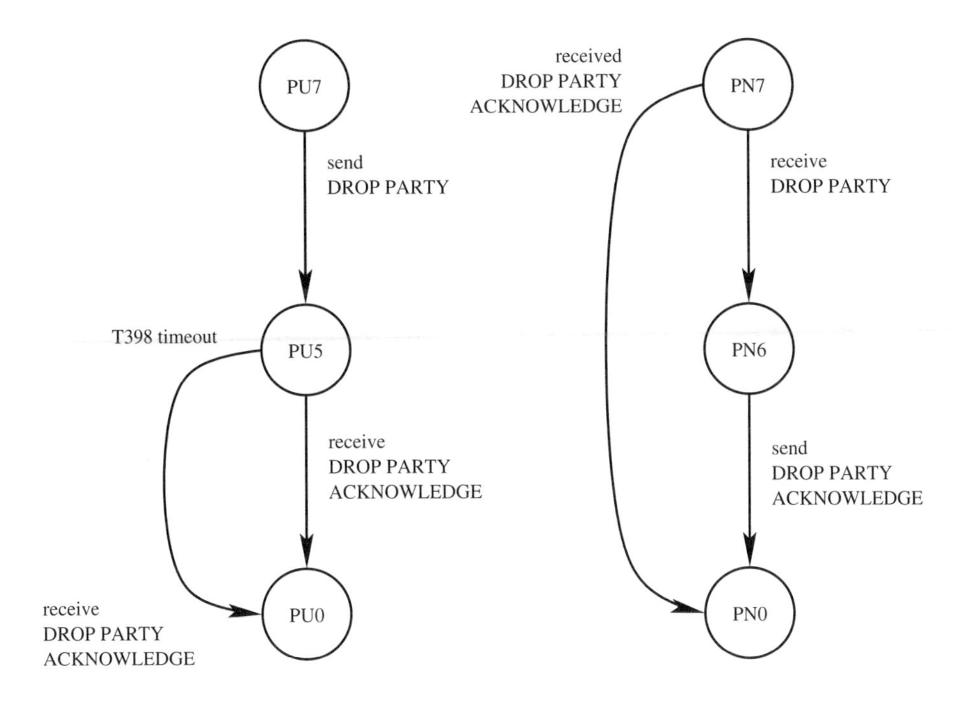

Figure 3.50: State machine for dropping a leaf

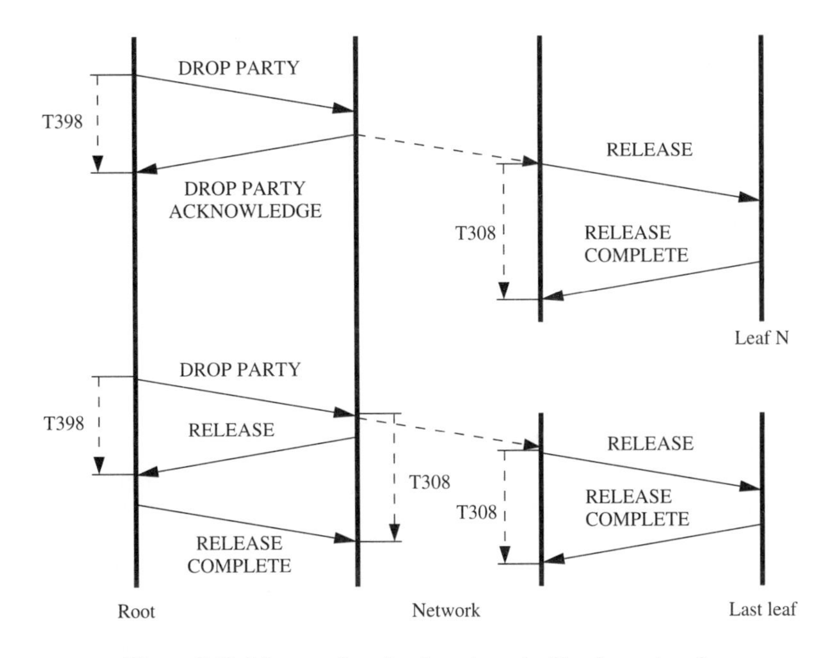

Figure 3.51: Message flow for dropping a leaf by the root node

PARTY to the network to drop the last leaf. The network initiates remote dropping towards the leaf and determines that this is the last leaf in this call. It then starts call clearing on the root interface by sending a RELEASE message to the root node. The root node in turn stops timer T398 which it started when sending the DROP PARTY, releases all local resources for the call and responds with a RELEASE COMPLETE. At this point the entire call including the call reference has been released.

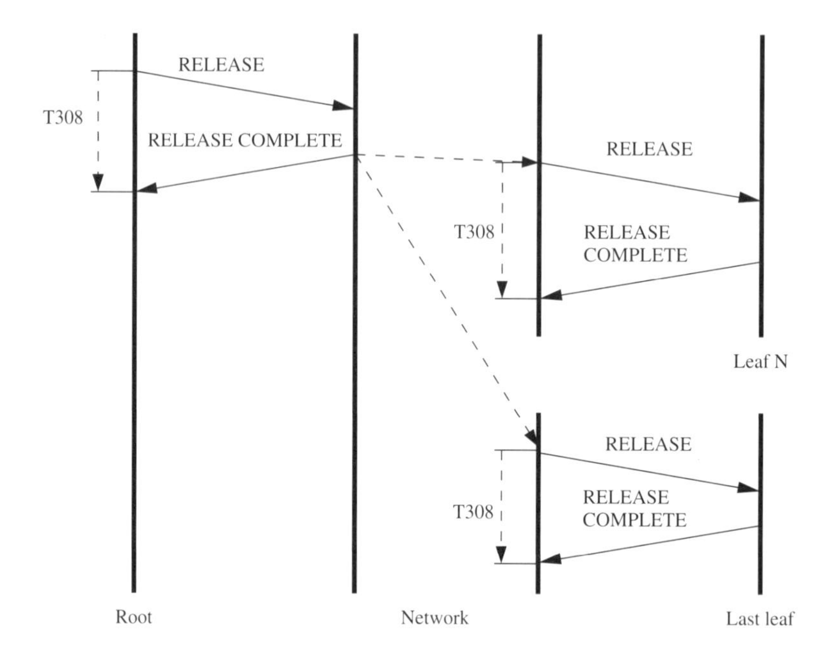

Figure 3.52: Dropping all parties from a call

A root node can also drop all parties at the same time and release all resources by sending a RELEASE message to the network (Figure 3.52). The network, when receiving this message, initiates party dropping towards all parties, releases resources and answers the root node with a RELEASE COMPLETE.

A leaf can also drop itself by sending a RELEASE to the network. The network forwards this message towards the root node, transforming it to a DROP PARTY on the way and responds to the leaf with a RELEASE COMPLETE (see Figure 3.53).

Let us look at some examples. The first trace shows a point-to-multipoint call with three leaves. The first two leaves are dropped by the root node and then the entire call is released:

```
1  E ⇒ S   0.000 uni cref={you,34} mtype=drop_party mlen=13
2                cause={loc=user,
3                   cvalue=normal,_unspecified,class=normal_event}
4                epref={type=local,flag=me,idval=2}
5  E ⇐ S   0.046 uni cref={me,34} mtype=drop_party_ack mlen=13
6                cause={loc=private_network_serving_local_user,
7                   cvalue=normal,_unspecified,class=normal_event}
```

```
 8                    epref={type=local,flag=notme,idval=2}
 9   E ⇒ S    8.224   uni cref={you,34} mtype=drop_party mlen=13
10                    cause={loc=user,
11                       cvalue=normal,_unspecified,class=normal_event}
12                    epref={type=local,flag=me,idval=3}
13   E ⇐ S    8.310   uni cref={me,34} mtype=drop_party_ack mlen=13
14                    cause={loc=private_network_serving_local_user,
15                       cvalue=normal,_unspecified,class=normal_event}
16                    epref={type=local,flag=notme,idval=3}
17   E ⇒ S    14.712  uni cref={you,34} mtype=release mlen=6
18                    cause={loc=user,
19                       cvalue=normal,_unspecified,class=normal_event}
20   E ⇐ S    14.806  uni cref={me,34} mtype=rel_compl mlen=6
21                    cause={loc=private_network_serving_local_user,
22                       cvalue=normal,_unspecified,class=normal_event}
```

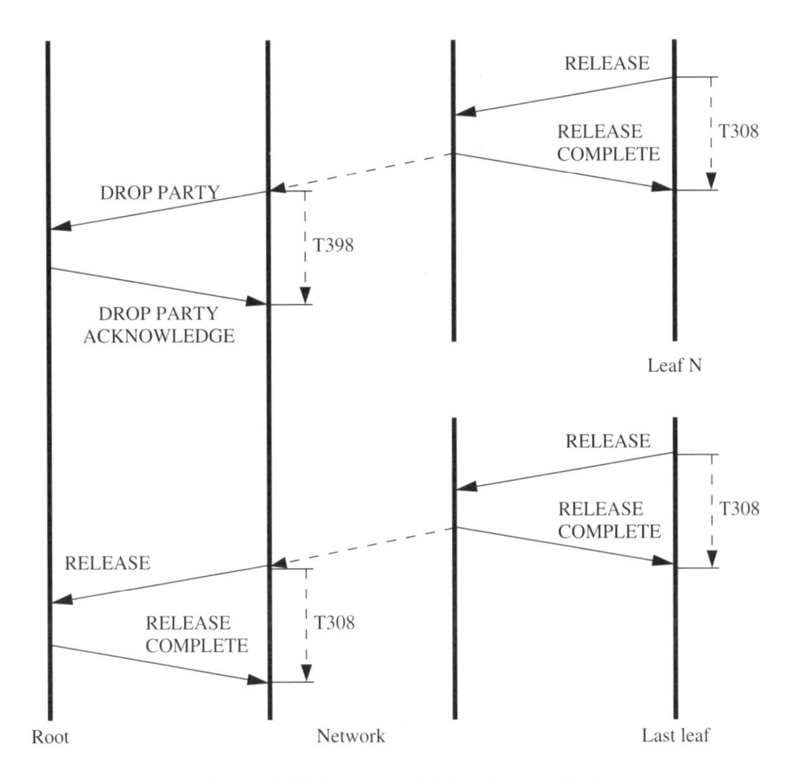

Figure 3.53: Dropping initiated by the leaf

The first DROP PARTY is sent by the end system in lines 1–4. The answer appears in lines 5–8. Then the party with endpoint reference 3 is dropped in lines 9–16. Finally the end system decides to release the entire call in lines 17–19 by sending a RELEASE. This is answered by the switch in lines 20–22.

The following trace shows dropping initiated by the leaves.

```
 1  E ⇐ S   0.000 uni cref={me,42} mtype=drop_party mlen=6
 2                cause={loc=private_network_serving_remote_user,
 3                   cvalue=destination_out_of_order,class=normal_event}
 4  E ⇒ S   0.212 uni cref={you,42} mtype=drop_party_ack mlen=6
 5                cause={loc=user,
 6                   cvalue=normal,_unspecified,class=normal_event}
 7  E ⇐ S  20.903 uni cref={me,42} mtype=drop_party mlen=6
 8                cause={loc=user,
 9                   cvalue=normal,_unspecified,class=normal_event}
10  E ⇐ S  25.026 uni cref={me,42} mtype=drop_party_ack mlen=15
11                cause={loc=user,
12                   cvalue=normal,_unspecified,class=normal_event}
13                cause={loc=private_network_serving_local_user,
14                   cvalue=recovery_on_timer_expiry,class=protocol_error,timer=T398}
15  E ⇐ S  42.402 uni cref={me,42} mtype=release mlen=6
16                cause={loc=private_network_serving_local_user,
17                   cvalue=normal,_unspecified,class=normal_event}
18  E ⇒ S  42.872 uni cref={you,42} mtype=rel_compl mlen=6
19                cause={loc=private_network_serving_local_user,
20                   cvalue=normal,_unspecified,class=normal_event}
```

Dropping the first leaf was generated by unplugging the cable from the switch. After the SAAL times out and cannot be re-established, leaf dropping is initiated towards the root. At the root this appears as a DROP PARTY message with cause number 27 ("Destination out of order") in lines 1–3. The acknowledge from the end system comes in lines 4–6. The second leaf drops itself normally. In lines 7–9 the DROP PARTY is received by the end system. This time the end system fails to respond, so timer T398 expires after approximately 4 seconds and a DROP PARTY ACKNOWLEDGE is sent by the switch. When the last leaf releases its branch of the call, the network clears the entire call towards the root by transmitting a RELEASE in lines 15–17. The end system acknowledges the RELEASE with a RELEASE COMPLETE in lines 18–20.

3.6.6 *Party Status Enquiry Procedure*

In Section 3.5.7 we saw how a UNI protocol instance can check the state of its peer entity with regard to a given call. In the same way it is possible to enquire the state of a point-to-multipoint call and the state of each party of such a call.

The call state of a point-to-multipoint call is questioned by sending a STATUS ENQUIRY over the UNI indicating the call reference of the call. The peer will answer with a STATUS message containing cause code 30 ("Response to STATUS ENQUIRY") and a call state information element. This is no different from the point-to-point case.

To enquire the state of a party the sender of the STATUS ENQUIRY must include an endpoint reference information element in the STATUS ENQUIRY message. The responding side will then include an endpoint state information element in addition to the call state information element in the STATUS message.

One problem, however, is that timer T322, which is used to detect lost STATUS ENQUIRY and STATUS messages, is a call timer, not a party timer. This means that at any given time only one STATUS ENQUIRY message can be outstanding. This poses a problem if the state of multiple parties is to be checked. This happens, for example, if the SAAL has failed and could be re-established. In this case the state of all parties must be checked with the peer. This is done by maintaining a queue of pending status enquiry requests for each call in a UNI instance. As soon as a STATUS message is received in response to a STATUS ENQUIRY the next STATUS ENQUIRY may be sent.

3.6.7 Leaf-Initiated Join

We have seen in the previous sections that in a point-to-multipoint call new leaves are always added by the root node. This works well as long as there are not too many leaves to handle and all the leaves are known. There are, however, situations where this may not be the case. An example of this are broadcasting services, where not all the receivers are known in advance, nor can they all be handled by one node because of their number. What is needed in this case is the ability for a leaf to say: I want to be connected to that call. This feature is the intention of the leaf-initiated join (LIJ) capability that was added to UNI4.0.

This feature uses two additional messages: LEAF SETUP REQUEST and LEAF SETUP FAILURE. Additionally three new information elements are used: the LIJ call identifier, the LIJ parameter IE and the leaf sequence number IE.

There are two types of LIJ calls:

1. *Root-prompted join.* In this case the leaf sends a request via UNI to the root node of the call indicating that it wants to be added to the call. The root adds the leaf by means of the normal point-to-multipoint procedures. This procedure allows the root to check the authorisation of the leaf to be added to the call. It also puts the burden of signalling on the root if there are many leaves in the call.
2. *Leaf-prompted join without root notification.* In this mode of operation the leaf sends a request over the UNI to the network to be added to a point-to-multipoint call. The adding of the leaf is entirely handled by the network—the root is not notified. When the call does not yet exist, the request is forwarded to the root node which can establish the call to the requesting leaf as the first leaf.

The user side of the UNI uses an FSM to handle the LIJ requests, which is different from the normal call state machine. The LIJ machine has two states:

- *Null.* The state machine does not exist.
- *Leaf-Setup initiated.* A LEAF SETUP REQUEST has been sent and the leaf is waiting for a SETUP, ADD PARTY, LEAF SETUP FAILURE or timer T331 to expire.

The state machine uses an additional timer: T331. This has a default value of 60 seconds and can optionally be restarted on expiry (with a retransmission of the LEAF SETUP REQUEST). Figure 3.54 shows the state machine.

When the UNI user requests to be added as a leaf to a point-to-multipoint connection a FSM is created. It sends a LEAF SETUP REQUEST, starts timer T331 and enters state LIJ1. Then one of four things can happen:

- **SETUP receipt.** If a SETUP is received it means that the network or root a going to add the user to the requested point-to-multipoint call. Usually SETUP procedures apply (see Section 3.6). The timer can be stopped and the FSM for the LEAF SETUP REQUEST can be destroyed.

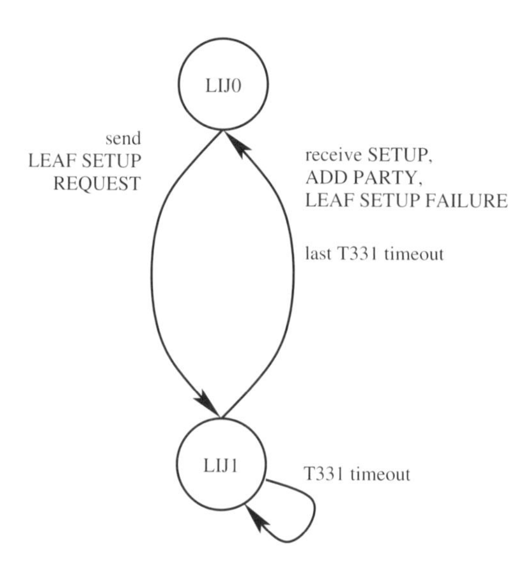

Figure 3.54: Leaf-initiated join FSM

- **ADD PARTY receipt.** This means that the network or root is going to add the user to the point-to-multipoint call and the new leaf is not the first one of the given call in this end system or switch. Usual ADD PARTY procedures apply. The timer can be stopped and the FSM destroyed.
- **LEAF SETUP FAILURE receipt.** The leaf cannot be added to the specified call. There are many reasons why this could happen. Besides the usual reasons why a call can fail there are additional cases: the feature may not be implemented, the call could not be found, the user is not authorised to be added to the call, etc.
- **T331 expiry.** In this case the LEAF SETUP REQUEST may be retransmitted several times. On final expiry the FSM is destroyed and the failure is reported to the user.

The LEAF SETUP REQUEST sent by the user includes the following information elements:

- Transit network selection.[1]
- Calling party number.
- Calling party subaddress.[1]
- Called party number.
- Called party subaddress.[1]
- LIJ call identifier.
- LIJ sequence number.

The transit network selection (TNS) and called party information elements are used as usual to locate the root node of the call in the network. Because a given root node may offer more that one point-to-multipoint call for which add requests can be received, an additional

[1] This is optional.

identifier for calls for a given root node is needed. This is the *LIJ call identifier* which is a 32-bit number that uniquely identifies the call on the given root.

When a node receives a SETUP, ADD PARTY or LEAF SETUP FAILURE message it needs to locate the LIJ FSM that handles the add request (given that the SETUP or ADD PARTY was triggered by a LEAF SETUP REQUEST). This is done by including a *leaf sequence number* into the LEAF SETUP REQUEST. This information element is echoed back by the network or root node in the ADD PARTY or SETUP message and enables the leaf node to correlate these messages to the original request. The leaf sequence number is a 32-bit integer that is unique on the given leaf.

When connecting the initial leaf of a point-to-multipoint call for which leaf join requests are expected, the root node must include at least the LIJ call identifier IE in the SETUP message. If the desired mode of joining is "network-prompted", then the root must also include an LIJ parameter IE. This IE has only one field (screening indication) for which one value, "Network Join Without Root Notification", is defined. If the SETUP is the response to a LEAF SETUP REQUEST, the invoker's leaf sequence number must also be included. (Note that the initial party may also be a normal party, joined by the root.)

Unfortunately we were not be able to verify the operation of the LIJ feature because the switch software of our switches does not yet support this feature. Because of this the following paragraphs show how LIJ *should* work.

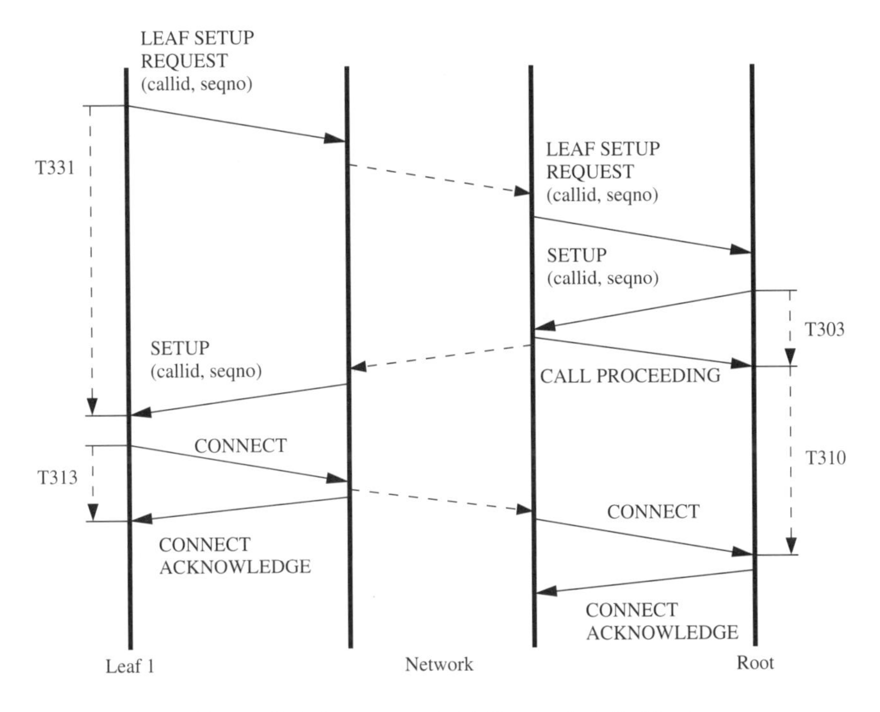

Figure 3.55: Root-prompted LIJ example, initial leaf

We first look at a call in root-prompted mode (Figure 3.55). Let us assume that the call does not yet exist when the first leaf requests the join. The first leaf sends a LEAF SETUP

REQUEST with the LIJ call identifier and the sequence number to the network. The network routes the message to the root node because it knows nothing about that call. The root node, upon receipt of the request, checks the message (whether the call identifier is legal and whether the leaf is authorised for this call). If everything is acceptable the root node initiates a point-to-multipoint call in the usual way with the first leaf being the leaf that sent the request (this is not necessarily so—the root could also set up the connection to any other address and add the requesting leaf in analogy to the second leaf). The call identifier and the invoker's sequence number must be included in the SETUP. The network proceeds with the establishment in the usual way and the leaf will receive a SETUP message with the sequence number and, optionally, the call identifier (for example, if this is an interface to a private network). It locates the FSM handling the LEAF SETUP REQUEST, stops the timer and destroys that FSM. It then continues to handle the point-to-multipoint call SETUP in the usual way. The CONNECT message will be sent back to the root node and at this point the connection to the first leaf is fully established.

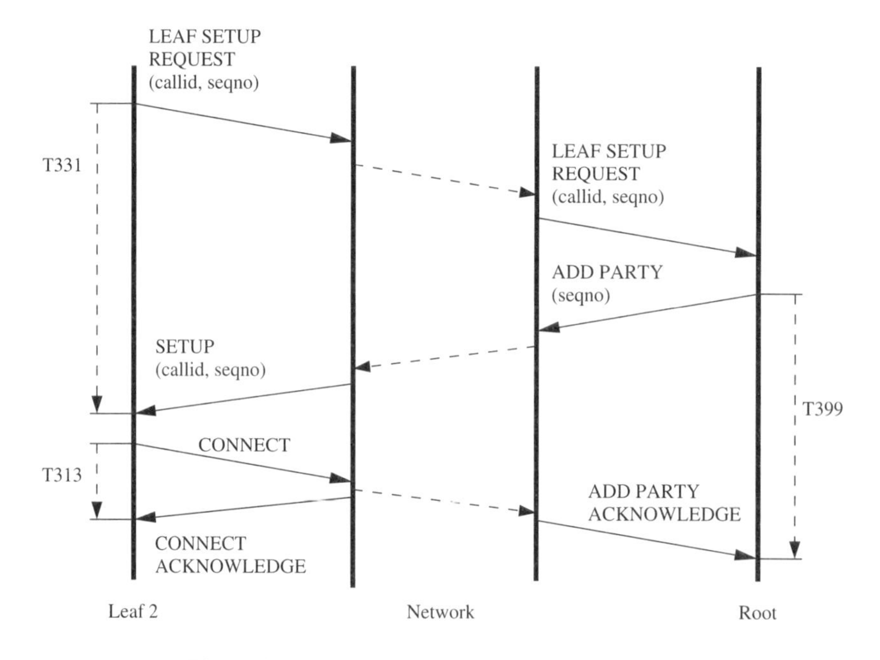

Figure 3.56: Root-prompted LIJ example, second leaf

Now the second leaf (residing on another ATM end system) requests to join the call with a LEAF SETUP REQUEST including the same call identifier and a sequence number (Figure 3.56). This request is forwarded through the network. Because the call is now known to the network it will check in which mode the call was established. Because the mode is "root-prompted" the LEAF SETUP REQUEST will be sent to the root node. The root node checks authorisation and whatever it needs to check and, if everything is fine, it sends an ADD PARTY message to the network to add the new leaf to the call. This message is forwarded through the network to the requesting node and pops out there as a SETUP containing the original sequence number. The leaf locates the LIJ FSM and destroys it after stopping the timer. It responds to the SETUP with a CONNECT, which is transferred to the ROOT as an ADD PARTY ACKNOWLEDGE. At this point the second leaf is fully added.

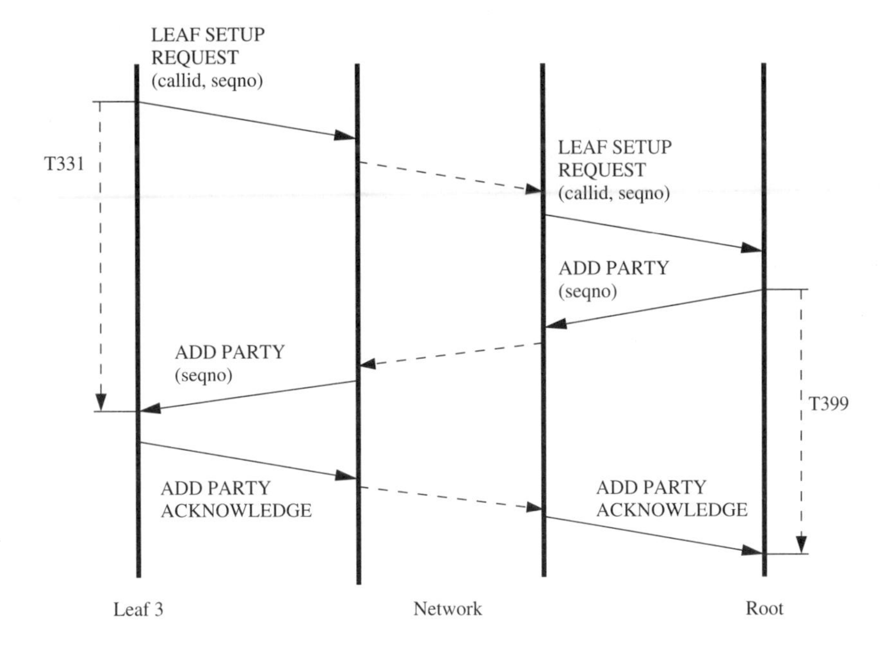

Figure 3.57: Root-prompted LIJ example, third leaf

Now let us suppose a third leaf is added (Figure 3.57), but now a branch of the point-to-multipoint call exists already through the node (this may be the case if the node is the border switch of a private network and this is the second leaf to join in that network). Everything starts exactly as in the previous case, only this time the setup request sent by the root appears as an ADD PARTY at the UNI. This time the leaf responds with an ADD PARTY ACKNOWLEDGE which gets forwarded to the root to finish joining the third leaf.

In the case of a network-prompted join the establishment of the first party happens in exactly the same way as for a root-prompted call. The only difference is that the root will include an LIJ parameter IE in the SETUP identifying the call as a network join call.

When the second leaf is set up (see Figure 3.58) it sends a LEAF SETUP REQUEST as usual. This request gets forwarded through the network until it finds a switch which knows the call to be joined to. When it identifies the call as a network-join call then instead of forwarding the LEAF SETUP REQUEST further to the root, it generates a SETUP message, includes the sequence number and optionally the call identifier from the request and sends this SETUP back to the leaf. The leaf accepts the SETUP by means of a CONNECT which gets forwarded to the point where the SETUP was originated. At this point the new leaf is fully added to the call. The root gets no indication that an additional leaf has appeared.

Note that for a network LIJ call the root itself may also add leaves. The dropping of leaves from such a call is handled by the root or the leaf for leaves added by the root, and the network

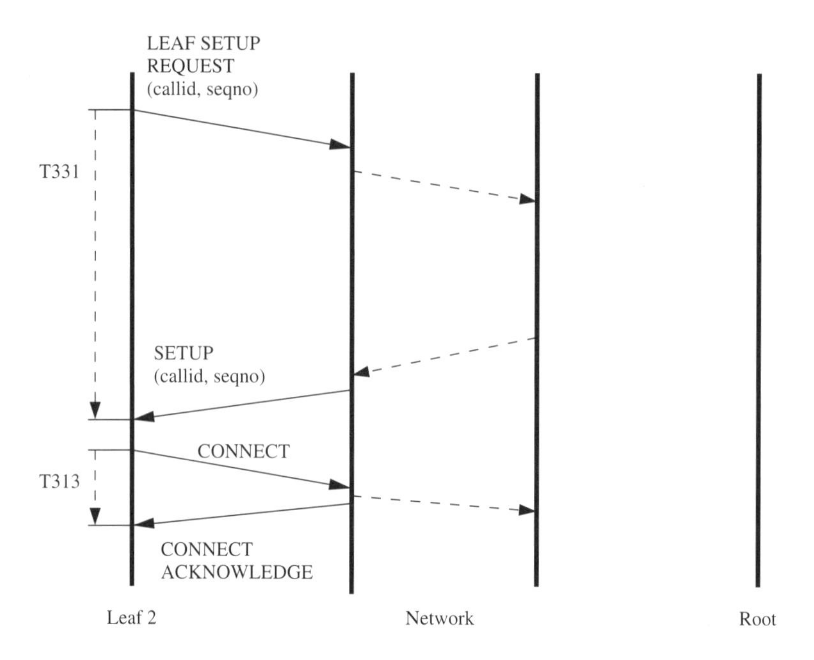

Figure 3.58: Network-prompted LIJ example, second leaf

and the leaf for automatically added leaves. As an exception to Q.2971, the root node should *not* release the call if it drops the last party, because there may be network-added leaves still active. In this case call clearing is handled by the network.

3.7 Restart Procedure

The restart procedure can be invoked on either side of a UNI to return one or all virtual channels or paths to a known state. This procedure is invoked via a request from the UNI user (call control in the switch or applications in the end system). Typical causes to invoke a restart are: no response to other messages from the peer, failure during call clearing (no RELEASE COMPLETE), and local failures (ATM board failure). When the procedure is finished all resources that were associated with the restarted channels and paths have been freed, including call references. This means that calls that use these channels and paths are cleared. When this happens on a switch, call clearing towards the remote user is also initiated.

The restart procedure is associated with the global call reference (all bits zero). It can be invoked from both sides of a UNI—the procedures are independent and the messages can be discriminated by looking at the call reference flag.

Two messages are used for this procedure: RESTART and RESTART ACKNOWLEDGE. Both messages always use the global call reference (all zeros). The call reference flag must be zero in the RESTART message (because it is always sent by the side invoking the procedure). In the RESTART ACKNOWLEDGE message the flag should be 1, but if the flag is 0, the message is interpreted as a RESTART message. Both messages can contain the same information elements: the restart indicator IE (see Figure 3.59) and the connection identifier IE (see Figure 3.26).

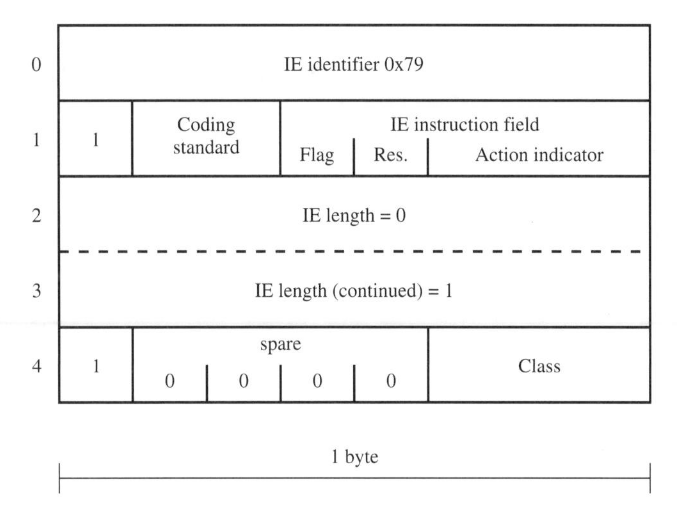

Figure 3.59: Restart indicator IE

Table 3.12: Restart classes

Code	Class
0	indicated virtual channel
1	all virtual channels in the indicated virtual path or the indicated virtual path that are controlled by the given UNI protocol instance
2	all virtual channels and paths that are controlled by the given UNI

The restart indicator describes the set of virtual channels and paths that are to be reset (Table 3.12).

Note that only *switched* channels and paths can be reset– semi-permanent or permanent channels and paths are ignored by this procedure.

If code 0 or 1 is used, the connection identifier IE must be included in the RESTART message. It indicates the channel or path that is to be restarted. For code 1 the action depends on whether the indicated virtual path is a switched path or not. If it is, it is restarted. If it is not, then all channels in this path are restarted.

For code 2 (restart all channels and paths) no connection identifier should be present in the message.

The state diagram for both sides of a UNI in the case of a restart is shown in Figure 3.60. Each side of a UNI always runs FSMs which go through both diagrams. The left FSM talks with the right FSM in the peer and vice versa.

The restart procedure is invoked by the UNI user by means of a Restart.request. When this signal is received, a RESTART message is sent to the peer with the global call reference and the call reference flag set to zero, which contains the appropriate restart indicator and connection identifier IEs.

After sending the RESTART timer T316 is started, which ensures that the peer completes the restart procedure in a limited time frame. This timer is about 2 minutes by default. The FSM moves to state REST1 (restart request).

The peer, upon receipt of the RESTART, informs its user with a Restart.indication, sets timer T317 and goes to state REST2 (restart). Timer T317 should be less than the peer's T316 to ensure that the UNI user completes the restart faster than the peer can time out. Now the application process on top of the UNI (either a real application or the call control stack in the case of the switch) should release all resources corresponding to the restarted channels and paths (including the call reference).

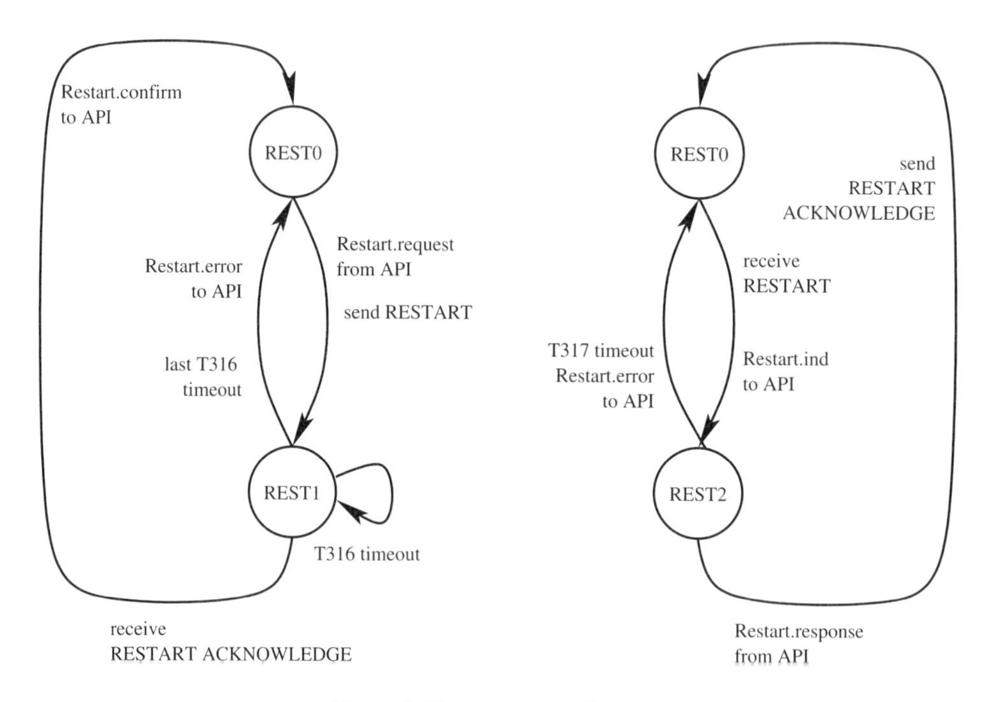

Figure 3.60: Restart state diagram

When the UNI user has finished this process, it sends a Restart.response signal to the UNI. The UNI in turn stops T317, sends a RESTART ACKNOWLEDGE, and enters state REST0.

The invoker of the restart procedure upon receipt of the RESTART ACKNOWLEDGE stops its timer T316, informs the user about the successful restart via a Restart.confirm and enters REST0. At this point the procedure is finished.

If timer T316 expires because the peer fails to answer the restart request, the RESTART message may be retransmitted and the timer restarted several times (default is 2). If no response is received even after all retransmissions a Restart.error indication is sent to the UNI user and the REST0 state is entered. The channels and paths that could not be restarted should be put in a state where they are not used.

If timer T317 expires because the UNI user fails to respond to the Restart.indication, a Restart.error indication is sent to the UNI user and the FSM returns to state REST0.

The following is a trace of a successful restart of channel 56 on path 0:

```
1  E ⇒ S   0.000 uni cref={you,global} mtype=restart mlen=14
2                connid={vpass=explicit,pex=exclusive_vpci_vci,vpci=0,vci=56}
3                restart={class=vc}
4  E ⇐ S   0.087 uni cref={me,global} mtype=res_ack mlen=14
5                connid={vpass=explicit,pex=exclusive_vpci_vci,vpci=0,vci=56}
6                restart={class=vc}
```

And the following is another trace for a restart of the whole UNI:

```
1  E ⇒ S   0.000 uni cref={you,global} mtype=restart mlen=5
2                restart={class=allvc_layer3}
3  E ⇐ S   0.278 uni cref={me,global} mtype=res_ack mlen=5
4                restart={class=allvc_layer3}
```

In both cases a RESTART message is sent in line 1. To restart only a channel, it contains a connection identifier IE for channel 0/56. To restart the entire UNI, no connection identifier IE is contained. The RESTART ACKNOWLEDGE should contain the same restart class and connection identification as the RESTART. This is checked by the end system that initiated the procedure. If a mismatch is detected, the RESTART ACKNOWLEDGE is ignored.

3.8 Interface to the SAAL

UNI as well as PNNI signalling uses a Signalling ATM Adaptation Layer (SAAL) consisting of the Service Specific Connection Oriented Protocol (SSCOP) with a UNI Service Specific Coordination Function (SSCF) on top. Below the SSCOP is a standard AAL5. Chapter 5 provides an in-depth description of the SAAL.

From the SAAL signals at the upper layer of the SSCF only the following are used:

`AAL-ESTABLISH.indication`
: This indicates to the UNI that either the peer has established a new SAAL connection, or the SAAL connection had to be re-initialised due to a protocol error.

`AAL-ESTABLISH.request`
: This is invoked by the UNI to start the SAAL. Although [Q.2931] Structured Description Language (SDL) diagrams show that an SAAL connection should be established, when a `Setup.request` signal is received from the UNI user, most implementations try to start the SAAL as soon as they are started. This reduces the setup time for the first call. It also makes error reporting better, because a broken UNI link is not detected when one tries to establish a connection, but when the terminal is switched on.

`AAL-ESTABLISH.confirm`
: This reports the successful establishment of the SAAL underlying SAAL connection.

`AAL-RELEASE.indication`
: This reports that the SAAL connection was lost either because of an error or because of a peer release request. Most implementations initiate a new SAAL establishment when they receive this indication.

`AAL-RELEASE.request`
: Releasing the SAAL may be requested from the UNI user. This seems to make sense only for access by management operations.

`AAL-DATA.request`
: This request is used by the UNI to send messages to the peer UNI.

`AAL-DATA.indication`
: This signal is invoked by the SAAL to inform the UNI about the arrival of a message from the peer UNI.

The unassured data transfer feature as well as user data in SSCOP connection control messages is not used.

Which channel is used for the SAAL depends on the configuration of the UNI (see Section 3.2). Normal configuration is to use VPI= 0 and VCI= 5 to control the entire link (non-associated signalling). [Q.2931] allows for associated signalling (in this case other VPs on the link can carry their own signalling channel in VCI= 5). [Q.2120] has support for the dynamic allocation of signalling channels. However, [UNI4.0] only allows the standard configuration. The only exception is proxy signalling or virtual UNI support. The VCI number and signalling mode is usually a configuration option in switches and end systems.

[UNI4.0] specifies also the characteristics of the signalling VC: service category, traffic contract, cell loss ratios and end-to-end transit delay.

The state of the entire UNI stack is described by the so-called coordinator state. This state can have the following values (see also Figure 3.61):

CU0—AAL Connection Released
: In this state no SAAL connection exists. No messages can be received or sent. An SAAL connection is established either by the peer, or by the UNI user by means of a `Restart.request` to execute a restart procedure, a `Setup.request` to establish an ATM connection or a `Link-Establish.request` to establish the SAAL connection. In the last three cases the coordinator issues an `AAL-ESTABLISH.request`, starts T309 and enters state CU1. In the first case it goes directly to CU3.

CU1—AAL Establish Request
: The coordinator has requested the establishment of the SAAL connection. If the connection is successfully established, the coordinator receives an `AAL-ESTABLISH.confirm`, stops T309 and enters CU3. If the establishment fails, either an `AAL-RELEASE.indication` is received or T309 times out. In both cases the coordinator goes back to CU0.

CU2—AAL Awaiting Release
: The UNI user has requested the release of the SAAL connection. The coordinator waits for the `AAL-RELEASE.confirm` signal. If it is invoked, CU0 is entered.

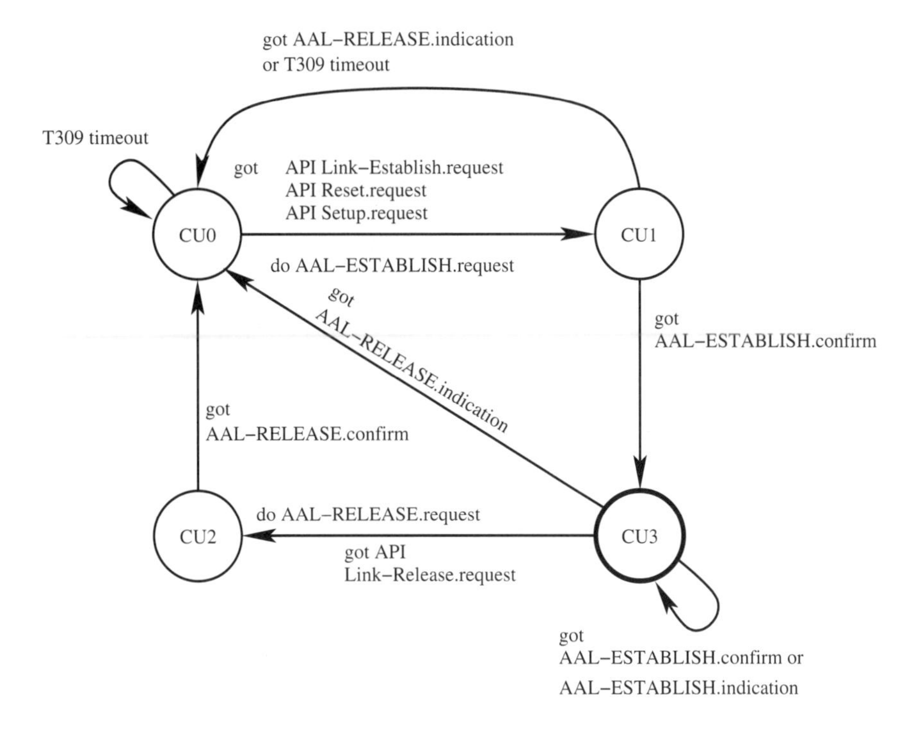

Figure 3.61: Coordinator states

CU3—AAL Connection Established

The SAAL connection is up and running. Messages can be transmitted and received.

The corresponding states on the network side UNI are called CN0, CN1, CN2 and CN3.

Normally the SAAL connection is established as soon as the interface is configured to be up. On successful establishment the coordinator remains in state CU3 or CN3 all the time. If there are problems on the link `AAL-ESTABLISH` or `AAL-RELEASE` signals can be received from the SAAL because of either error recovery or connection release by the peer or by one of the SSCOP instances. The coordinator translates these signals to four different signals that are then sent to all call FSMs:

Link-Establish.indication

This means that there was a problem on the link and the SAAL connection was either successfully re-established or the error recovery procedure was invoked and finished. If a call FSM receives this signal the action of the FSM depends on the current state. For calls in the release states (U11 and U12) no action is taken. For calls in the establishment states (U1–U9) the status enquiry procedure may be invoked. If the call is in state U10, then either the call status enquiry is invoked or the party status enquiry for all active parties for point-to-multipoint calls. Parties in the drop states are ignored and the state of parties in the add states may optionally be enquired.

Link-Establish.confirm

This means that there was a problem on the link and the SAAL connection was successfully re-established. Because there was a Link-Release.indication received previously only active calls and parties exist at this point. For all calls and parties the status enquiry procedure is invoked.

Link-Release.indication

This means that the link was released either by the peer or by the SSCOP because of extensive errors. All calls that are not in state U10 are locally released. For point-to-multipoint calls all parties not in the active party state are released. This means that only active calls and active parties remain. Additionally the re-establishment of the SAAL connection is requested if any call is found in state U10. Normal implementation always request SAAL establishment, even if no call is active.

Link-Release.confirm

This happens only if the request to release the SAAL connection was invoked from the UNI user. It means that the link is now released.

There is also a Link-Establish.error indication. This is invoked by the coordinator if the SAAL (re-)establishment fails. In this case all calls and parties are locally cleared.

(Re-)establishment of the SAAL is controlled by timer T309. This timer is started whenever an `AAL-ESTABLISH.request` signal is sent to the SAAL (provided that it is not running already). If the timer expires, an error is assumed and all calls are cleared. The timer normally has a values of 10 seconds which is far more than the SAAL should need to establish the connection (the SSCF at the UNI (see Section 5.2) specifies four attempts with an interval of 1 second).

3.9 Exception Handling

In such a complex protocol like the UNI with such a complex message structure many exceptional situations can arise. An additional problem is the existence of different protocol variants (Q.2931, Q.2971, UNI3.1, UNI4.0) and the existence of many optional features and enhancements (for example, bandwidth modification of existing connections [Q.2963.1]–[Q.2963.3]). In the UNI protocol even the error handling is complex.

There are four basic kinds of errors that can occur:

- Wrong message encoding. This can happen because of implementation errors or misinterpretations of the standard. Features that are implemented on one side of a UNI, but not on the other, also lead to coding errors, because non-implemented information elements or messages are handled like coding errors.
- Status errors. In this class are errors like: receiving a message not expected in the given state, information elements that should be in the message for the current state but are missing, or information elements that should not be contained in the message but are found. As in the previous case this can be the result of a non-implemented optional feature. An example is the receipt of a CONNECT message in response to an outgoing SETUP without an intervening CALL PROCEEDING. According to Q.2931 this is an optional feature (see Annex H of [Q.2931]). If an end system has not implemented this features, but the access switch has been configured to not send a CALL PROCEEDING, the CONNECT or ALERTING that is received later will be treated as an error.

- Semantic errors in the messages. These errors are quite hard to spot if they occur. As already noted, there are many interdependencies between the information elements of a message, and even between different messages for one call. If, for example, the bearer capability information element specifies a point-to-multipoint connection, the backward cell rates in the traffic descriptor must be zero. Because a switch need not to implemented all legal bearer types, traffic classes, QoS classes and traffic characteristics and a call typically travels through many switches in the network, all possibly with different implementations and configurations, these kinds of errors can happen at any stage in the call and it is difficult for the user to located, where the error really occurred. To make it even more complicated, if the network uses dynamic routing (PNNI), one call make succeed but a later call with the same characteristics may fail.
- Not enough resources. Even if everything is fine, the network may lack the resources needed for the call.

If an error situation occurs a UNI can react in different ways:

- It can release the call. This is probably the most drastic action, but in a case like, for example, the lack of resources, there is nothing the network can do about the call.
- It can report the problem by means of a STATUS, ignore the offending information elements and try to continue.
- It can report the problem with a STATUS message, ignore the entire message that had an error and remain in the state it was prior to receipt of that message.
- It can ignore erroneous information elements and try to continue.
- It can ignore the offending message and remain in its state.

The standard specifies a default action for all errors.[2] This default error handling can be overridden on a per-information-element or per-message basis by means of the action indicator (see Figure 3.11 and Table 3.5). In addition the PNNI has a Pass Along Request bit in the message and the information element header which, if set, entirely inhibits error processing. If this bit is set, the information element is forwarded to the next link as it is. This feature allows the support of new information elements and messages in PNNI even if not all switches implement the new feature.

If an error is not entirely ignored, the call is either released or a STATUS message is sent back to the originator of the offending message. In any case one of three message types is generated: RELEASE, RELEASE COMPLETE, or STATUS. RELEASE COMPLETE is used either in "very" exceptional situations, when the call is to be released very quickly, or when a call attempt has to be rejected. The normal action to release a call is by means of a RELEASE message. If exception handling specifies that the call should not be cleared, a STATUS message is emitted. All three messages contain a cause information element (see Figure 3.36) that tells the peer UNI what the problem is. In the case of an error during the release procedure, the RELEASE message can even include two of these causes (this happens when, during a release, timer T308 times out).

Let us look at some common error handling examples in the following traces.

When a message is received the highest priority check is for the correctness of the message header. If the protocol discriminator is wrong, the call reference has a wrong format (reserved

---

[2] Well, not for all, but for most. Some kinds of errors are simply not mentioned in the standards and the implementor must choose a sensible way of handling them.

bits not zero or length not 3) or the message is shorter than 9 bytes (so that the length is missing), then the message is totally ignored. This type of error is not shown in the following examples.

The first two traces shows the reaction of the network to a message with a wrong call reference.

```
1  E ⇒ S   0.000 uni cref={me,27} mtype=connect mlen=0
2  E ⇐ S   0.078 uni cref={you,27} mtype=rel_compl mlen=6
3                cause={loc=private_network_serving_local_user,
4                   cvalue=invalid_call_reference_value,class=invalid_message}
```

Here a CONNECT message is sent with a call reference that is not used yet. The network answers with a RELEASE COMPLETE. This reaction seems sensible, because the CONNECT message could be the result of the user side thinking that the call exists. The RELEASE COMPLETE lets it release all resources for that call, so it is now non-existent on both sides of the interface.

```
1  E ⇒ S   0.000 uni cref={you,global} mtype=setup mlen=51
2                traffic={fpcr01=0,bpcr01=0,be}
3                bearer={class=bcob-x,traffic=noind,timing=noind,clip=not,user=p2p}
4                called={type=unknown,plan=aesa,addr=plan=aesa,addr=spock}
5                qos={forw=class0/unspecified,back=class0/unspecified}
6  E ⇐ S   0.091 uni cref={me,global} mtype=status mlen=11
7                callstate={0-Null}
8                cause={loc=private_network_serving_local_user,
9                   cvalue=invalid_call_reference_value,class=invalid_message}
```

In this example a SETUP is sent with the global call reference. This is illegal so the network rejects this with a STATUS message indicating that the call reference is invalid. Upon receipt of this message the UNI stack does compatibility checking of the call states and finds a mismatch—the call state on the user side is U1 (call initiated), but the response from the network indicates U0 (not existent). This leads to a local clearing of the call.

The protocol reactions to these kinds of errors are specified in this way to synchronise the state of the UNI protocol instances on both ends of the interface.

The following two examples show the default reaction to messages received in the wrong state.

```
1  E ⇒ S   0.000 uni cref={you,35} mtype=stat_enq mlen=0
2  E ⇐ S   0.101 uni cref={me,35} mtype=status mlen=11
3                callstate={10-Active}
4                cause={loc=private_network_serving_local_user,
5                   cvalue=response_to_status_enquiry,class=normal_event}
6  E ⇒ S   5.027 uni cref={you,35} mtype=connect mlen=0
7  E ⇐ S   5.090 uni cref={me,35} mtype=status mlen=11
8                callstate={10-Active}
9                cause={loc=private_network_serving_local_user,
10                  cvalue=message_not_compatible_with_call_state,
11                  class=protocol_error,mtype=0x07}
```

In this example we have a call in the active state. This is verified by invoking a status enquiry in line 1. Lines 2–5 show the STATUS message from the switch which indicates state N10. In line 6 a CONNECT message is sent, which is clearly wrong in that state. The switch responds with a STATUS. The call state indicates that it has effectively ignored the message (the state did not change). The cause value shows that the last message was received in the wrong state.

```
1   E ⇒ S   0.000 uni cref={you,41} mtype=stat_enq mlen=0
2   E ⇐ S   0.137 uni cref={me,41} mtype=status mlen=11
3                 callstate={10-Active}
4                 cause={loc=private_network_serving_local_user,
5                    cvalue=response_to_status_enquiry,class=normal_event}
6   E ⇒ S   3.870 uni cref={you,41} mtype=!0x88
7                    unparsed={88:80:00:0c:59:80:00:08:84:00:07:d0:85:00:07:
8                    d0:00:00:00:00:00}
9   E ⇐ S   3.923 uni cref={me,41} mtype=status mlen=12
10                callstate={10-Active}
11                cause={loc=private_network_serving_local_user,
12                   cvalue=message_type_non-existent_or_not_implemented,
13                   class=protocol_error,mtype=0x88}
```

This trace shows the invocation of an unimplemented feature. The user sends a MODIFY REQUEST message to change the peak cell rate of a CBR connection (the call state is again checked by means of a status enquiry procedure). This feature is not implemented in the switch, so it responds with a status message. This time the error code shows that the message was not completely understood. We see also another problem: the `sigdump` tool that is used to trace the message also does not understand the MODIFY REQUEST and prints its contents as hex numbers. Before we try to invoke the modify procedure we verify that the call is in state 10.

Errors in single information elements are classified into two groups:

- errors in mandatory information elements;
- errors in non-mandatory information elements.

Whether an information element is mandatory in a given type of message or not may depend on the state of the call and on previous messages. If, for example, an ALERTING message is sent by the network to the calling end system, then the connection identifier is mandatory if no CALL PROCEEDING was sent previously, and non-mandatory if a CALL PROCEEDING was sent. Another example would be if the called user has received an end-to-end transit delay information element in the SETUP, he must include an end-to-end transit delay IE in the CONNECT.

When a mandatory information element is missing or has a content error, the message is usually ignored and a STATUS message sent. Exceptions are: a SETUP is rejected by means of a RELEASE COMPLETE; an invalid or missing cause in a RELEASE is interpreted as a normal cause; and an error in the cause of a RELEASE COMPLETE is ignored. The following traces show these actions.

```
1  E ⇒ S  0.000 uni cref={you,27} mtype=setup mlen=26
2               traffic={fpcr01=0,bpcr01=0,be}
3               bearer={class=bcob-x,traffic=noind,timing=noind,clip=not,user=p2p}
4               qos={forw=class0/unspecified,back=class0/unspecified}
5  E ⇐ S  0.056 uni cref={me,27} mtype=rel_compl mlen=7
6               cause={loc=private_network_serving_local_user,
7                  cvalue=mandatory_information_element_is_missing,
8                  class=protocol_error,ie=called}
```

In this example the user sends a SETUP message without a called party number information element (lines 1–4. This is rejected by the network with an appropriate RELEASE COMPLETE indicating the missing IE (lines 5–8).

Now an example of a missing cause in a RELEASE:

```
1  E ⇒ S  0.000 uni cref={you,28} mtype=release mlen=0
2  E ⇐ S  0.137 uni cref={me,28} mtype=rel_compl mlen=7
3               cause={loc=private_network_serving_local_user,
4                  cvalue=mandatory_information_element_is_missing,
5                  class=protocol_error,ie=cause}
```

The RELEASE message in line 1 is missing the cause information element. The switch clears the call anyway and responds with a RELEASE COMPLETE indicating the missing cause in lines 2–5.

The behaviour for information elements with errors is the same as for missing IEs, only error code 100 ("Invalid information element contents") is used instead of 81.

For errors in non-mandatory information elements the actions of the UNI stack are different—normally a STATUS message is sent after handling the message. The information element with an error is ignored. This may of course lead to the establishment of a connection with the wrong characteristics. This may be hard to detect. If, for example, the calling user wants to negotiate AAL parameters (see Section 3.5.3) he includes an AAL parameters IE into the SETUP. The called user should adjust the parameters in this IE and return it in the CONNECT message. If he makes an error in the coding of this IE, the network sends a STATUS message to the called user and forwards the CONNECT without the AAL parameters to the calling user. He in turn will falsely assume that the called user has accepted the settings in the AAL parameters IE. In such a case the called user should probably release the call.

Let us look at an example. The calling user sends a SETUP message with an error in the AAL parameters information element:

```
1  E ⇒ S  0.000 uni cref={you,12} mtype=setup mlen=85
2               aal={!illegal_aaltype6}
3               traffic={fpcr01=0,bpcr01=0,be}
4               bearer={class=bcob-x,traffic=noind,timing=noind,clip=not,user=p2p}
5               called={type=unknown,plan=aesa,addr=plan=aesa,addr=spock}
6               calling={type=unknown,plan=aesa,addr=kirk}
7               qos={forw=class0/unspecified,back=class0/unspecified}
8  E ⇐ S  0.141 uni cref={me,12} mtype=status mlen=12
9               callstate={1-Call_initiated}
```

```
10                      cause={loc=private_network_serving_local_user,
11                        cvalue=access_information_discarded,class=resource_unavailable,
12                        ie=aal}
13   E ⇐ S     0.167  uni cref={you,12} mtype=call_proc mlen=9
14                      connid={vpass=explicit,pex=exclusive_vpci_vci,vpci=0,vci=223}
```

The calling user sends a SETUP with an invalid AAL type in the AAL parameters information element. The network discards this information element and proceeds the call (line 11). Before the CALL PROCEEDING is sent, a STATUS message reports the error (line 8). Note that in the case of the BLLI, BHLI, AAL parameters and sub-address information elements the error is 43 ("Access information discarded") while for other information elements the error would be 100 ("Invalid information element contents").

If there is an entirely unknown information element in a message a STATUS with cause code 99 ("Information element non-existent or not implemented") is sent with a diagnostic indicating the information element. This may be used by the sending entity to detect that a certain optional feature is missing.

The previous examples have shown the action of the UNI stack in the case of default error handling. These rules can be overwritten by the specification of explicit action indicators in single information elements or for the entire message. The message action indicator is inspected when the message type is unknown or the message is unexpected in the given state. The information element action indicator is used when the message itself is acceptable in this state, but the information element is either in error or unknown.

In the following example we have a call established and in the active state. Then the calling user sends a CONNECT message with the action indicator set to "clear call". Of course, the CONNECT in state 10 is totally unexpected by the network and it invokes its error procedures:

```
1    E ⇒ S     0.000  uni cref={you,51} mtype=stat_enq mlen=0
2    E ⇐ S     0.027  uni cref={me,51} mtype=status mlen=11
3                       callstate={10-Active}
4                       cause={loc=private_network_serving_local_user,
5                         cvalue=response_to_status_enquiry,class=normal_event}
6    E ⇒ S     4.711  uni cref={you,51} mtype=connect mlen=0
7    E ⇐ S     4.807  uni cref={me,51} mtype=release mlen=7
8                       cause={loc=private_network_serving_local_user,
9                           cvalue=message_not_compatible_with_call_state,
10                          class=protocol_error,mtype=0x7}
11   E ⇒ S     5.197  uni cref={you,51} mtype=rel_compl mlen=6
12                      cause={loc=user,cvalue=normal,_unspecified,class=normal_event}
```

In lines 1–5 the end system enquires the state from the switch to verify that the call is in state U10. It then sends a CONNECT message with the action indicator set to "clear call". Unfortunately `sigdump` is not able to show the action indicator. The switch responds to the unexpected message with a RELEASE showing cause code 101 ("Message not compatible with call state"). The end system in turn finishes call clearing with a RELEASE COMPLETE.

Error handling in UNI is quite complex. The rules in the standards often are not very clear and do not handle all situations. The possibility of fine-grained error handling by means of action indicators makes error handling even more complex, but allows the step-by-step addition of new features to a network without the need to update all switches and end systems.

3.10 The Structure of a UNI Protocol Instance

Figure 3.62 shows the structure of a UNI. This structure is usually described as a set of processes, although the implementation need not be multi-threaded. The FreeUNI implementation uses only one thread of execution for the entire UNI plus the SAAL (see [Beg]).

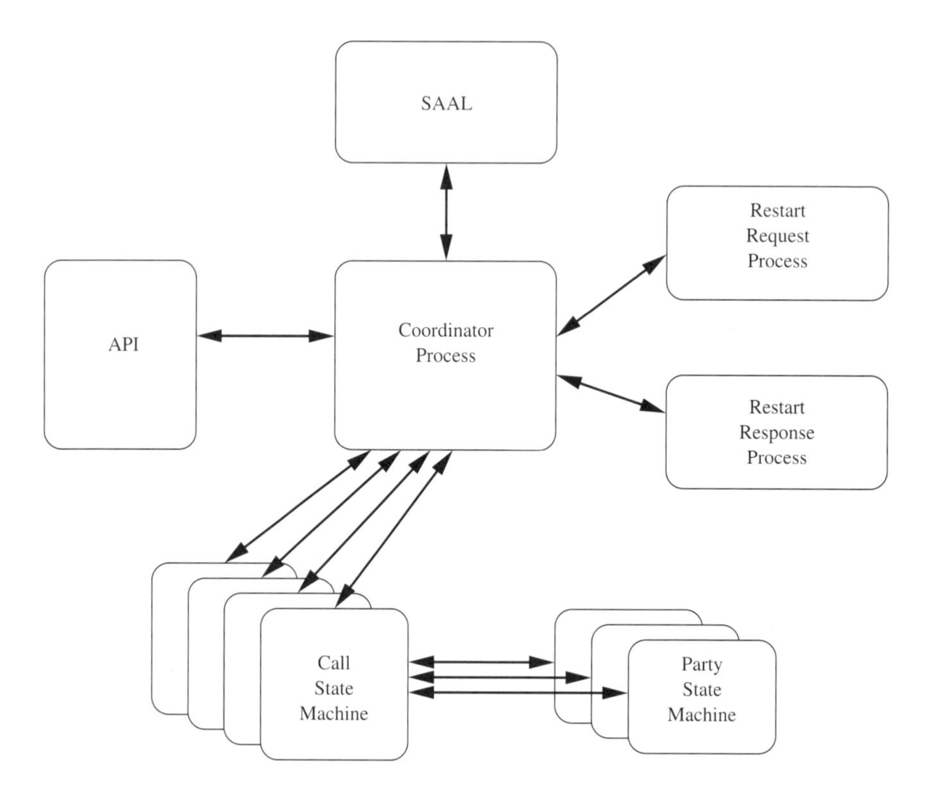

Figure 3.62: The structure of a UNI protocol instance

The central part is the coordinator process called Co-ord-N and Co-ord-U on the network and the user sides, respectively. This process controls the SAAL connection, sends messages to the SAAL, dispatches received messages to other processes, communicates with the Application Programming Interface (API) and creates UNI state machines for new calls. The coordinator is a Finite State Machine (FSM) operating on the coordinator state (see Section 3.8).

Two additional processes are always present: the restart requestor (Restart-Start-N and Restart-Start-U) and the restart responder (Restart-Respond-N and Restart-Respond-U). They are FSMs—the requestor operates on the call state of the global call reference with the call reference value 0; the responder works on the global call reference with flag 1. This means that the requestor communicates with the peer UNI's responder and vice versa. RESTART and RESTART ACKNOWLEDGE messages, as well as STATUS and STATUS ENQUIRY messages with a global call reference are handled by these processes. There is also a set of signals by which the API can trigger the restart process or gets an indication about a peer-initiated restart.

The coordinator instantiates a new call FSM whenever an outgoing call is requested from the API via a Setup.request signal when an incoming call is received, i.e. a SETUP message arrives. Messages carrying an existing call reference are then routed to the appropriate FSM. A call FSM is destroyed when a releasing procedure is executed, a restart of the entire UNI or the channel that handles the FSM is requested, or when local call clearing occurs in certain exceptional situations.

For point-to-multipoint calls each call instantiates a party FSM for each party. This instantiation is done at the root node when an Add-party.request signal is received from the API or an ADD PARTY message is received at a leaf node. Party-related messages and API signals are routed from the coordinator to the call FSM and from the call FSM to the party FSM. Outgoing signals and messages go the opposite way. A party FSM is destroyed when the party is finally dropped or the parent call FSM is destroyed.

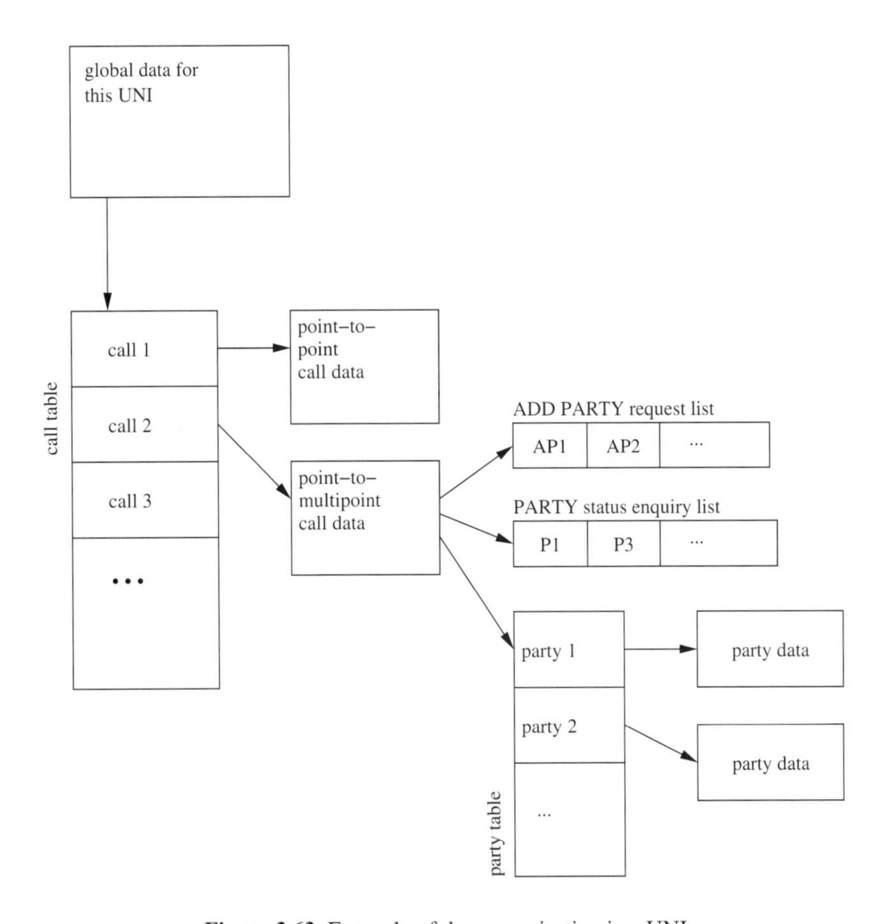

Figure 3.63: Example of data organisation in a UNI

Figure 3.63 shows the principal organisation of data in a UNI. There is global data associated with the UNI like:[3]

- initial timer values;
- initial values for timer restart counters;
- the VPI/VPCI mapping table;
- the list of VPCIs controlled by this UNI;
- the call state associated with the restart processes;
- the coordinator state;
- timers T316 and T317;
- SAAL associated data.

The call list contains a data structure for each active call FSM which contains data like:[4]

- the call reference including an ownership flag;
- the call type: point-to-point, point-to-multipoint root, point-to-multipoint leaf, Connectionless Bearer-Independent (COBI) or leaf-initiated join;
- the current call state;
- timers T301, T303, T308, T310, T322 and their repeat counts (if applicable);
- the connection identifier;
- various saved messages and information elements for retransmission.

If the call is a point-to-multipoint there are additional data fields for a call:

- a list of pending ADD PARTY requests (only at the network side);
- a list of pending party status enquiry requests;
- a party list.

The party list in turn contains a data structure for each active party FSM for a given point-to-multipoint call. Among this data are:[5]

- the endpoint reference including an ownership flag;
- the endpoint state;
- timers T397, T398 and T399.

Timers in a UNI are usually configurable. Table 3.13 shows the default values (in seconds) as specified in the standards. Some of the timers have optional restart counters.

---

[3] This list is not exhaustive.
[4] This list is not exhaustive.
[5] This list is not exhaustive.

Table 3.13: UNI timers

Timer	Restarts	User	Network	Usage
Point-to-point connection control				
T301	0	$\geq$ 3 min	$\geq$ 3 min	Limits the length of the peer alerting phase.
T303	1[c]	4 s	4 s	Protects against lost SETUP messages.
T308	1	30 s	30 s	Retransmits RELEASE in case it is lost.
T310	0	30–120 s	10 s	Limits the time a SETUP can travel to the peer.
T313	0	4 s	—	Protects against losing the CONNECT at the destination interface.
T322	N[d]	4 s	4 s	Status enquiry procedure.
Restart procedure				
T316	N[d]	2 min	2 min	Ensures that the peer answers to a restart request.
T317	0	$<$ peer T316	$<$ peer T316	Ensures that the layer above UNI answers to a restart request from the peer.
SAAL control				
T309	0	10 s	10 s	Re-establishes SAAL in case it fails.
Point-to-multipoint connection control				
T397	0	$\geq$ 3 min	$\geq$ 3 min	Restricts the time the remote party is alerted.[a]
T398	0	4 s	4 s	Locally drops a party in the case when the DROP PARTY or DROP PARTY ACKNOWLEDGE is lost.
T399	0	34–124 s	34–124 s	Protects against loss of ADD PARTY or the answer messages to it.[b]
Leaf-initiated join				
T331	1[c]	60 s	—	Ensures LEAF SETUP REQUEST gets answered.

[a]Same value as T301
[b]Sum of T303 and T310
[c]Restart is optional
[d]Restart is optional. The number of restarts is an implementation option.

4

ATM Addresses

4.1 The semantics of Addresses

Addresses in networks may have three semantics: they can be simple identifiers, they can be locators or, as in ISDN and telephone networks, an address may also be a service selector (for example, numbers starting with 800 and 900).

If addresses are identifiers they do not carry any information that can be used to locate the addressed entity. Examples of such addresses are Ethernet Multiple Access Control (MAC) addresses. These addresses are usually attached to the network card by the manufacturer and if the card is moved from one computer to another, the address moves with the card. This means that these addresses cannot directly be used to route data to the addressed entity.

Addresses used as locators need to have an internal structure. The address itself can be used to compute the route (or at last parts of it) to the addressed entity. ATM addresses are locators. They indicate the location of an ATM interface in the network through which traffic enters or exits the given routing domain.

Addresses can also be service request identifiers. Certain telephone number prefixes, rather than addressing another telephone, invoke a service that can, for example, provide address translation or special billing variants.

ATM addresses as defined by the ATM-Forum in [ADDR1.0], [ADDRR] and [ADDRUNI] are always locators—their intention is to address an ATM interface in the given routing domain and to aid in the computing of the route to this interface.

4.2 Called and Calling Party Numbers

ATM addresses are used in several information elements (IEs) in signalling message as well as in ILMI messages. There are two groups of addressing IEs: party numbers and party subaddresses. Party numbers are used for route computations, whereas subaddresses are used either by the end system or by network ingress and egress nodes for the tunnelling of address information through intermediate networks (for example, to carry an AESA between parts of a private ATM network which are connected through a public ATM network that supports only E.164 addressing).

The IEs containing party numbers are:

- **Called party number.** (see Figure 4.1) This information element is mandatory in each SETUP sent by a user to the network. On the basis of this information element the route to the called user is computed. The information element is delivered to the called user which can optionally perform a check of the number or use it to address services or applications in

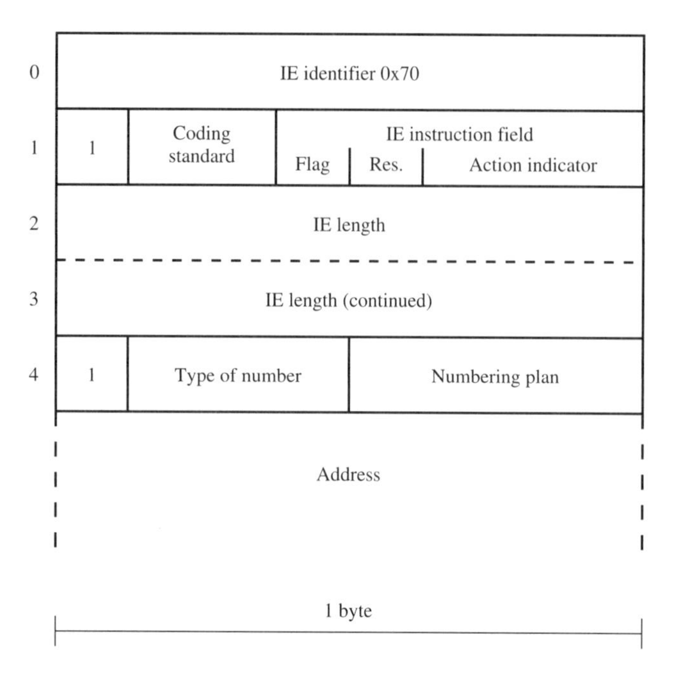

Figure 4.1: Called party number IE

the user equipment (if, for example, more than one address is assigned to the called user).

- **Calling party number.** This information can be included by the calling user in the SETUP message. Its contents need not be verified by the network and may be delivered to the called user. This delivery is subject to subscription restrictions of the calling user.
- **Connected number.** This information maybe included by the called user in the CONNECT he sends to the network. In this case its contents are verified by the network to contain one of the user's addresses. If the user did not provide the IE, it is inserted by the network on a per-subscription basis of the user. The contained address provides a unique identification of the ATM interface that is connected during this call. This may be different from the calling party number in the case of re-directions or service invocations (800 or 900 numbers). The presentation of the information element to the calling user may be restricted on a per-subscription basis by the called user (for example, for emergency services).

The subaddress information elements generally cannot be verified by the network and are transparently transferred from one endpoint of the connection to the other. The network nodes provide only syntactic checks of the information elements:

- **Called party subaddress.** (see Figure 4.2) This information element contains an NSAP address or an AESA which is used by the called end system to address sub-entities. One major application of this IE is the tunnelling of private addressing information through public networks. The private border switch in the calling network maps the private AESA to an E.164 public address and moves the private AESA into a subaddress IE. The border switch of the called network then throws away the called party number (after verification), moves the AESA from the subaddress IE to the called party number IE and continues to

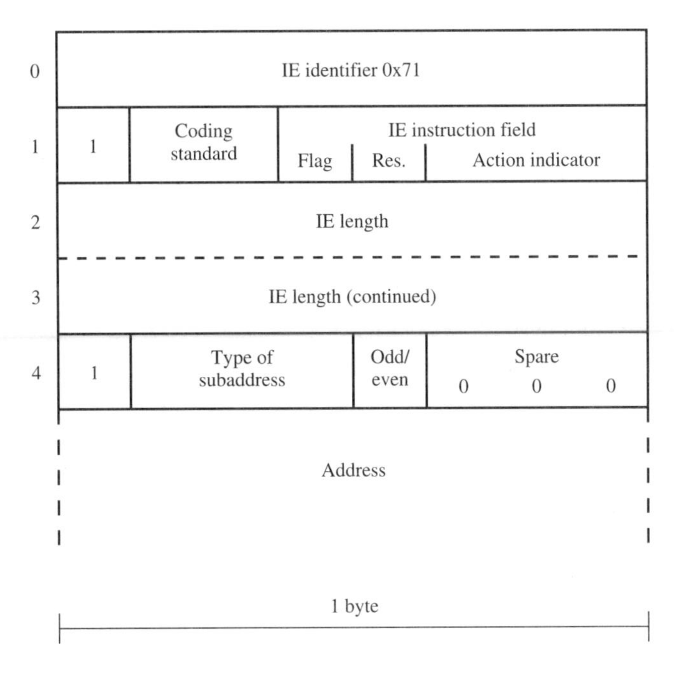

Figure 4.2: Called party subaddress IE

route the call in the private network according to the AESA. This technique is also the reason for the support of two subaddress IEs in [UNI4.0] instead of one as in [Q.2931]. This allows subaddress information to be used also in tunnelled private ATM networks.

- **Calling party subaddress.** This IE can be provided by the calling user to indicate the subentity of the end system that originated the call. In the case of tunnelling private networks through public networks, this IE may contain the original calling party number while the SETUP moves through the public network.
- **Connected subaddress.** This IE can be provided by the called user to indicated the subentity of the end system that handles the call.

There are three kinds of addresses used in ATM networks, where the second group is a sub-group of the third one:

- **E.164 addresses.** These addresses are also known as telephone numbers. They consist of up to 15 digits and may be of several types: international, national, network specific, subscriber number, abbreviated number. E.164 is the name of the ITU-T recommendation that defines the format of these numbers (see [E.164]). UNI4.0 allows only international numbers and defines the use of these numbers only for addressing purposes, i.e. the use of service request identifiers like 800 or 900 is not defined.
- **ATM End System Address (AESA).** This is a subclass of NSAP addresses as defined in [X.213] and [ISO8348]. They are always 20 bytes in length and are the major class of addresses used in UNI and PNNI.
- **Network Service Access Point (NSAP).** These can be used in the subaddress information elements.

Private ATM networks that conform to ATM-Forum standards *must* support AESAs and *may* support native E.164 addresses. Public ATM networks should support either one of them or both.

4.3 AESA: ATM End System Address

ATM End System Addresses (AESA) are a subclass of NSAP addresses as defined in [ISO8348]. They identify the location of one or more ATM interface in a given routing domain. There are two groups of these addresses: individual and group. Individual addresses identify a single ATM interface, whereas group addresses can be associated with more than one interface (see Section 4.5). There are different formats of AESA addresses (see Figure 4.3).

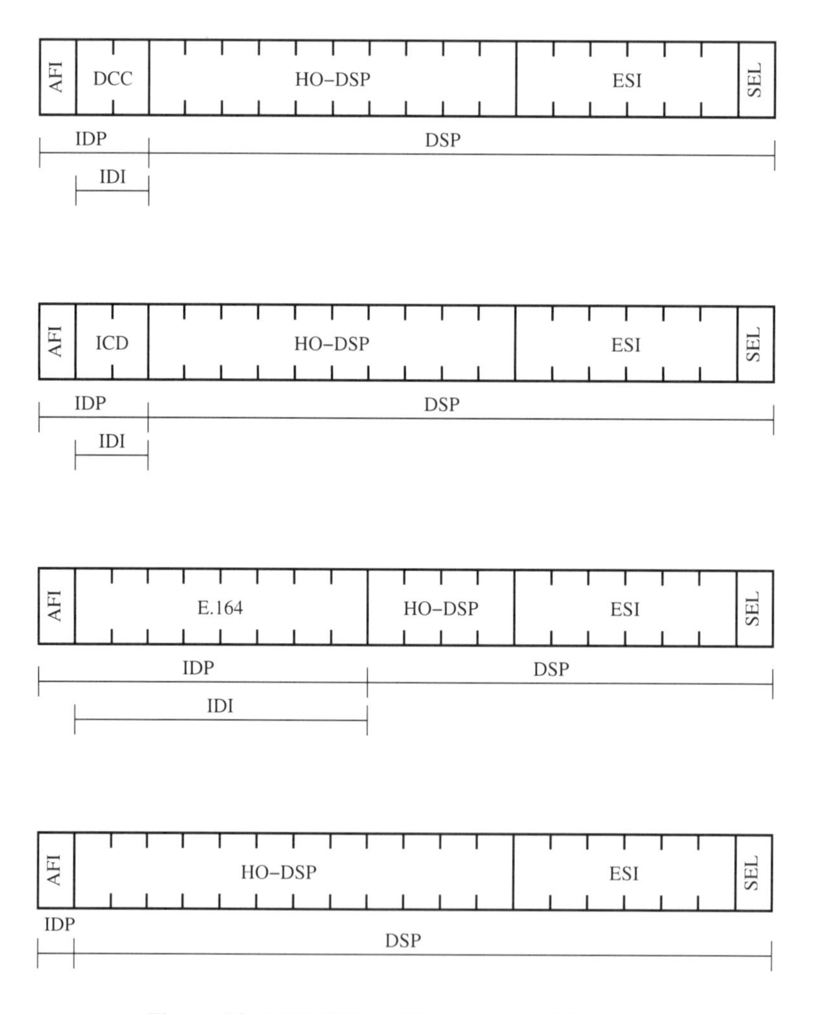

Figure 4.3: DCC, ICD, E.164 and Local AESA format

An AESA is always 20 bytes in length. The first byte is the Authority and Format Identifier (AFI). The AFI specifies the following:

- the format of the Initial Domain Identifier (IDI);
- the network addressing authority that is responsible for the allocation of the IDI values;
- whether or not leading zeros in the IDI are significant;
- the syntax of the Domain Specific Part (DSP).

The ATM-Forum specifies the values (Table 4.1) to be used for the allocation of AESA. It allows also for the use of other AFIs.

Table 4.1: AFI values for ATM End System Addresses

Individual AFI	Group AFI	Format
`0x39`	`0xBD`	DCC format
`0x45`	`0xC3`	E.164 format
`0x47`	`0xC5`	ICD format
`0x49`	`0xC7`	Local format

The Initial Domain Identifier (IDI) part of the address specifies the network addressing authority that is responsible for the allocation of the Domain Specific Part (DSP) values and the domain from which these values are allocated. The AFI and the IDI together form the Initial Domain Part (IDP) that uniquely identifies the Administrative Authority (AA) responsible for assignment of DSP values.

The IDI of a Data Country Code (DCC) address specifies the country that registered the address. These codes are given in [ISO3166] and are always three digits long. They are Binary Coded Decimal (BCD) encoded and padded to the right with the valued `0xF`. This results in an IDI length of two bytes.

The IDI of an International Code Designator (ICD) address specifies the authority that is responsible for the allocation of DSP values. It is a four-digit value that is BCD encoded into two bytes.

The IDI of an E.164 AESA is the Integrated Services Digital Network (ISDN) number of the addressed ATM interface. E.164 AESA include telephone numbers. Only the international format of these numbers is allowed. The number is padded to 15 digits with leading zeros, BCD encoded and the value `0xF` appended for a total of eight bytes.

Local AESAs can be used to carry other non-OSI address formats. They have no extra IDI part. They can be used in private networks, but should never be seen outside such a network.

The Domain Specific Part (DSP) of an AESA consists of three parts: the High-Order DSP (HO-DSP), the End System Identifier (ESI) and the Selector (SEL).

The selector is the last byte of the AESA and is never used for routing purposes. It may be used to pass information from one end system to another (it could be used, for example, like the port number in IP packets to address different applications).

The End System Identifier (ESI) is a six-byte value and uniquely identifies an end system for a given value of IDP+HO-DSP. Usually this identifier is an IEEE MAC address, and thus globally unique.

The format and value of the HO-DSP field is defined by the authority identifier by the

IDP. This field can and should contain subfields for which the authority is delegated to sub-authorities. These subfields should be allocated in a way so that the resulting HO-DSP contains enough topology information to facilitate call routing. As an example Figure 4.4 shows the structure of a US DCC AESA.

Figure 4.4: US DCC AESA format

The authority for US AESAs is ANSI. The country code is `840` and the first byte of the HO-DSP is the Domain Specific Part Format Indicator (DFI), which currently has a single value assigned: `0x80`. The DFI defines the format of the remaining bytes in the HO-DSP. With DFI=`0x80` the next three bytes are the Organisation Name which is assigned by ANSI. The remaining bytes are then defined by the named organisation.

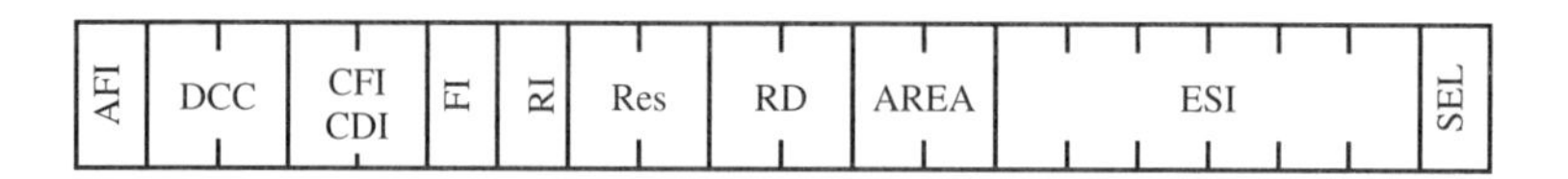

Figure 4.5: DFN DCC AESA format

Another example is the address format used by the Deutsches Forschungsnetzwerk (DFN) in Germany (see Figure 4.5). The country code for Germany is `0x276`. The two bytes following the country code are the Country Domain Part (CDP). This consists of a leading Country Format Identifier (CFI) which defines the format and length of the following Country Domain Identifier. The one-digit CFI in our example has the value 3 and the three-digit CDI has the value `0x100` (this is the code assigned to the organisation "DFN"). The next byte is the Format Identifier (FI); a value of `0x01` specifies that the rest of the address follows US-GOSIP Format II. The FI is followed by a four-digit Region Identifier (RI). There are values assigned to the various parts of Germany as well as to some large organisations. After two reserved bytes with the values `0x0000` follows a two-byte Routing Domain (RD) field. The values in this field are assigned to organisations within the given region. These organisations then define the format and values of the remaining two-byte Area field.

The AESAs are designed in such a way that subfields are allocated from left to right. This allows addresses to be grouped on subfield boundaries by using prefixes. During the routing process a prefix defines the route for all AESAs beginning with that prefix. In this way reachability information can be hierarchically grouped to reduce routing data base size and routing traffic.

4.4 Native E.164 Addresses

Native E.164 addresses are defined in ITU-T Recommendation [E.164]. Recommendation [E.191] then defines how these numbers are used in the B-ISDN context. The structure of an E.164 number is shown in Figure 4.6.

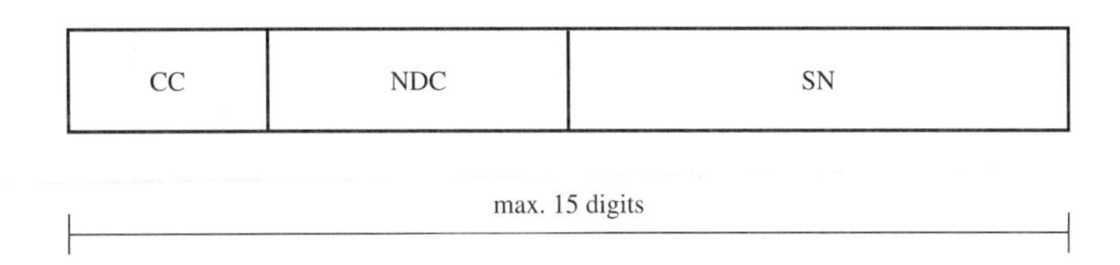

Figure 4.6: Native E.164 address format

An E.164 consists of a maximum of 15 digits. It starts with a country code (CC), which identifies either a country or geographical region, a global service or a network. By assigning a code to a country, a network or a region the responsibility for the assignment of values for the subsequent fields is delegated to an organisation of this country, region or network. The National Destination Code (NDC) identifies a trunk, an area in the given country, a national service (like freephone) or a network. The subscriber number (SN) identifies an addressable entity in the domain covered by the CC and the NDC.

E.164 cannot only be addresses, but also service identifiers. They can be abbreviated (there are national, local and other types of E.164 numbers). They can be prefixed with dialling plan digits or service selection escapes like "*" or "#". In the context of the ATM-Forum only the use of international E.164 as addresses is specified. Other uses are not precluded.

4.5 ATM Anycast

ATM Anycast is specified in [UNI4.0] as an optional feature that allows a user to establish a point-to-point connection to any member of a group of ATM end systems.

An anycast connection is requested by setting the called party number information element in the SETUP message to one of the ATM group addresses (see Table 4.1 for the AFIs of group addresses). The user can optionally include a connection scope information element which restricts the routing range used for the selection of the group member (see Table 4.2). If no scope is indicated by the user "localNetwork" is used. The network, upon receipt of an anycast SETUP, selects one of the end systems that is a member of the group and routes the SETUP to this system. The called end system can return its real address in the connected number and connected number subaddress information element of the CONNECT message. How the member is selected by the network is not specified by the standard.

The Anycast capability has several uses among which are:

- Enhanced reliability by duplication of critical servers, for example the ATM Address Resolution Protocol (ATMARP) server in CLIP (see Chapter 8) or the LAN Emulation Configuration Server (LECS) in LANE.
- Load balancing. This allows a call to be routed to a group of servers based on the current

load of each member of the group.

The ATM Anycast group membership is handled by ILMI. ILMI supports the registration of group addresses for an ATM interface as well as the registration of the scope of these addresses.

Table 4.2: Organisational connection scopes

Code	Meaning
0x00	reserved
0x01	local network
0x02	local network plus one
0x03	local network plus two
0x04	site minus one
0x05	intra-site
0x06	site plus one
0x07	organisation minus one
0x08	intra-organisation
0x09	organisation plus one
0x0a	community minus one
0x0b	intra-community
0x0c	community plus one
0x0d	regional
0x0e	inter-regional
0x0f	global

4.6 Address Aggregation

As was mentioned in previous sections, the fields of AESAs are allocated from left to right in a way to carry network topology information in the address. This allocation strategy facilitates route aggregation by means of address prefixes. An address prefix is the leading part of an AESA. A prefix is specified by a number L between 0 and 152 specifying the number of leading bits in the prefix (remember that the Selector byte is never used for routing, hence the maximum of 152), and an AESA with the rightmost $160 - L$ bits set to zero. A prefix length of zero aggregates all possible AESAs; a prefix length of 152 designates a single ATM interface.

Prefixes are aggregated in a hierarchical fashion with higher levels in the hierarchy using shorter prefixes than lower levels. This scheme is used in the PNNI routing protocol (see Chapter 6) to aggregate routing information in hierarchical networks. Figure 4.7 shows an example of hierarchical aggregation.

In this example an (US) ATM Service Provider (ASP) has obtained the DCC format AESA `39.840F.80.777888`. A German multi-site customer has in turn obtained three blocks of addresses from this service provider for its sites in Berlin, Frankfurt and Munich. These blocks are specified by using a two-byte (four-digit) field from the ASP-defined part of the AESA. The Berlin site has two facilities, so the address is further sub-divided by the customer by a one-byte field. Inside each facility the customer has different buildings for which he uses

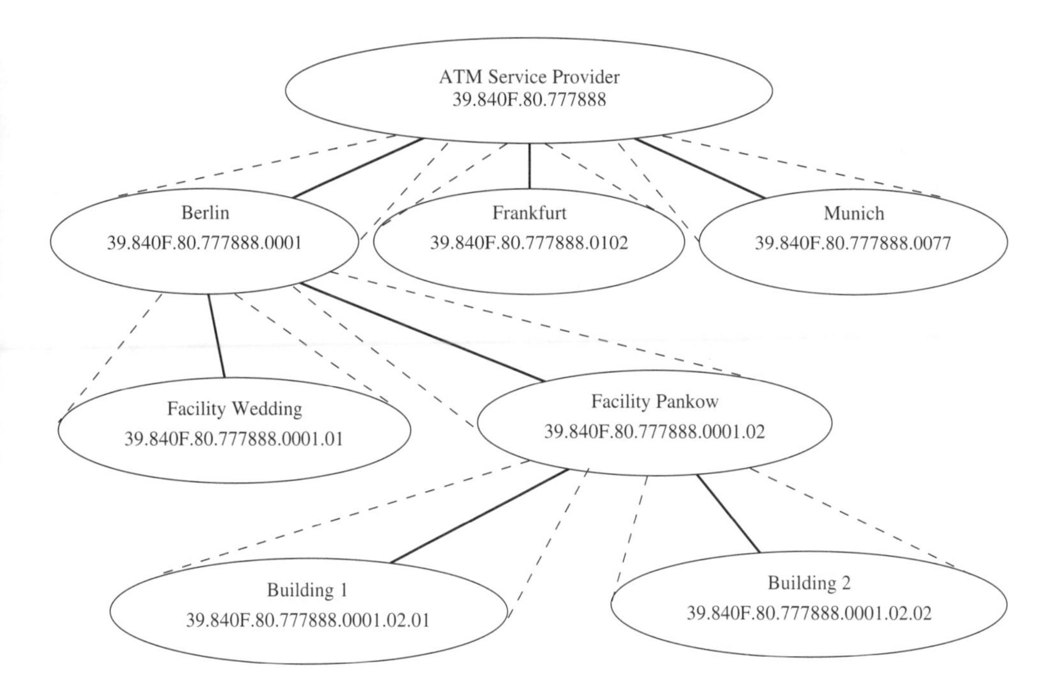

Figure 4.7: Example of an AESA hierarchy

another byte. This address organisation aggregates quite nicely:

- All ATM end systems in Building 1 in Berlin-Pankow can be reached through the address prefix `88/39.840F.80.777888.0001.02.01`.
- All systems in the Pankow Facility can be reached through the address prefix `80/39.840F.80.777888.0001.02`.
- All customer ATM systems in Berlin are reachable through the address prefix `72/39.840F.80.777888.0001`.
- To reach all end systems of this customer the service provider need only announce its own prefix `56/39.840F.80.777888` to other ASPs. This prefix aggregates all addresses allocated to this ASP (not only those of the example customer).

If the customer were to use an address owned by him, which cannot be aggregated with the ASP's addresses (for example, an ICD format address), the ASP would need to announce the customer's prefixes and carry the corresponding routes in all its nodes. In this case the routing tables grow linearly with the number of attachment points of such customers to the ASP.

Address prefixes are used for routing purposes. When routing a call the switch compares the destination address with all known prefixes in the routing table. Of course, there may be more than one match. Let us consider an example. A call to the customer from the previous example has to be routed by the ASP. The called number is

```
39.840F.80.777888.0001.02.02.0700000000.434f00.00
```

and the switch has the following routing table entries:

```
/0 (default route)
47.0081/24
39.840F.80.777999/56
39.840F.80.777888.0/60
39.840F.80.777888.1/60
39.840F.80.777888.0001/72
39.840F.80.777888.0002/72
```

In this case the entries in lines 1, 4 and 6 give a match. The route is selected based on the longest match, which is in line 6.

If there is more than one match with the same prefix length, the selection among these routes can be based on other information like trunk load or the traffic contract.

Hierarchical address aggregation is directly built into PNNI (although it allows other configurations). Nodes at lower levels of the hierarchy derive their address by appending a number of bits to the prefix of the next highest hierarchy. End systems compute their address by appending the ESI and SEL to the switch prefix. If a PNNI network is organised in this way, route aggregation is optimal.

4.7 Summary

In this chapter we have analysed which types of addresses are used in ATM networks. There are two types of addresses: telephone number-like addresses, called E.164 that are mainly used in public ATM networks, and ATM end system addresses (AESAs) that are a subset of NSAP addresses and are used in private ATM networks. A special feature of ATM addressing is the Anycast capability that can be used to build high-performance or high-reliability servers. An explanation of how works address aggregation and why it is useful closes the chapter.

5

SAAL: Signalling ATM Adaptation Layer

The Signalling ATM Adaptation Layer (SAAL) is the lowest layer in the control plane of the ATM that is different from the layers at the user plane [Q.2100]. Its main function is the reliable transfer of signalling messages between two signalling entities. It consists of several sublayers (see Figure 5.1).

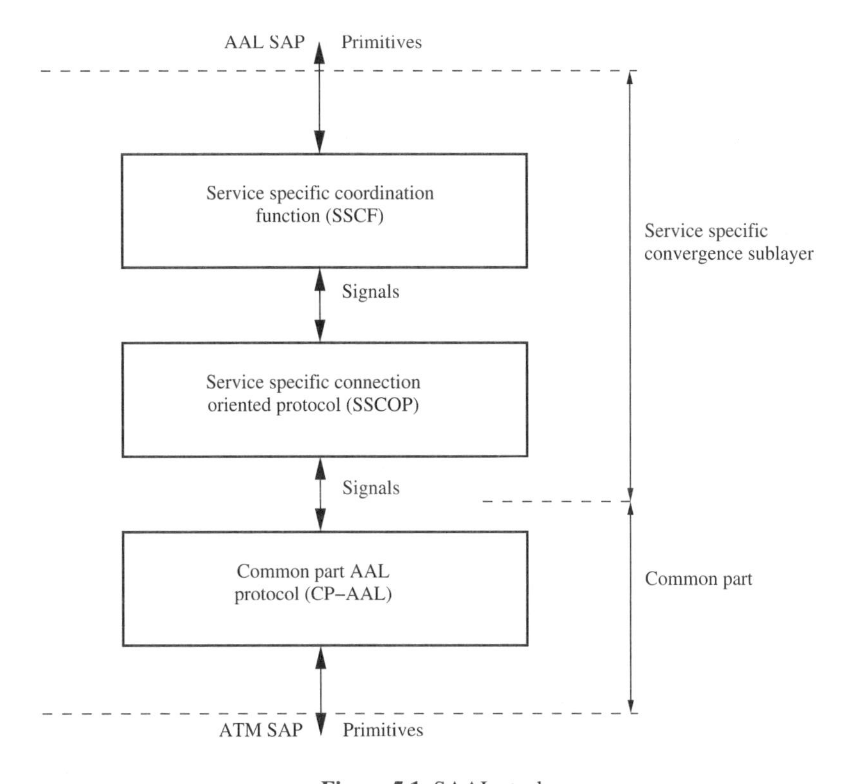

Figure 5.1: SAAL stack

The lower sublayer (CP-AAL) is usually AAL5 although AAL3/4 are also usable. In modern ATM interface cards this layer is implemented in the hardware for performance reasons. The CP-AAL is used by the SSCOP to send and receive AAL messages. No user-to-user information is carried in the AAL frames.

The Service Specific Connection Oriented Protocol (SSCOP) sublayer is the most interesting part of the SAAL. SSCOP is a transport protocol that provides guaranteed, in sequence delivery of messages to the layer above it. It includes also flow control, error reporting to the management plane and a keep-alive function.

The Service Specific Coordination Function (SSCF) actually comes in two types: the Service Specific Coordination Function (SSCF) at the User–Network Interface (UNI) and the SSCF at the Network–Node Interface (NNI). Whereas the SSCF at the UNI provides no additional functionality, just some simplification of the upper interface, the SSCF at the NNI constitutes a real protocol. This protocol is used to monitor the performance and quality of NNI links and helps in error situations. Because the SSCF at the NNI is used only in public networks which run the Broadband ISDN User Part (B-ISUP) stack, it is not of much interest for this book and we will focus on SSCOP and the UNI SSCF. A short description of the SSCF at the NNI shows how SSCOP can be tailored depending on the requirements of the application and how a more obscure feature of SSCOP—local message retrieval—can be used. For a more thorough treatment of the SSCF at the NNI the user should read [Q.2140] and [Q.2144].

5.1 SSCOP: Service Specific Connection Oriented Protocol

The SSCOP is defined in International Telecommunication Union (ITU-T) recommendation [Q.2110]. It provides the following functions to its user:

- *Reliable and in-sequence delivery of variable size messages.* Note that there is no error checking for the contents of the message in addition to what is done by the ATM Adaptation Layer Common Part (AAL-CP). Reliable just means that every message sent, will arrive. This is done by selective retransmission of lost or corrupted messages. The contents of message must, if necessary, protected by higher layers.
- *Unreliable transfer of data.* This may be used to send messages outside the reliable data stream (for example management or urgent data).
- *Flow control.*
- *Keep-alive.* This function ensures that the connection is alive even if no data is sent. This is checked by an exchange of status messages at regular intervals.
- *Local data retrieval.* This rather obscure function allows the SSCOP user to retrieve messages from the SSCOP that have not yet been sent. This is used by the NNI stack when it switches from a failed NNI link to a backup link. In this case unsent messages are retrieved from the failed SSCOP and sent on the backup link.

It must be noted that SSCOP seems to be overly complex for the simple task of signalling message transport at the UNI. Even careful handwritten C-code is at least in the order of 5000 lines of code. Signalling traffic on the UNI is typically very low and could use a simple Remote Procedure Call (RPC) scheme. Nevertheless SSCOP is now in wide use and is even used as a general-purpose transport protocol for ATM.

SSCOP is a trailer protocol—protocol information is *appended* to the user information. The user information part is padded to a multiple of four bytes and protocol information is always a multiple of four bytes long. This means that SSCOP can effectively be processed on modern computer architectures. Figure 5.2 shows how higher layer data is packed into SSCOP and AAL5 PDUs.

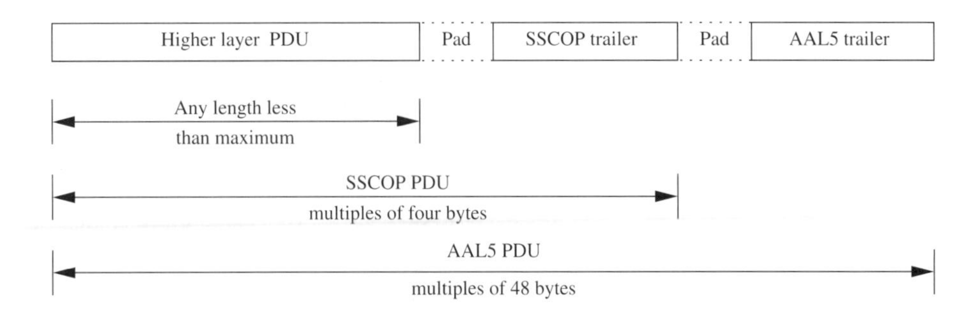

Figure 5.2: SAAL PDUs

SSCOP uses 15 different messages (see Figures 5.3–5.21 later in the chapter). The message fields have the following meaning:

User-to-user information
: This is additional information that can be transferred to the remote user with connection control messages during establishment, tear-down and resynchronisation. It is not used at the UNI. There is no guarantee that this information will arrive—if the SSCOP needs to retransmit a connection control message it may choose not to include the user information in the retransmitted message.

Padding
: This field is up to three bytes long and pads the size of user data to a multiple of four bytes. It should contain zeros. The length of the user-to-user information plus padding bytes is denoted by *Len* in Figures 5.3–5.21.

N(SQ)
: This is the sequence number of BGN, RS and ER messages and is used to identify retransmissions of these messages at the receiver. It is initialised to 0 and incremented after each transmission of these messages.

N(MR)
: This is one higher than the highest sequence number of an SD message that the receiver will accept (the upper bound of the receiver window). Note that the receiver may lower this variable, and even close the window.

N(S)
: This is the sequence number of the next new SD message.

PL
: This contains the number of padding bytes.

S
: This is set to 1 if the connection is released by SSCOP, and 0 if released by the SSCOP user.

—
: This denotes reserved fields, which should be set to zero.

The state diagram for SSCOP is quite complex and is not reproduced here. It can be found in [Q.2110].

An SSCOP connection is in one of 10 states (the names in parentheses will later be used in the communication traces):

`1(IDLE)`
: *Idle.* No connection exists.

`2(OUT_PEND)`
: *Outgoing connection pending.* SSCOP has sent a BGN message and is waiting for an acknowledgement or rejection.

`3(IN_PEND)`
: *Incoming connection pending.* SSCOP has received a BGN and informed the user. It waits for the user to decide whether to acknowledge or reject it.

`4(OUT_DIS_PEND)`
: *Outgoing disconnect pending.* SSCOP has sent a END message and waits for the acknowledgement.

`5(OUT_RESYNC_PEND)`
: *Outgoing resynchronisation pending.* An RS message has been sent on the user's request to resynchronise the states of the SSCOP entities. An RSAK is expected from the peer. The result of this operation is that the user knows that all messages have been acknowledged by the remote SSCOP entity.

`6(IN_RESYNC_PEND)`
: *Incoming resynchronisation pending.* An RS message has been received. SSCOP waits for the user to decide upon the acknowledgement.

`7(OUT_REC_PEND)`
: *Outgoing recovery pending.* SSCOP has detected a problem and has requested recovery from the peer by sending an ER message.

`8(REC_PEND)`
: *Recovery response pending.* SSCOP has completed recovery (an ERAK was received), has informed its user and waits for a response from the user to enter State 10.

`9(IN_REC_PEND)`
: *Incoming recovery pending.* SSCOP has recovered on request from the peer (which sent an ER message) and waits for a response from the user.

`10(READY)`
: *Data transfer ready.* SSCOP is ready to receive and transmit data.

Transitions between these states occur when SSCOP receives messages, signals or timers expire. The exact transitions can be found in the standard.

SSCOP needs five timers for processing:

CC
: This timer is used during connection control phases (establishment, tear-down, error recovery and resynchronisation). If no answer from the remote side is received when the timer expires, the message is retransmitted. This timer should be greater than the round trip time. The standard value is 1 second. The number of retransmissions before giving up is configurable; the standard value is four seconds.

POLL
: Each SSCOP sends its peer entity POLL messages at regular intervals to ensure that status information (data acknowledgements and the window size) are actual. The timer controls the interval between these messages with a standard value of 750 milliseconds. This timer is used only when data messages are to be sent or acknowledgements are outstanding.

KEEP-ALIVE
: If there are no data messages to transmit and there are no outstanding acknowledgements, SSCOP switches from timer POLL to timer KEEP-ALIVE. The normal value is 2 seconds.

IDLE
: If the connection is stable enough and there are no data messages to transmit and no outstanding acknowledgements, SSCOP switches from timer KEEP-ALIVE to timer IDLE. The standard value is 15 seconds.

NO-RESPONSE
: In parallel to timer POLL and timer KEEP-ALIVE the timer NO-RESPONSE is running. This timer determines the maximum time interval during which at least one STAT message must be received in response to a POLL. On expiry of this timer the connection is aborted. The standard value is 7 seconds.

The values of these timers are not given in the SSCOP standard but rather in the SSCF standards that reside on top of SSCOP.

5.1.1 SSCOP Interfaces

At the lower interface (the interface to the AAL5 CPCS) two signals are used:

CSCP-UNITDATA.invoke
: This is used by SSCOP to send an AAL frame to the remote user.

CSCP-UNITDATA.signal
: This is generated by the CPCS if an AAL frame from the remote user arrives.

The upper interface is more complex and uses many signals with different parameters which may have *request*, *response*, *indication* and *confirmation* forms.

AA-ESTABLISH
: This group of signals is used to establish a SSCOP connection. There are four variants of it (request, response, indication and confirmation).

AA-RELEASE
: This group of signals is used to either reject a connection request or to tear-down a connection. There is no `AA-RELEASE.response`. (It is not possible to reject a connection release.)

AA-DATA
: These signals are used to reliably send data or receive such data. There is a request and an indication.

AA-RESYNC
: This group is used to resynchronise the SSCOP connection. There are four forms as in `AA-ESTABLISH`.

AA-RECOVER
: The indication and response forms are used to recover from protocol errors. SSCOP invokes the recovery procedure for cases that are not covered by the retransmission of messages—if a message is received that has already been received or sequence numbers are outside the expected range. There are two forms: `AA-RECOVER.indication` and `AA-RECOVER.response`.

AA-UNITDATA
: The request can be used to send an unassured message to the remote user. The remote user will get an indication (if the message is not lost). Requests and indications exist for `AA-UNITDATA`.

AA-RETRIEVE
: By invoking this request the SSCOP user indicates that he wishes to get all messages out of SSCOP's send buffer that have not yet been transmitted. The user can retrieve either single messages by message number or all messages. For each retrieved message the SSCOP will invoke an indication. There is only a request and an indication form.

AA-RETRIEVE-COMPLETE
: When all requested messages have been retrieved from the SSCOP it invokes this indication. Only the indication form exists.

The interface to layer management uses two groups of signals:

MAA-ERROR
: If the SSCOP detects an error it invokes this indication. The indication contains a one-character code which will describe the error condition. The error codes are listed in Table 5.2 on page 142. There is only an indication form of this signal.

MAA-UNITDATA
: This pair of request and indication can be used by layer management to send messages to the remote management entity. Note, that this is an unassured service. The UNI does not use this feature. There is a request and an indication form.

5.1.2 Message Types

SSCOP uses 15 different message types as show in Table 5.1. The exact format of these messages will be shown in the following sections.

5.1.3 State Variables

SSCOP needs a number of state variables. They can roughly be divided into three groups: connection control variables, status enquiry variables and data flow variables.

The following variables are used to control a connection:

VT(SQ) *Transmitter Connection Sequence.* This variable is used to detect retransmissions of connection control messages: BGN, RS and ER. Each time one of these messages is transmitted *VT(SQ)* is incremented and inserted into the message. If the message is to be retransmitted the same value will be used.

Table 5.1: SSCOP PDU types

Function	Message name	Code	Description
Establishment	BGN BGAK BGREJ	0001 0010 0111	establish connection acknowledge connection establishment reject connection
Tear-down	END ENDAK	0011 0100	tear-down connection acknowledge tear-down
Resynchronisation	RS RSAK	0101 0110	start resynchronisation acknowledge resynchronisation
Error recovery	ER ERAK	1001 1111	start error recovery acknowledge error recovery
Assured data transfer	SD POLL STAT USTAT	1000 1010 1011 1100	sequenced data transmit and request state information solicited state information unsolicited state information
Unassured data transfer	UD	1101	unassured user data
Management data transfer	MD	1110	unassured management data

VT(CC) *Transmitter Connection Control State.* This indicates the number of outstanding BGN, END, ER or RS messages. This variable is set to zero if one of the associated procedures is started (establishment, tear-down, error recovery or resynchronisation) and incremented for each of the above messages. If a limit (the default is four) is reached without getting a response from the peer, the connection is aborted.

VR(SQ) *Receiver Connection Sequence.* This is the receiver side variable that corresponds to *VT(SQ)*. When a BGN, RS or ER message is seen, the current value of the variable is compared with the sequence number in the message. If they are equal, the message was caused by a re-transmission. After the comparison the variable is set to the value in the message.

For the status enquiry procedure the following variables are used:

VT(PS) *Poll Send State.* This is the current poll message sequence number which is incremented before each transmission of a POLL message and is inserted into this message.

VT(PA) *Poll Acknowledge State.* When SSCOP receives a POLL message it inserts the sequence number from the POLL message into the STAT message it is sending to acknowledge the POLL. *VT(PS)* is the lowest poll sequence number the sender expects to see in a STAT message. If a STAT message is received, the variable is set to the sequence number in the STAT message.

VT(PD) *Poll Data State.* This variable counts the number of SD (data) messages sent since the last POLL message was emitted. It is reset to zero when the next POLL message is sent.

VT(PS)@VT(PS) and *VT(PA)* form the window for legal sequence number values in received STAT messages. A STAT message is legal if its sequence number *SEQ* holds:

$$VT(PS) \leq SEQ \leq VT(PA)$$

The rest of the variables used to control the data flow are:

VT(S) *Send State.* This variable contains the sequence number of the next newly transmitted SD message and is incremented after each transmission, but not for retransmissions.

VT(A) *Acknowledge State.* This is the sequence number of the next in-sequence SD messages expected to be acknowledged. This is updated when the next in-sequence SD message is acknowledged.

VT(MS) *Maximum Send State.* This is the maximum sequence number plus one that the remote receiver will accept. This is updated on receipt of any of the messages that contain an *N(MR)* field.

VR(R) *Receive State.* This is the sequence number of the next in-sequence SD message that the receiver expects. It is incremented when that message is seen.

VR(H) *Highest Expected State.* This is the highest expected sequence number in an SD message. This may be updated from the next SD or POLL message and should roughly be equal to the peer's *VT(S)*.

VR(MR) *Maximum Acceptable Receive State.* This is the highest SD sequence number plus one that the receiver will accept. How this is updated is implementation dependent.

VR(R) and *VR(MR)* form the window for legal SD sequence numbers on the receiver side. An SD message is accepted when its sequence number *SEQ* holds:

$$VR(R) \leq SEQ < VR(MR)$$

5.1.4 Connection Establishment

In the next sections the operation of SSCOP will be shown by means of communication traces. These traces have been taken by connecting two SSCOPs with a bi-directional pipe. Between these two SSCOPs a program is used (`stee`) that records all messages with timestamps in a packet stream file. These files were then analysed with the `sscopdump` program.

The reader should imagine two SSCOPs—one to the left and one to the right. There is also an SSCOP user to the left of the left SSCOP and one to the right of the right SSCOP. There are three colums after the starting “SSCOP” in each trace: the first stands for the upper interface of the left SSCOP, the middle for the communication between the SSCOPs and the right for the upper interface of the left SSCOP. A right arrow in the left column means a signal sent from the left SSCOP user to the left SSCOP; a left arrow in that column means a signal from the left SSCOP to the left SSCOP user. The right column is the other way around: a right arrow denotes a signal to the user; a left arrow from the user. The second column may contain one of the following symbols:

$\Rightarrow\times$	Denotes that this message from the left to the right SSCOP was lost.
$\times\Leftarrow$	Denotes that this message from the right to the left SSCOP was lost.
$\Rightarrow$	In the left column this means that the left SSCOP user has sent a signal to the SSCOP; in the right column this means that the right SSCOP has sent a signal to its user.
$\Leftarrow$	In the right column this means that the right SSCOP user has sent a signal to the SSCOP; in the left column this means that the left SSCOP has sent a signal to its user.
↴	This message, sent by the left SSCOP, will be delayed on the link.
↙	This message, sent by the right SSCOP, will be delayed on the link.
↳	The delayed message from the left SSCOP arrives.
↲	The delayed message from the right SSCOP arrives.

The delay of messages is sometimes necessary to show what happens if both protocol instances do something simultaneously. Note that a delayed message is shown twice: once when it is sent and once when it is received (both times with the sending SSCOP).

For signals, the rest of the line is the signal name and the state change, which is the result of the signal. For messages, the next column shows the time relative to the first message in seconds and the rest of the line describes the different fields of the message.

Before user data can be transmitted over SSCOP, the connection must be established. For signalling connections on the UNI this is usually initiated by the switch by sending a BGN message (see Figure 5.3) and starting the CC timer. The other side answers with either a BGREJ (see Figure 5.5) to reject the establishment of the connection (this is unusual for signalling connections) or BGAK (Figure 5.4). If the timer CC expires and no message is received, the BGN message is retransmitted. To identify duplicate messages, BNG messages have an eight-bit sequence number which is incremented each time a BGN is transmitted. The number of retransmissions, which is counted in the variable *VR(CC)*, is configurable—the UNI SSCF defines four retransmissions. If the maximum number of retransmissions is exceeded, the SSCOP gives up, sends an END message and remains in the idle state.

It should be noted that most ATM switches behave differently, i.e. they restart connection establishment after sending the END message.

Let us take a look at a successful connection establishment. The following is a trace of the communication between two SSCOP entities. It shows what happens if the first BGAK is lost.

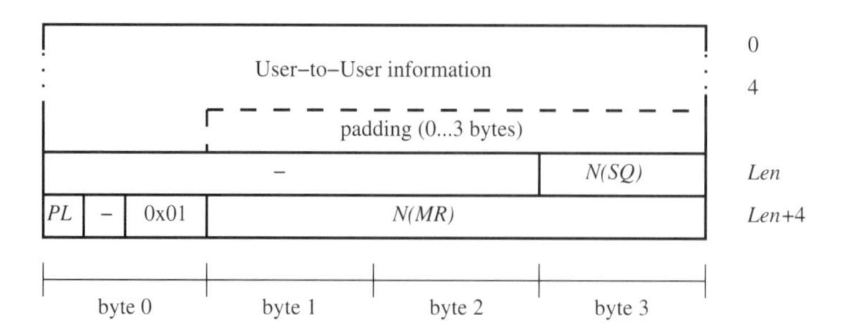

Figure 5.3: SAAL BGN message

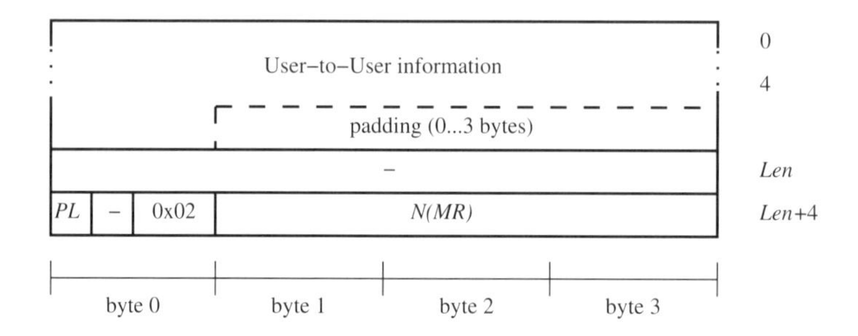

Figure 5.4: SAAL BGAK message

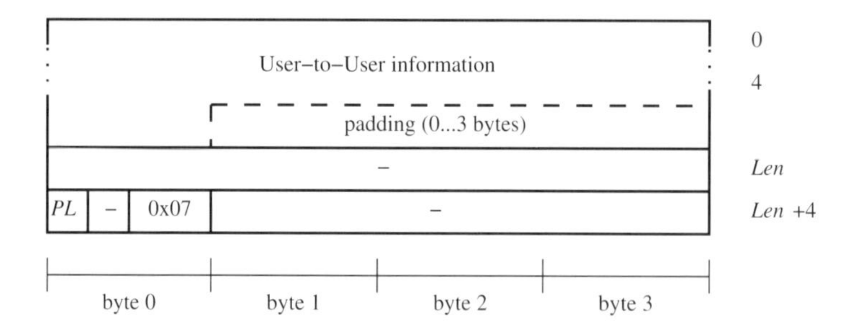

Figure 5.5: SAAL BGREJ message

```
1  SSCOP ⇒ . .          AA-ESTABLISH.request in state S_IDLE
2  SSCOP . ⇒ .   0.002  begin mr=128 sq=1
3  SSCOP . . S          state S_IDLE -> S_IN_PEND
4  SSCOP . . ⇒          AA-ESTABLISH.indication in state S_IN_PEND
5  SSCOP S . .          state S_IDLE -> S_OUT_PEND
6  SSCOP . . ⇐          AA-ESTABLISH.response in state S_IN_PEND
7  SSCOP . ×⇐ .  0.657  bgak mr=128
8  SSCOP . . S          state S_IN_PEND -> S_READY
```

```
 9  SSCOP  .  ⇒  .   1.016  begin mr=128 sq=1
10  SSCOP  .  ⇐  .   1.018  bgak mr=128
11  SSCOP  ⇐  .  .          AA-ESTABLISH.confirm in state S_OUT_PEND
12  SSCOP  S  .  .          state S_OUT_PEND -> S_READY
13  SSCOP  .  ⇐  .   1.416  poll s=0 ps=1
14  SSCOP  .  ⇒  .   1.419  stat r=0 mr=128 ps=1 list={}
```

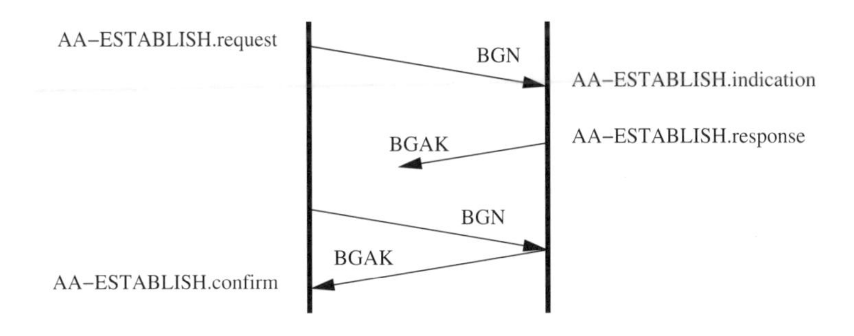

Figure 5.6: Losing a BGAK message

The initiating protocol instance sends a BGN (lines 1 and 2), starts timer CC and enters state 2. Upon receiving this BGN, the destination protocol instance informs its user (line 3) which in turn tells the SSCOP to accept the connection (line 4). The SSCOP sends a BGAK (line 5), starts timer POLL and timer NO-RESPONSE, enters state 10 and is thus ready to transfer data. This first BGAK is lost, so the first SSCOP times out on timer CC and retransmits the BGN (it contains the same sequence number, so it is a retransmission) (line 6). Because the peer is already in state 10 (data transfer ready) it can answer directly, without user intervention, by retransmitting the BGAK (line 7). After receiving the BGAK the initiating protocol instance stops timer CC, starts the timers POLL and NO-RESPONSE and enters state 10 (line 8). Now both SSCOPs are in the ready state (this can be seen by the exchange of POLL and STAT messages).

Another interesting question is: what happens if both SSCOPs happen to send a BGN simultaneously? The following trace shows this scenario:

```
 1  SSCOP  ⇒  .  .          AA-ESTABLISH.request in state S_IDLE
 2  SSCOP  S  .  .          state S_IDLE -> S_OUT_PEND
 3  SSCOP  .  ⇁  .   0.002  begin mr=128 sq=1
 4  SSCOP  .  .  ⇐          AA-ESTABLISH.request in state S_IDLE
 5  SSCOP  .  .  S          state S_IDLE -> S_OUT_PEND
 6  SSCOP  .  ⇐  .   0.510  begin mr=128 sq=1
 7  SSCOP  ⇐  .  .          AA-ESTABLISH.confirm in state S_OUT_PEND
 8  SSCOP  S  .  .          state S_OUT_PEND -> S_READY
 9  SSCOP  .  ↳  .   1.011  begin mr=128 sq=1
10  SSCOP  .  .  ⇒          AA-ESTABLISH.confirm in state S_OUT_PEND
11  SSCOP  .  .  S          state S_OUT_PEND -> S_READY
12  SSCOP  .  ⇒  .   1.012  bgak mr=128
13  SSCOP  .  ⇐  .   1.012  bgak mr=128
```

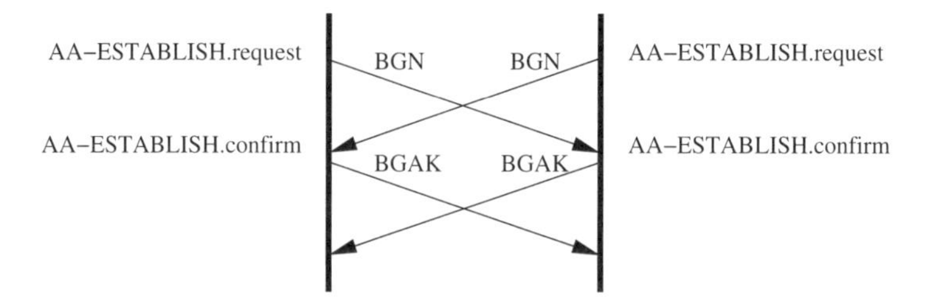

Figure 5.7: Simultaneous BGN message

To do this experiment the first BGN message from SSCOP 1 is delayed for half a second so that the peer can also send a BGN (lines 2 and 4). On line 5 the BGN is delivered to SSCOP 2. As one can see, upon receiving the BGN in state 2 (outgoing connection pending), the first SSCOP sends a BGAK, enters state 10 (ready), stops timer CC and starts the data transfer timers (POLL and NO-RESPONSE). The same holds for its peer. The BGAK message arriving in state 10 (ready) is ignored.

A connection request from the remote SSCOP can be rejected by responding to the BGN message with a BGREJ:

```
1  SSCOP ⇒ . .          AA-ESTABLISH.request in state S_IDLE
2  SSCOP S . .          state S_IDLE -> S_OUT_PEND
3  SSCOP . ⇒ .   0.001  begin mr=128 sq=1
4  SSCOP . . ⇒          AA-ESTABLISH.indication in state S_IN_PEND
5  SSCOP . . S          state S_IDLE -> S_IN_PEND
6  SSCOP . . ⇐          AA-RELEASE.request in state S_IN_PEND
7  SSCOP . . S          state S_IN_PEND -> S_IDLE
8  SSCOP . ⇐ .   0.629  bgrej
9  SSCOP ⇐ . .          AA-RELEASE.indication in state S_OUT_PEND
10 SSCOP S . .          state S_OUT_PEND -> S_IDLE
```

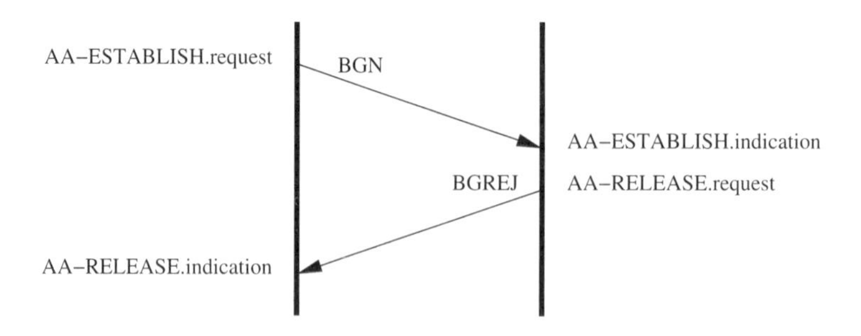

Figure 5.8: Reject connection

An unsuccessful connection establishment because of not getting any answer is shown in the following trace. The maximum number of connection control message retransmissions is (as per default) four:

```
1   SSCOP ⇒ .  .          AA-ESTABLISH.request in state S_IDLE
2   SSCOP S .  .          state S_IDLE -> S_OUT_PEND
3   SSCOP . ⇒  .   0.002  begin mr=128 sq=1
4   SSCOP . .  ⇒          AA-ESTABLISH.indication in state S_IN_PEND
5   SSCOP . .  S          state S_IDLE -> S_IN_PEND
6   SSCOP . ⇒  .   1.008  begin mr=128 sq=1
7   SSCOP . ⇒  .   2.018  begin mr=128 sq=1
8   SSCOP . ⇒  .   3.028  begin mr=128 sq=1
9   SSCOP ⇐ .  .          AA-MERROR.indication in state S_OUT_PEND
10  SSCOP ⇐ .  .          AA-RELEASE.indication in state S_OUT_PEND
11  SSCOP S .  .          state S_OUT_PEND -> S_IDLE
12  SSCOP . ⇒  .   4.039  end reason=sscop
13  SSCOP . .  ⇒          AA-RELEASE.indication in state S_IN_PEND
14  SSCOP . .  S          state S_IN_PEND -> S_IDLE
15  SSCOP . ⇐  .   4.040  endak
```

It can be seen that the remote SSCOP user fails to respond to the `AA-establish.request`. The BGN message is retransmitted four times after the expiry of timer CC. The trace also shows that the failure to establish the connection is reported to layer management by means of an `AA-ERROR.indication`.

Note that it is possible to send additional information to the user of the remote SSCOP. BGN, BGAK and BGREJ messages can contain user information. The UNI does not use this feature.

5.1.5 Connection Tear-down

Each side of an SSCOP connection can tear down the connection by sending a END message (see Figure 5.9). This can be initiated by either the user or the protocol itself in the case of fatal protocol errors; a bit in the protocol trailer indicates who released the connection. The other SSCOP responds with a ENDAK message (see Figure 5.10).

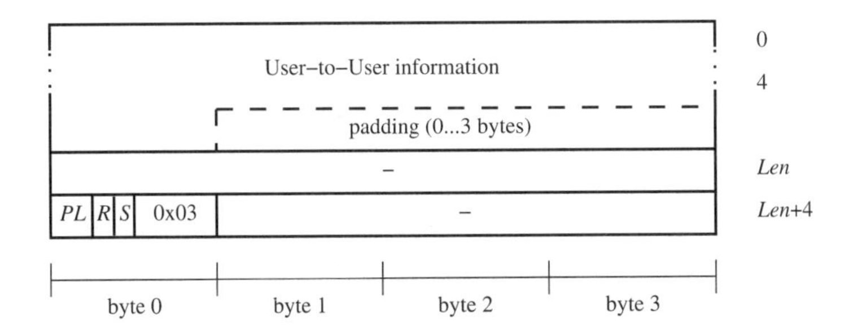

Figure 5.9: SAAL END message

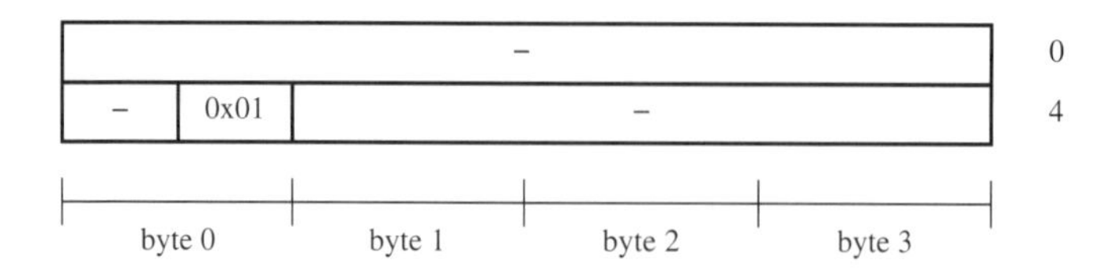

Figure 5.10: SAAL ENDAK message

The SSCOP user initiates connection tear-down by sending the `AA-RELEASE.request` signal. The user of the remote SSCOP will receive an `AA-RELEASE.indication`, which has not be to answered (it is not possible to reject a release). The remote SSCOP answers the END with an ENDAK, which generates an `AA-RELEASE.confirmation` to the releasing user:

```
1  SSCOP ⇒ . .          AA-RELEASE.request in state S_READY
2  SSCOP S . .          state S_READY -> S_OUT_DIS_PEND
3  SSCOP . ⇒ .   4.051  end reason=user
4  SSCOP . . ⇒          AA-RELEASE.indication in state S_READY
5  SSCOP ⇐ . .          AA-RELEASE.confirm in state S_OUT_DIS_PEND
6  SSCOP . . S          state S_READY -> S_IDLE
7  SSCOP . ⇐ .   4.052  endak
8  SSCOP S . .          state S_OUT_DIS_PEND -> S_IDLE
```

Of course, simultaneous release also works:

```
1  SSCOP ⇒ . .          AA-RELEASE.request in state S_READY
2  SSCOP S . .          state S_READY -> S_OUT_DIS_PEND
3  SSCOP . ⤵ .   2.119  end reason=user
4  SSCOP . . ⇐          AA-RELEASE.request in state S_READY
5  SSCOP . . S          state S_READY -> S_OUT_DIS_PEND
6  SSCOP . ⇐ .   2.912  end reason=user
7  SSCOP ⇐ . .          AA-RELEASE.confirm in state S_OUT_DIS_PEND
8  SSCOP S . .          state S_OUT_DIS_PEND -> S_IDLE
9  SSCOP . ↳ .   3.123  end reason=user
10 SSCOP . . ⇒          AA-RELEASE.confirm in state S_OUT_DIS_PEND
11 SSCOP . . S          state S_OUT_DIS_PEND -> S_IDLE
12 SSCOP . ⇐ .   3.124  endak
13 SSCOP . ⇒ .   3.124  endak
```

In this case the ENDAK messages are effectively ignored.

Note that a receiving an END message is legal in all states except state 2 (outgoing connection pending)—a connection request is rejected with a BGREJ.

5.1.6 Assured Data Transfer and Keep-Alive

Data are sent to the remote SSCOP by means of SD messages (see Figure 5.11). Each SD message contains a 24-bit sequence number that is taken from the variable *VT(S)*, which is incremented after sending a new message. This sequence number identifies the data message.

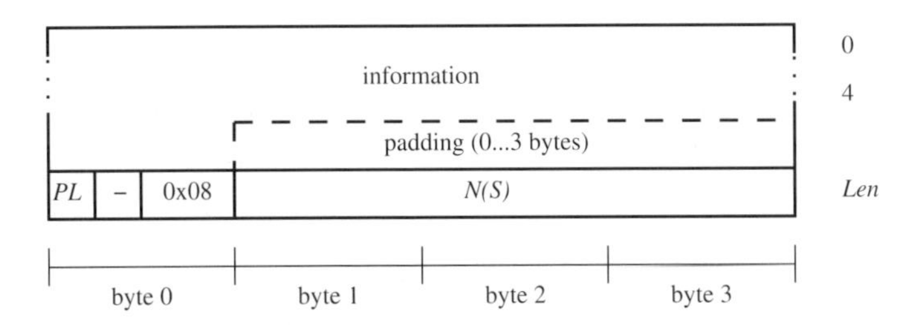

Figure 5.11: SAAL SD message

If a message needs to be resent, the same number is used again. (Messages are held in a queue until they are acknowledged, so it is easy to retransmit the message with the same sequence number.) The number of bytes that can be transported by an SD message is specified by the parameter *k* of the protocol. For signalling purposes this parameter is set to 4096 as required by ITU-T recommendation [Q.2130]. Note that an SD message can also be empty.

Acknowledgements and flow control information are transported in STAT messages (see Figure 5.12). These messages are not emitted automatically, but rather requested by sending a POLL message (Figure 5.13).

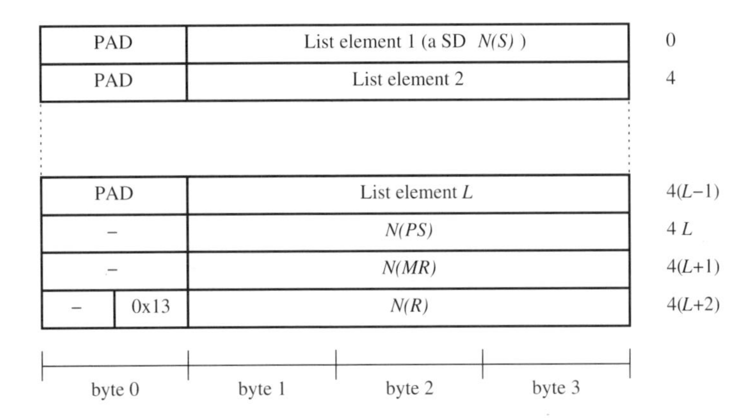

Figure 5.12: SAAL STAT message

POLL messages contain the current value *N(S)* of the *VT(S)* variable (e.g. the sequence number of the next fresh SD message) and the poll sequence number *N(PS)* from the variable *VT(PS)*. The first of these values is needed to detect lost messages. If, for example, SSCOP 0 has just emitted the SD message with the sequence number 56, the following POLL message will contain the sequence number 57 (remember that *VT(S)* is incremented after sending the message.) If message 56 is lost, SSCOP 1 will detect this loss, because it expects to see sequence number 56 in the POLL.

The STAT message that is constructed in response to a POLL contains at least three numbers and a variable sized list of SD message sequence numbers. This list specifies ranges of SD

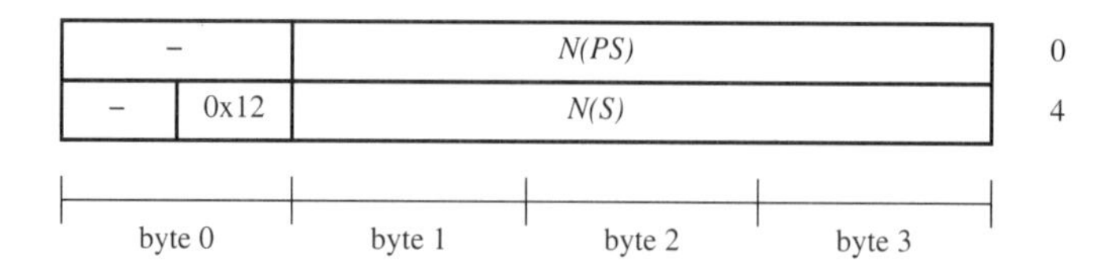

Figure 5.13: SAAL POLL message

messages that were apparently lost and may be empty. *N(R)* is the current value of *VR(R)*, which describes the sequence number of the next in-sequence SD message that the receiver expects. *N(MR)* (the current value of *VR(MR)*) specifies the current receive window and specifies one higher than the maximum sequence number the receiver is willing to accept. *N(PS)* is the poll sequence number of the POLL message that triggered this STAT message. The relationship between the different SD sequence numbers is shown in Figure 5.14.

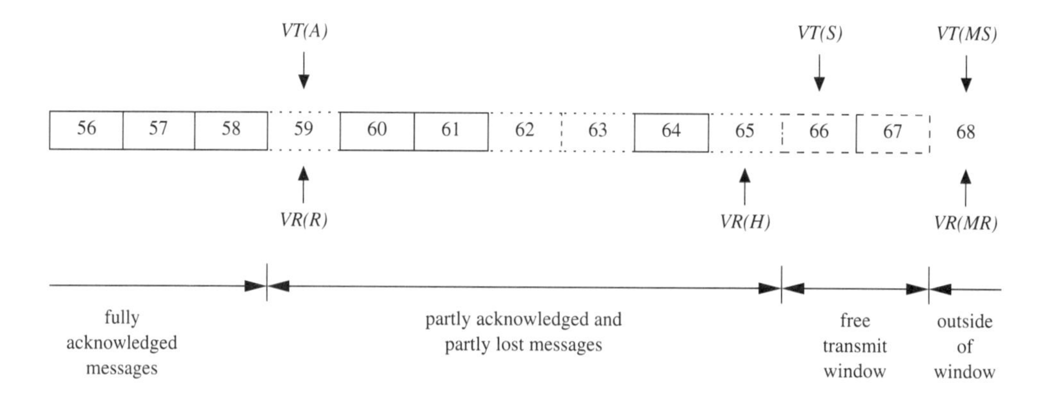

Figure 5.14: SSCOP sequence number relationship

In this figure messages up to number 58 are successfully received and also acknowledged. This means that both *VT(A)* and *VR(R)* have the same value, namely 59: *VT(A)* because the next expected in-sequence acknowledgement belongs to message 59; *VR(R)* because the next expected in-sequence message is message 59. Messages 59, 62, 63 and 65 have been lost. *VT(S)* is 66—the next new SD message will carry this sequence number. *VR(H)* is 65 because the highest SD sequence number seen so far was 64. The receive window continues up to message 67; 68 would already be outside the window so both *VT(MS)* and *VR(MR)* are 68 (given that a STAT message was received since the last update of these variables).

The POLL message and the resulting STAT message for the situation depicted in Figure 5.14 is shown in Figure 5.15.

The POLL message says that the transmitting side has sent messages with sequence numbers up to 65; 66 would be the next. The STAT message says that the receiver expects the next in-sequence message to be message number 59 and that it is willing to receive messages up to (but not including) number 68. It also contains a list of sequence numbers. These

byte 0		bytes 1–3	offset
–		*N(PS)*=XXX	0
–	0x12	*N(S)*=66	4

byte 0		bytes 1–3	offset
0		66	0
0		65	4
0		64	8
0		62	12
0		60	16
0		59	20
–		*N(PS)* =XXX (from POLL)	24
–		*N(MR)*=68	28
–	0x13	*N(R)*=59	32

byte 0 byte 1 byte 2 byte 3

Figure 5.15: POLL message and STAT message with lost messages list

sequence numbers must be interpreted in pairs N, M, each pair N, M tells the other side that messages with numbers from N up to $M - 1$ have been lost and should be retransmitted. In the example this means that the receiver has not received messages 59, 62, 63 and 65.

Data can be sent only if the SSCOP is in state 10 (ready). Actually there are three sub-states of the ready state:

active The protocol entity is in this state if there are any SD messages to be transmitted or when there are outstanding acknowledgements. In this state the timers POLL and NO-RESPONSE are running in parallel. Timer POLL ensures that POLL messages are transmitted (these are needed to get STAT messages that contain acknowledgements and update the send window (see Section 5.1.7)).

The protocol does not insist on a STAT message in response to each POLL. During the interval of timer NO-RESPONSE at least one STAT message must be received. In this case the timer is restarted. If the timer expires, the SSCOP connection is aborted.

transient When all queued SD PDUs are sent, there are no outstanding acknowledgements and timer POLL expires SSCOP enters the active sub-state. Instead of restarting timer POLL, timer KEEP-ALIVE is started, which is considerably longer. As in the active state, loss of POLL or STAT messages is protected by the NO-RESPONSE timer.

The state changes back to active whenever new data is to be transmitted.

Timer KEEP-ALIVE is greater than timer POLL and greater than the round trip delay. This means that viewer POLL messages are sent.

idle When a STAT message is received and timer KEEP-ALIVE is still running, both timers KEEP-ALIVE and NO-RESPONSE are stopped and timer IDLE started instead. While timer IDLE is running no POLL messages are sent. When the timer expires, the transient state is entered again. Because timer KEEP-ALIVE expires only when POLL or STAT PDUs are lost, the switch to the idle state occurs only when the connection seems stable enough. Timer IDLE is considerably greater than timer KEEP-ALIVE.

The following trace shows how SSCOP switches between these three sub-states.

```
 1  SSCOP ⇒ .  .         AA-ESTABLISH.request in state S_IDLE
 2  SSCOP S .  .         state S_IDLE -> S_OUT_PEND
 3  SSCOP . ⇒  .  0.004  begin mr=128 sq=1
 4  SSCOP . .  S         state S_IDLE -> S_IN_PEND
 5  SSCOP . .  ⇒         AA-ESTABLISH.indication in state S_IN_PEND
 6  SSCOP . ⇒  .  1.014  begin mr=128 sq=1
 7  SSCOP . .  ⇐         AA-ESTABLISH.response in state S_IN_PEND
 8  SSCOP . .  S         state S_IN_PEND -> S_READY
 9  SSCOP . ⇐  .  1.144  bgak mr=128
10  SSCOP ⇐ .  .         AA-ESTABLISH.confirm in state S_OUT_PEND
11  SSCOP S .  .         state S_OUT_PEND -> S_READY
12  SSCOP . ⇐  .  1.907  poll s=0 ps=1
13  SSCOP . ⇒  .  1.911  poll s=0 ps=1
14  SSCOP . ⇒  .  1.915  stat r=0 mr=128 ps=1 list={}
15  SSCOP . ⇐  .  1.917  stat r=0 mr=128 ps=1 list={}
16  SSCOP . ⇐  . 16.925  poll s=0 ps=2
17  SSCOP . ⇒  . 16.927  stat r=0 mr=128 ps=2 list={}
18  SSCOP . ⇒  . 16.929  poll s=0 ps=2
19  SSCOP . ⇐  . 16.931  stat r=0 mr=128 ps=2 list={}
20  SSCOP . ⇒  . 31.936  poll s=0 ps=3
21  SSCOP . ⇐  . 31.938  poll s=0 ps=3
22  SSCOP . ⇐  . 31.940  stat r=0 mr=128 ps=3 list={}
23  SSCOP . ⇒  . 31.941  stat r=0 mr=128 ps=3 list={}
24  SSCOP ⇒ .  .         AA-DATA.request in state S_READY
25  SSCOP . ⇒  . 39.539  sd s=0 k(4)=67:75:67:75
26  SSCOP . .  ⇒         AA-DATA.indication in state S_READY
27  SSCOP . ⇒  . 40.294  poll s=1 ps=4
28  SSCOP . ⇐  . 40.296  stat r=1 mr=129 ps=4 list={}
29  SSCOP . ⇒  . 41.055  poll s=1 ps=5
30  SSCOP . ⇐  . 41.056  stat r=1 mr=129 ps=5 list={}
31  SSCOP . ⇐  . 46.956  poll s=0 ps=4
32  SSCOP . ⇒  . 46.957  stat r=0 mr=128 ps=4 list={}
33  SSCOP . ⇒  . 56.075  poll s=1 ps=6
34  SSCOP . ⇐  . 56.076  stat r=1 mr=129 ps=6 list={}
35  SSCOP . ⇐  . 61.975  poll s=0 ps=5
36  SSCOP . ⇒  . 61.977  stat r=0 mr=128 ps=5 list={}
```

After a connection has been established the active phase is entered by starting timer POLL in lines 8 and 11. At expiry of this timer POLL messages are sent (lines 12 and 13) and, because the transmit buffer is empty and no acknowledgements are expected (i.e. *VT(S)* is equal to *VT(A)*), timer KEEP-ALIVE is started and thus the transient phase entered. In both phases timer NO-RESPONSE is running in parallel (this is not shown). In lines 14 and 15 both SSCOPs receive STAT messages in response to their POLL messages. Because at that time the KEEP-ALIVE timers are still running, the idle phase is entered by stopping timers KEEP-ALIVE and NO-RESPONSE and starting timer IDLE. These timers expire after 15 seconds and POLL messages are transmitted (lines 16 and 18). The transient phase is entered again (to wait for the STAT messages) by starting timers NO-RESPONSE (to protect against losing connectivity) and KEEP-ALIVE. At lines 17 and 19 responses are received and both SSCOPs switch to the idle phase again (lasting until lines 20 and 21) . In this manner the SSCOPs instances toggle between the transient phase to wait for a STAT and the idle phase.

If in the transient phase a STAT message (or the requesting POLL) is lost, a new POLL is transmitted and the SSCOP remains in the transient phase, i.e. starts timer KEEP-ALIVE. This may be repeated a number of times, which depends on the relation of the values of timers KEEP-ALIVE and NO-RESPONSE. With the standard values (2 and 7 seconds, respectively) three retransmissions of the POLL would be done before aborting the connection at expiry of timer NO-RESPONSE.

If a message is to be transmitted the SSCOP leaves the transient or idle phase and enters the active phase (see lines 24 and 25). Timer POLL is started and timer NO-RESPONSE started or restarted and, at expiry of timer POLL (line 27), a POLL message is sent. The STAT message in line 28 carries the acknowledgement of the data message, so the SSCOP can enter the transient phase again.

Note that there is a maximum number of SD messages, that can be sent without an intervening POLL. The standard value for this parameter is 25, which means that after transmitting 25 SD messages without expiry of timer POLL, a POLL message is sent in any case. This prevents the receiving SSCOP from getting too busy processing data messages and not being able to acknowledge these messages. If this were to happen, the sending SSCOP could use up the available send window and would then stop sending, which in turn would drain the pipe and reduce performance.

Upon receipt of an SD message the SSCOP remains in the state it was—only in line 31 is a new POLL sent, after timer IDLE has expired.

To speed up the recovery process in the case of lost messages, the receiving SSCOP has the ability to report problems immediately to the sending SSCOP by means of USTAT messages (Figure 5.16). Such messages are transmitted in two cases upon receipt of an SD message: when the SD message falls outside the receive window (this could happen, for example, if the receiver has reduced the window) and when an SD message loss is detected.

The first case is detected when the sequence number in the received SD message is not less than *VR(MR)*. If this happens for the first time, an USTAT is transmitted, containing *VR(H)* (the next highest SD PDU expected) and *VR(MR)* (the first out-of-window SD PDU). Upon sending this message, *VR(H)* is set equal to *VR(MR)*. If on receipt of another out-of-the-window message these variables are found equal, that SD PDU is simply ignored. The following trace shows this behaviour:

```
1  SSCOP . ⇒ .  1.521  poll s=0 ps=1
2  SSCOP . ⇐ .  1.524  poll s=0 ps=1
3  SSCOP . ⇒ .  1.525  stat r=0 mr=128 ps=1 list={}
4  SSCOP . ⇐ .  1.527  stat r=0 mr=128 ps=1 list={}
```

```
 5  SSCOP ⇒ .  .          AA-DATA.request in state S_READY
 6  SSCOP .  . ⇒          AA-DATA.indication in state S_READY
 7  SSCOP .  ⇒ .   3.279  sd s=0 k(1)=31
 8  SSCOP .  ⇒ .   4.040  poll s=1 ps=2
 9  SSCOP .  ⇐ .   4.042  stat r=1 mr=129 ps=2 list={}
10  SSCOP ⇒ .  .          AA-DATA.request in state S_READY
11  SSCOP .  ⇒ .   4.145  sd s=131 k(1)=32
12  SSCOP . ×⇐ .   4.147  ustat r=1 mr=129 list={1,129}
13  SSCOP . ⇒× .   4.800  poll s=2 ps=3
14  SSCOP ⇒ .  .          AA-DATA.request in state S_READY
15  SSCOP .  ⇒ .   5.298  sd s=132 k(1)=33
16  SSCOP . ⇒× .   5.570  poll s=3 ps=4
```

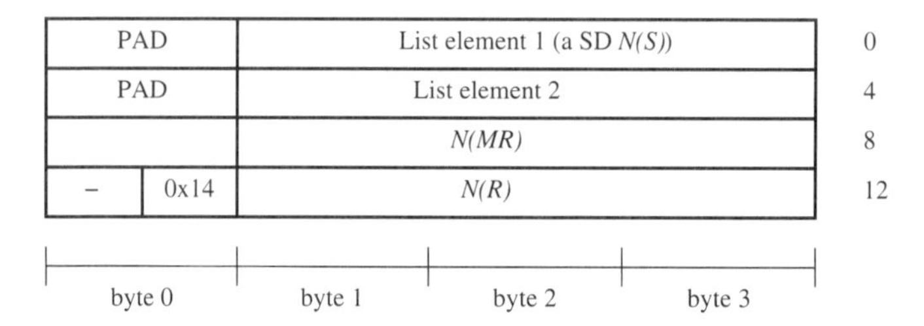

Figure 5.16: SAAL USTAT message

SSCOP 1 announces a window of 128 in line 4 and SSCOP 0 sends an SD message with sequence number 0 in line 7. In line 9 the receiver announces the new window 129. However, in line 11 SSCOP 0 sends an SD message with sequence number 131, which is outside the window and SSCOP 1 drops that message and sends a USTAT. This USTAT says that the next in-sequence SD PDU should be number 1 (`r=1`), that messages up to and not including sequence number 129 are OK to send and that the highest sequence number seen so far was 1 and the window is 129.

The second case for transmission of a USTAT is detected when the received SD message is not the next in sequence SD message (that is, its sequence number is not equal to *VR(R)*) and it is not the next expected highest SD message, but a later one (i.e. its sequence number is not equal to *VR(H)*, but higher). If the SD message's sequence number is equal to *VR(H)*, the new message is simply put into the buffer. If the new SD message has a sequence number higher than *MR(H)*, messages have been lost. In this case a USTAT is generated containing the current *VR(H)* and the received sequence number, thus requesting the immediate retransmission of all messages in between. This behaviour can be seen in the following trace:

```
1  SSCOP .  ⇒ .   1.510  poll s=0 ps=1
2  SSCOP .  ⇐ .   1.510  poll s=0 ps=1
3  SSCOP .  ⇒ .   1.513  stat r=0 mr=128 ps=1 list={}
4  SSCOP .  ⇐ .   1.514  stat r=0 mr=128 ps=1 list={}
5  SSCOP ⇒ .  .          AA-DATA.request in state S_READY
6  SSCOP .  ⇒ .   2.363  sd s=0 k(1)=31
```

```
7   SSCOP  .  .  ⇒            AA-DATA.indication in state S_READY
8   SSCOP  .  ⇒  .    3.120   poll s=1 ps=2
9   SSCOP  .  ⇐  .    3.121   stat r=1 mr=129 ps=2 list={}
10  SSCOP  .  ⇒  .    3.880   poll s=1 ps=3
11  SSCOP  .  ⇐  .    3.882   stat r=1 mr=129 ps=3 list={}
12  SSCOP  ⇒  .  .            AA-DATA.request in state S_READY
13  SSCOP  .  ⇒×  .   4.349   sd s=1 k(1)=32
14  SSCOP  ⇒  .  .            AA-DATA.request in state S_READY
15  SSCOP  .  ⇒  .    4.653   sd s=2 k(1)=33
16  SSCOP  .  ⇐  .    4.654   ustat r=1 mr=129 list={1,2}
17  SSCOP  ⇐  .  .            AA-MERROR.indication in state S_READY
18  SSCOP  .  ⇒  .    4.658   sd s=1 k(1)=32
19  SSCOP  .  .  ⇒            AA-DATA.indication in state S_READY
20  SSCOP  .  .  ⇒            AA-DATA.indication in state S_READY
21  SSCOP  .  ⇒  .    5.111   poll s=3 ps=4
22  SSCOP  .  ⇐  .    5.113   stat r=3 mr=131 ps=4 list={}
```

The SD message with sequence number 1 sent in line 13 is lost. When the next message with sequence number 2 is received in line 15, the receiver detects that its sequence number is higher than *VR(H)*. *VR(H)* at that point is 1, which means that the next highest message expected is 1. When the receiver sees sequence number 2 it deduces, that message number 1 must have been lost and sends a USTAT message requesting immediate retransmission of message 1. Upon receipt of this message the transmitter retransmits message 1 in line 18 and the receiver can deliver both messages 1 and 2 to the SSCOP user (lines 19 and 20).

Upon receipt of a USTAT message, SSCOP removes all acknowledged messages from its buffers, updates *VT(A)* and *VT(NS)* (the send window) accordingly and moves all messages between the receivers *VR(H)* and *VR(MR)* (which were reported in the USTAT) from the transmission buffer, where they wait for acknowledgement, to the retransmission buffer, where they will be held until the window allows them to be sent. If the window is open, they will be sent immediately.

In the case, when no USTAT message is generated, but messages where lost, the next STAT message will contain ranges of message numbers to be retransmitted. When these messages are received, all messages that are in-sequence are delivered to the SSCOP user and deleted from the receive buffer. The next STAT or USTAT message then acknowledges these messages and moves the window. In the following traces messages 2, 3 and 4 are lost and then later retransmitted:

```
1   SSCOP  .  ⇒  .    1.585   poll s=0 ps=1
2   SSCOP  .  ⇐  .    1.587   poll s=0 ps=1
3   SSCOP  .  ⇒  .    1.588   stat r=0 mr=128 ps=1 list={}
4   SSCOP  .  ⇐  .    1.590   stat r=0 mr=128 ps=1 list={}
5   SSCOP  ⇒  .  .            AA-DATA.request in state S_READY
6   SSCOP  .  ⇒  .    2.604   sd s=0 k(1)=31
7   SSCOP  .  .  ⇒            AA-DATA.indication in state S_READY
8   SSCOP  .  ⇒  .    3.364   poll s=1 ps=2
9   SSCOP  .  ⇐  .    3.366   stat r=1 mr=129 ps=2 list={}
10  SSCOP  .  ⇒  .    4.124   poll s=1 ps=3
11  SSCOP  .  ⇐  .    4.126   stat r=1 mr=129 ps=3 list={}
12  SSCOP  ⇒  .  .            AA-DATA.request in state S_READY
13  SSCOP  .  ⇒×  .   5.171   sd s=1 k(1)=32
14  SSCOP  ⇒  .  .            AA-DATA.request in state S_READY
```

```
15  SSCOP  . ⇒× .   5.411  sd s=2 k(1)=32
16  SSCOP ⇒ .  .           AA-DATA.request in state S_READY
17  SSCOP  . ⇒× .   5.664  sd s=3 k(1)=32
18  SSCOP ⇒ .  .           AA-DATA.request in state S_READY
19  SSCOP  . ⇒  .   5.938  sd s=4 k(1)=32
20  SSCOP  . ⇒  .   5.940  poll s=5 ps=4
21  SSCOP  . ⇐  .   5.942  ustat r=1 mr=129 list={1,4}
22  SSCOP ⇐ .  .           AA-MERROR.indication in state S_READY
23  SSCOP  . ⇒× .   5.947  sd s=1 k(1)=32
24  SSCOP ⇐ .  .           AA-MERROR.indication in state S_READY
25  SSCOP  . ⇐  .   5.948  stat r=1 mr=129 ps=4 list={1-4,5}
26  SSCOP  . ⇒  .   5.950  sd s=2 k(1)=32
27  SSCOP  . ⇒  .   5.952  sd s=3 k(1)=32
28  SSCOP ⇒ .  .           AA-DATA.request in state S_READY
29  SSCOP  . ⇒  .   6.554  sd s=5 k(1)=32
30  SSCOP  . ⇒  .   6.704  poll s=6 ps=5
31  SSCOP  . ⇐  .   6.706  stat r=1 mr=129 ps=5 list={1-2,6}
32  SSCOP ⇐ .  .           AA-MERROR.indication in state S_READY
33  SSCOP  . ⇒  .   6.709  sd s=1 k(1)=32
34  SSCOP  .  . ⇒          AA-DATA.indication in state S_READY
35  SSCOP  .  . ⇒          AA-DATA.indication in state S_READY
36  SSCOP  .  . ⇒          AA-DATA.indication in state S_READY
37  SSCOP  .  . ⇒          AA-DATA.indication in state S_READY
38  SSCOP  .  . ⇒          AA-DATA.indication in state S_READY
39  SSCOP  . ⇒  .   7.464  poll s=6 ps=6
40  SSCOP  . ⇐  .   7.466  stat r=6 mr=134 ps=6 list={}
```

This example shows how both mechanisms, STAT and USTAT, work. The sender's SSCOP tries to send five messages. The first one (sequence number 0) is received and delivered to the receiver's user (lines 5–7). The next three messages are lost (lines 13, 15 and 17). The receiver detects this loss when receiving message five in line 19 (because its *VR(H)* is still 1) and sends a USTAT to request retransmission of messages 1 to 3. The transmitter resends these, but message 2 is lost again (line 23). This time the loss is not detected by the receiver until the transmitter requests a status report. The emitted STAT message in line 31 tells the transmitter SSCOP that message 1 is still missing. After successful retransmission and receipt of this message all buffered messages 1–5 can be delivered to the user.

This example shows that in complex loss scenarios retransmission is triggered solely by the exchange of POLL and STAT messages. If timer POLL is too large, this can slow down protocol performance. To overcome this problem Q.2110 offers an implementation option called "Poll after retransmission". If this option is implemented a POLL message is emitted after the last retransmission. Each time an SD message is retransmitted, the retransmit buffer is checked and if it is found to be empty, the POLL message is sent. In this way the receiver will report still missing messages immediately and they can be retransmitted as fast as possible.

5.1.7 Flow Control

As already seen in previous paragraphs, SSCOP includes a window scheme for flow control. The receiver maintains a variable *VR(MR)* which contains the sequence number of the first SD message that the receiver will not accept. SD messages with numbers equal to or higher than *VR(MR)* will be discarded by the receiver. In certain circumstances the receiver will emit a USTAT message in this case.

VR(MR) is sent to the transmitter in the *N(MR)* field of BGN, BGAK, RS, RSAK, ER, ERAK, STAT and USTAT messages. The transmitter copies the value from this field into its *VT(MS)* variable. It uses this variable to decide whether or not another SD message can be transmitted, when the transmit buffer is not empty.

The window on the transceiver side is bound by *VT(A)* at the lower end and *VT(MS)-1* at the upper end. SD messages with sequence numbers between these values may be transmitted and the receiver should be able to provide enough buffers in its receive queue to buffer all these messages. If there is not enough buffer space in the receiver it may discard incoming messages (which means that for performance reasons in large bandwidth-delay product situations the receiver may allow a larger window than it is able to buffer), but may never discard already acknowledged messages and must always be able to buffer at least the message with sequence number *VR(R)*, unless the window is completely closed (*VR(R)* = *VR(H)* = *VR(MR)*).

The receiver is allowed to reduce the upper end of the window, but only to a minimum of *VR(H)*. This means that the window cannot be reduced below the highest currently acknowledged message. The lower end of the window is maintained automatically by *VT(A)* and *VR(R)*; the upper end is maintained through unspecified procedures in *VT(MS)* and *VR(MR)*. Because of the conditions

$$VT(A) \leq VT(S) \leq VT(MS)$$

and

$$VR(R) \leq VR(H) \leq VR(MR)$$

the window can be completely closed only when all messages are acknowledged by the receiver.

Sequence number arithmetic in SSCOP is done modulo 2^{24}. This limits the operating window of the protocol to $2^{24} - 1$, but, because in contrast to TCP/IP the sequence numbers are message and not byte numbers, this limits the amount of outstanding bytes in the pipe to $(2^{24} - 1)k$, where k is the maximum number of bytes in the user part of SD, MD and UD messages. If this is set to the maximum allowed by AAL5 this amounts to $(2^{24} - 1)65535 = 1{,}099{,}494{,}785{,}025$, which is about 1 terabyte and should be enough even for high-bandwidth satellite links.

The algorithm for choosing the initial window *VR(MR)* as well as the method of updating the window is not specified in Q.2110, but left to the implementation. The window should be initialised to a value that is computed based on the round-trip delay of the connection, the bandwidth and the value of the various timers. It must be updated each time a data receive event occurs. Appendix IV of [Q.2110] provides a formula for computing the default window size:

$$w = 2 + (2T_{\text{Poll}} + 6T_{\text{Delay}})\frac{BW}{8Len}$$

where

w = the window size in messages,

T_{Poll} = the timer POLL value of the peer in seconds,

T_{Delay} = the end-to-end transit delay in seconds,

BW = the bandwidth of the connection in bits per second, and

Len = the message length in bytes.

SSCOP events like message reception and signal processing are normally processed in the order in which they occur. However, in the event of congestion, SSCOP status events have priority over data transfer. When congestion at layers below SSCOP is detected (this may happen, for example, if the underlying ATM connection is traffic-shaped, in which case a queueing delay may be introduced and even blocking), the SSCOP entity may choose to suspend servicing data requests from the upper layer and retransmissions. During suspension messages are held in three different queues, for ordinary, management and unassured data, until the lower layer gets decongested again. The interface between the SSCOP and layer management or lower layers to detect congestion is not specified in the standard.

The following trace illustrates how the peer-to-peer flow control algorithm works:

```
1   SSCOP  . ⇒ .   1.683  poll s=0 ps=1
2   SSCOP  . ⇐ .   1.686  stat r=0 mr=3 ps=1 list={}
3   SSCOP ⇒ . .           AA-DATA.request in state S_READY
4   SSCOP  . . ⇒          AA-DATA.indication in state S_READY
5   SSCOP  . ⇒ .   3.065  sd s=0 k(1)=31
6   SSCOP ⇒ . .           AA-DATA.request in state S_READY
7   SSCOP  . ⇒ .   3.728  sd s=1 k(1)=32
8   SSCOP  . . ⇒          AA-DATA.indication in state S_READY
9   SSCOP  . ⇒ .   3.824  poll s=2 ps=2
10  SSCOP  . ⇐ .   3.825  stat r=2 mr=3 ps=2 list={}
11  SSCOP ⇒ . .           AA-DATA.request in state S_READY
12  SSCOP  . ⇒ .   4.252  sd s=2 k(1)=33
13  SSCOP  . . ⇒          AA-DATA.indication in state S_READY
14  SSCOP  . ⇒ .   4.584  poll s=3 ps=3
15  SSCOP  . ⇐ .   4.585  stat r=3 mr=3 ps=3 list={}
16  SSCOP ⇐ . .           AA-MERROR.indication in state S_READY
17  SSCOP ⇒ . .           AA-DATA.request in state S_READY
18  SSCOP  . ⇒ .   5.354  poll s=3 ps=4
19  SSCOP  . ⇐ .   5.355  stat r=3 mr=3 ps=4 list={}
20  SSCOP ⇒ . .           AA-DATA.request in state S_READY
21  SSCOP  . ⇒ .   6.114  poll s=3 ps=5
22  SSCOP  . ⇐ .   6.115  stat r=3 mr=3 ps=5 list={}
23  SSCOP  . ⇒ .   6.874  poll s=3 ps=6
24  SSCOP  . ⇐ .   6.875  stat r=3 mr=3 ps=6 list={}
25  SSCOP  . ⇒ .   7.634  poll s=3 ps=7
26  SSCOP  . ⇐ .   7.635  stat r=3 mr=3 ps=7 list={}
27  SSCOP  . ⇒ .   8.394  poll s=3 ps=8
28  SSCOP  . ⇐ .   8.396  stat r=3 mr=3 ps=8 list={}
29  SSCOP  . ⇒ .   9.155  poll s=3 ps=9
30  SSCOP  . ⇐ .   9.157  stat r=3 mr=3 ps=9 list={}
31  SSCOP  . ⇒ .   9.914  poll s=3 ps=10
32  SSCOP  . ⇐ .   9.915  stat r=3 mr=6 ps=10 list={}
33  SSCOP ⇐ . .           AA-MERROR.indication in state S_READY
34  SSCOP  . ⇒ .   9.918  sd s=3 k(1)=34
35  SSCOP  . . ⇒          AA-DATA.indication in state S_READY
36  SSCOP  . . ⇒          AA-DATA.indication in state S_READY
37  SSCOP  . ⇒ .   9.921  sd s=4 k(1)=35
38  SSCOP  . ⇒ .   10.674 poll s=5 ps=11
39  SSCOP  . ⇐ .   10.675 stat r=5 mr=6 ps=11 list={}
```

In line 2, the receiving SSCOP protocol instance announces that it is willing to accept messages with a sequence number up to, but not including, 3. The sending SSCOP receives requests from its user to send five messages in lines 3, 6, 11, 17 and 20. Three of these are actually sent and delivered to the receiver's user in lines 4, 8 and 13. Then the sender detects that the window is closed, polls a STAT message and, on receiving this message (which announces a closed window) in line 15, issues a management error signal with code "W" (no credit) in line 16. The sender now remains in the active phase and keeps polling the receiver in the hope that the window will open. The active state is maintained because timer POLL is much less than timer IDLE, so the opening of the window can be detected faster. In line 32 a STAT message is finally received, which announces a new *VR(MR)* of 6 and the sender sends the two queued messages.

5.1.8 Recovery from Protocol Errors

Two cases showing how the protocol recovers from transmission errors have already been described in the previous section. By means of STAT and USTAT messages the receiver tells the transmitter which messages seem to be missing. The transmitter retransmits these messages until they are acknowledged.

Situations may occur, however, when the peer entities get confused about the state of each other. This may happen, for example, when bit errors occur that are undetected by the AAL5 layer below (this is very unlikely) or, more likely, in the case of implementation errors of the peer entity. For such cases the protocol includes a mechanism for explicit error recovery, which is initialised automatically each time an error is detected. On the peer side the error recovery process is reported to the upper layer and must be confirmed from that layer, so appropriate actions for upper layer protocol integrity can be done.

Error recovery is started in state 10 (Data Transfer Ready) in the following cases:

- an SD PDU is received that is already in the receive buffer;
- a POLL message indicates a send state (i.e. the number of the next new SD message the transmitter will send) that is lower than the maximum SD sequence number seen so far;
- the receiver requests the retransmission of an SD message it has already been acknowledged;
- the receiver requests the retransmission of an SD message outside the window;
- a STAT or USTAT message contains bad sequence numbers (first number of the range larger than the second one);
- a STAT message contains a POLL sequence number outside the POLL sequence window.

Figure 5.17 shows the signals, messages and state changes during a normal error recovery.

If one of the above conditions is detected, the SSCOP stops all timers, reports an error to the management entity, resets the flow control window to an initial value, empties the retransmission queue and, if allowed, empties the transmission queue and buffer (see Section 5.1.11). It then emits an ER message (see Figures 5.18 and 5.19), starts timer CC and enters state 7 (Outgoing Recovery Pending).

On receipt of the ER PDU the peer SSCOP stops all timers, initialises the send window from the ER PDU, empties the retransmission queue and, if allowed, empties the transmission queue and buffer, clears the receiver buffer (after delivering all in-sequence data to its user) and sends an `AA-RECOVER.indication` to the SSCOP user. It then enters state 9 (Incoming Recovery Pending).

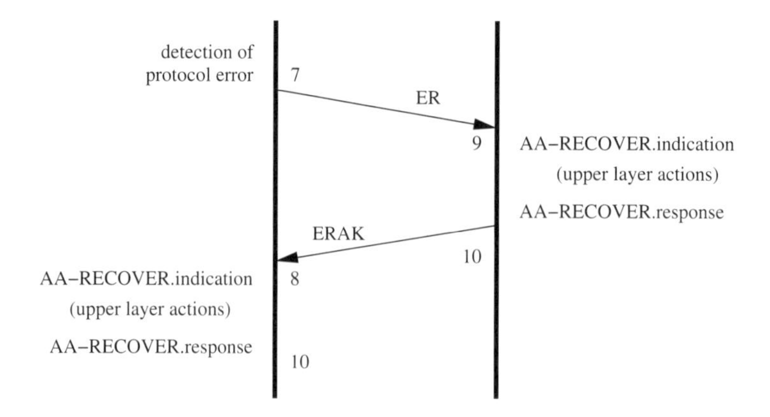

Figure 5.17: SSCOP error recovery

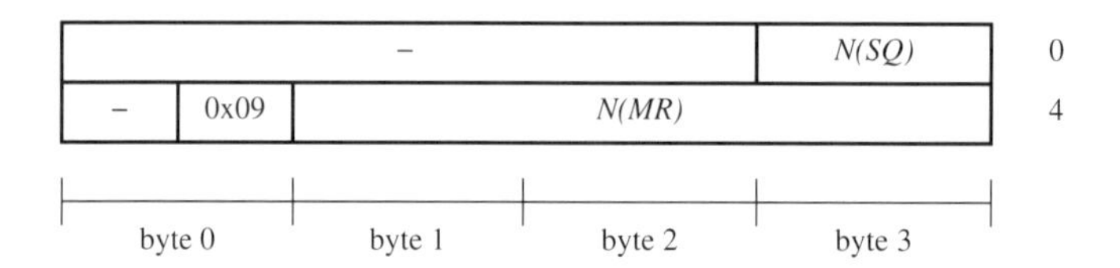

Figure 5.18: SAAL ER message

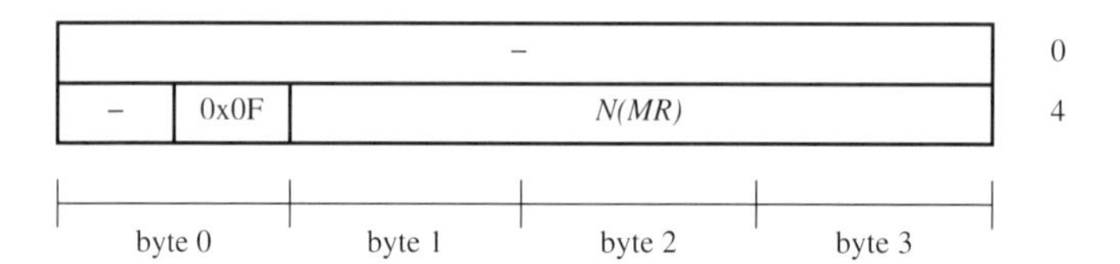

Figure 5.19: SAAL ERAK message

In this state the SSCOP user can take the appropriate actions to guarantee its integrity (error recovery is initiated in cases where the protocol cannot guarantee integrity) and then invokes an `AA-RECOVER.response`. Upon receipt of this signal SSCOP the empties the transmission buffer (if it was not allowed previously), reinitialises all data transfer state variables, transmits an ERAK message and enters the active phase of the Data Transfer Ready state (state 10).

When the original invoker of the error recovery procedure receives the ERAK message, it initialises its send window from the value in the ERAK, delivers any in-sequence data to its user, clears the receive buffer, sends an `AA-RECOVER.indication` to its user and enters state 8 (Recovery Response Pending). Then the user has the chance of doing everything to guarantee its integrity and answers with an `AA-RECOVER.response`. When the SSCOP receives this response it clears the transmit buffers (if it was not allowed to do this earlier), initialises all send state variables and enters the active phase of state 10 (Data Transfer Ready). At this point both SSCOP entities are ready to transfer data.

The following trace shows an actual error recovery. This trace was obtained by patching a POLL message to contain a wrong value of *N(S)* (the sender's *VT(MS)*).

```
 1  SSCOP  .  ⇒  .   5.225  poll s=2 ps=3
 2  SSCOP  .  ⇐  .   5.227  stat r=2 mr=5 ps=3 list={}
 3  SSCOP  ⇒  .  .          AA-DATA.request in state S_READY
 4  SSCOP  .  ⇒  .   5.449  sd s=2 k(1)=33
 5  SSCOP  .  .  ⇒          AA-DATA.indication in state S_READY
 6  SSCOP  .  ⇒  .   5.985  poll s=2 ps=4
 7  SSCOP  .  .  ⇒          AA-MERROR.indication in state S_READY
 8  SSCOP  .  .  S          state S_READY -> S_OUT_REC_PEND
 9  SSCOP  .  ⇐  .   5.988  er mr=3 sq=1
10  SSCOP  ⇐  .  .          AA-RECOVER.indication in state S_READY
11  SSCOP  S  .  .          state S_READY -> S_IN_REC_PEND
12  SSCOP  ⇒  .  .          AA-RECOVER.response in state S_IN_REC_PEND
13  SSCOP  .  ⇒  .   6.923  erak mr=3
14  SSCOP  .  .  ⇒          AA-RECOVER.indication in state S_OUT_REC_PEND
15  SSCOP  S  .  .          state S_IN_REC_PEND -> S_READY
16  SSCOP  .  .  S          state S_OUT_REC_PEND -> S_REC_PEND
17  SSCOP  .  ⇒  .   7.685  poll s=0 ps=1
18  SSCOP  .  .  ⇐          AA-RECOVER.response in state S_REC_PEND
19  SSCOP  .  .  S          state S_REC_PEND -> S_READY
20  SSCOP  .  ⇐  .   8.505  poll s=0 ps=1
21  SSCOP  .  ⇒  .   8.507  stat r=0 mr=3 ps=1 list={}
22  SSCOP  .  ⇒  .   9.695  poll s=0 ps=2
23  SSCOP  .  ⇐  .   9.697  stat r=0 mr=3 ps=2 list={}
```

In line 4 the sender emits SD message number 2 and a short time later in line 6 it generates a POLL message with *N(S)* set to 2. This is clearly wrong, because the next SD message number (this is what is conveyed in *N(S)*) will be 3. The receiver detects this error and starts error recovery in line 9, announcing its window. The sender exchanges `AA-RECOVER` signals with its user and acknowledges the error procedure with the ERAK in line 13. At this point the sender has reset all its state variables, reinitialised the window and is back in state 10. The receiver now exchanges `AA-RECOVER` signals with its user and is back in state 10 in line 19. The POLL and STAT messages in the last four lines show that the state in both instances is reinitialised and synchronised.

5.1.9 Resynchronisation

The SSCOP includes an additional procedure to initialise synchronisation of the peer on request from the protocol's user. This is called resynchronisation.

Resynchronisation can be requested in either state 10 (Data Transfer Ready) or in one of the error states 7, 8 or 9 by issuing an `AA-RESYNC.request`. That is, the resynchronisation procedure takes precedence over the recovery procedure.

Upon receipt of the resynchronisation request from its user, the SSCOP stops all timers, reinitialises the window, sends an RS message (see Figure 5.20) and clears all buffers and queues. It then starts the connection control timer (timer CC) and enters state 5 (Outgoing Resynchronisation Pending).

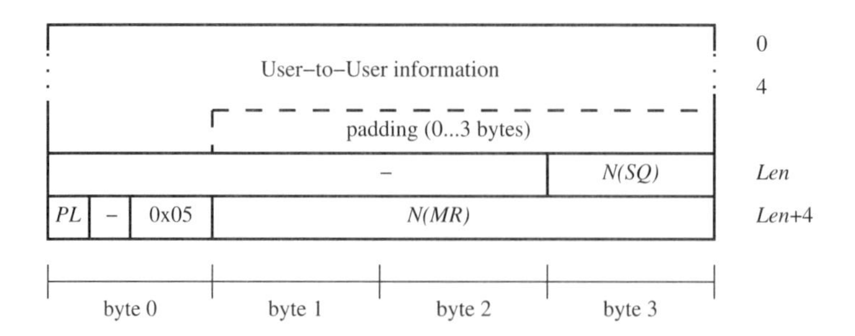

Figure 5.20: SAAL RS message

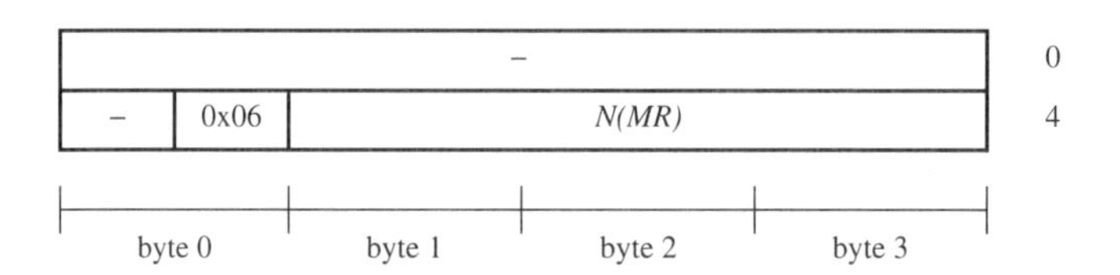

Figure 5.21: SAAL RSAK message

The peer, upon receipt of the RS PDU, stops its data transfer timers, initialises the send window from the value carried in the RS PDU, informs its user by means of an `AA-RESYNC.indication` and clears queues and buffers if it is allowed to do so. It then enters state 6 (Incoming Resynchronisation Pending) awaiting a response from the user.

The user answers with an `AA-RESYNC.reponse` and the SSCOP reinitialises the receive window, sends an RSAK PDU (see Figure 5.21), clears buffers and queues it was previously not allowed to, initialises the data transfer state variables and enters the active phase of state 10. At this point the resynchronisation ends for the peer entity.

The invoker of the procedure, upon receipt of the RSAK, initialises the send window from the value in the RSAK PDU, confirms the resynchronisation to its user, initialises the data transfer state and enters the active phase of state 10.

At this point both instances are guaranteed to be synchronised. The following trace shows an example of a resynchronisation:

```
1   SSCOP  . ⇒ .   1.683  poll s=0 ps=1
2   SSCOP  . ⇐ .   1.684  poll s=0 ps=1
3   SSCOP  . ⇒ .   1.686  stat r=0 mr=128 ps=1 list={}
4   SSCOP  . ⇐ .   1.687  stat r=0 mr=128 ps=1 list={}
5   SSCOP  ⇒ . .          AA-DATA.request in state S_READY
6   SSCOP  . ⇒ .   2.473  sd s=0 k(1)=31
7   SSCOP  . . ⇒          AA-DATA.indication in state S_READY
8   SSCOP  . ⇒ .   3.232  poll s=1 ps=2
9   SSCOP  . ⇐ .   3.234  stat r=1 mr=129 ps=2 list={}
10  SSCOP  ⇒ . .          AA-RESYNC.request in state S_READY
11  SSCOP  . ⇒ .   3.447  rs mr=128 sq=2
12  SSCOP  . . ⇒          AA-RESYNC.indication in state S_READY
```

```
13 SSCOP S  .  .          state S_READY -> S_OUT_RESYNC_PEND
14 SSCOP .  .  S          state S_READY -> S_IN_RESYNC_PEND
15 SSCOP .  ⇒  .   4.452  rs mr=128 sq=2
16 SSCOP .  .  ⇐          AA-RESYNC.response in state S_IN_RESYNC_PEND
17 SSCOP .  ⇐  .   4.557  rsak mr=128
18 SSCOP ⇐  .  .          AA-RESYNC.confirm in state S_OUT_RESYNC_PEND
19 SSCOP S  .  .          state S_OUT_RESYNC_PEND -> S_READY
20 SSCOP .  .  S          state S_IN_RESYNC_PEND -> S_READY
21 SSCOP .  ⇒  .   5.313  poll s=0 ps=1
22 SSCOP .  ⇐  .   5.314  poll s=0 ps=1
23 SSCOP .  ⇒  .   5.316  stat r=0 mr=128 ps=1 list={}
24 SSCOP .  ⇐  .   5.317  stat r=0 mr=128 ps=1 list={}
25 SSCOP ⇒  .  .          AA-DATA.request in state S_READY
26 SSCOP .  ⇒  .   7.165  sd s=0 k(1)=32
27 SSCOP .  .  ⇒          AA-DATA.indication in state S_READY
28 SSCOP .  ⇒  .   7.922  poll s=1 ps=2
29 SSCOP .  ⇐  .   7.924  stat r=1 mr=129 ps=2 list={}
30 SSCOP .  ⇒  .   8.682  poll s=1 ps=3
31 SSCOP .  ⇐  .   8.684  stat r=1 mr=129 ps=3 list={}
```

In lines 5–9 of the trace an SD message is sent and the window and acknowledge numbers updated accordingly. In line 10 the user of the left SSCOP invokes the resynchronisation procedure by sending an `AA-RESYNC.request`. An RS message is emitted and the SSCOP goes into states 5 (Outgoing Resynchronisation Pending) and 6 (Incoming Resynchronisation Pending). The user of the right SSCOP, after receiving the indication about the resynchronisation, sends the `AA-RESYNC.response` and the SSCOP can answer its peer with an RSAK. When this message is received by the left SSCOP both protocol instances are ready again. The POLL and STAT messages show that the window and sequence numbers are reset to their respective start values.

5.1.10 Unassured Data Transfer

The SSCOP provides two channels for the unassured transfer of data messages. One of these is reserved for use by the management plane and one can be used by the SSCOP user. This data transfer is done via UD (unassured data) and MD (management data) messages (see Figures 5.22 and 5.23).

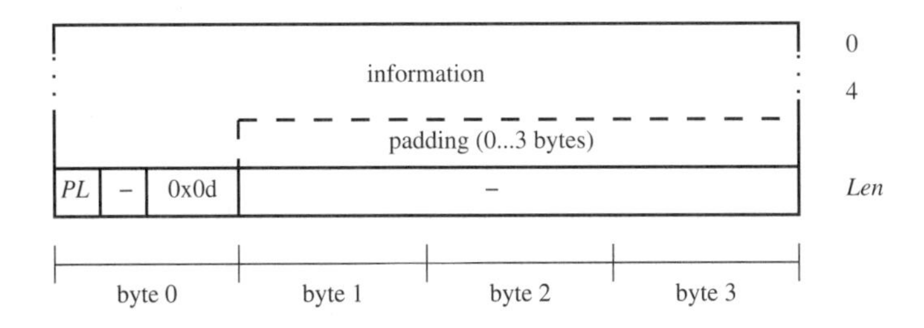

Figure 5.22: SAAL UD message

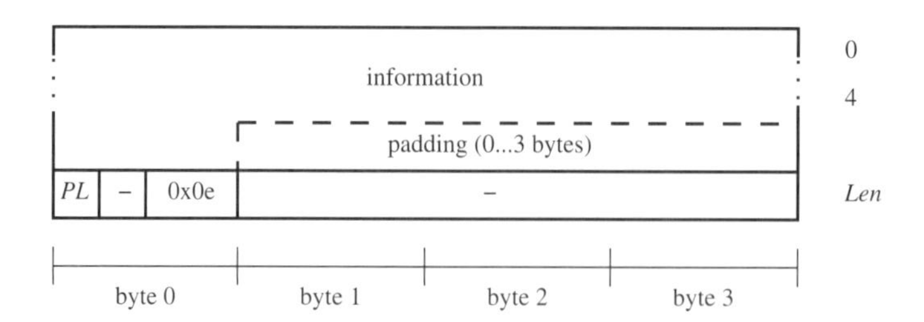

Figure 5.23: SAAL MD message

Messages of this kind are not acknowledged and are not subject to flow control except for the lower layer busy case. If a congestion in the layers below the SSCOP is detected, the messages are queued until they can be sent. MD and UD messages can be sent in any state, with or without an established connection, even when the SSCOP is in the idle state (state 0).

5.1.11 Message Retrieval and Buffer Management

The SSCOP uses a number of conceptual queues and buffers. Conceptual means that an actual implementation can choose to use fewer queues or other mechanisms, but must provide the same behaviour as in the standard. Queues and buffers can store messages. The number of messages that can be stored may be restricted. Queues and buffers differ in the way messages inside it can be accessed: messages in queues are accessed in a strict first-in, first-out manner; messages in buffers can be retrieved by message number. Figure 5.24 shows the SSCOP buffers and queues and their relationship.

The MD and UD queues are used to hold MD and UD messages while the lower layer is congested. As soon as the congestion disappears, these messages are sent to the peer SSCOP. The SD queue holds SD messages while either the lower layer is congested, the send window is closed or there are messages waiting in the retransmission queue. All three queues are fed from the user of the SSCOP. Whereas messages are blocked in the SD queue by lower layer congestion and a closed window, the retransmission queue is blocked only by congestion. When sending an SD message, the retransmission queue takes precedence over the SD queue. If an SD message is sent, it is put into the transmission buffer. This buffer holds SD messages until a positive (ACK) or negative (NACK) acknowledgement is received from the peer. A message for which an ACK is received leaves the SSCOP (it is destroyed); a message for which a NACK is received is put back into the retransmission queue (and held in the transmission buffer). In this way messages circulate between the transmission buffer and the retransmission queue until they are finally acknowledged by the peer SSCOP. Note that the transmission queue and the retransmission queue are different in the sense that the transmission queue holds messages, while the retransmission queue holds only links to messages in the transmission buffer.

On the receiving side the configuration is much simpler. Each message received from the peer is checked, whether it is the next in-sequence message or not. If so, it is directly delivered to the user; if not, it is put into the receive buffer. The message is held in the receive buffer until all messages with lower sequence numbers are received. At this moment the message can be removed from the buffer and delivered to the SSCOP user.

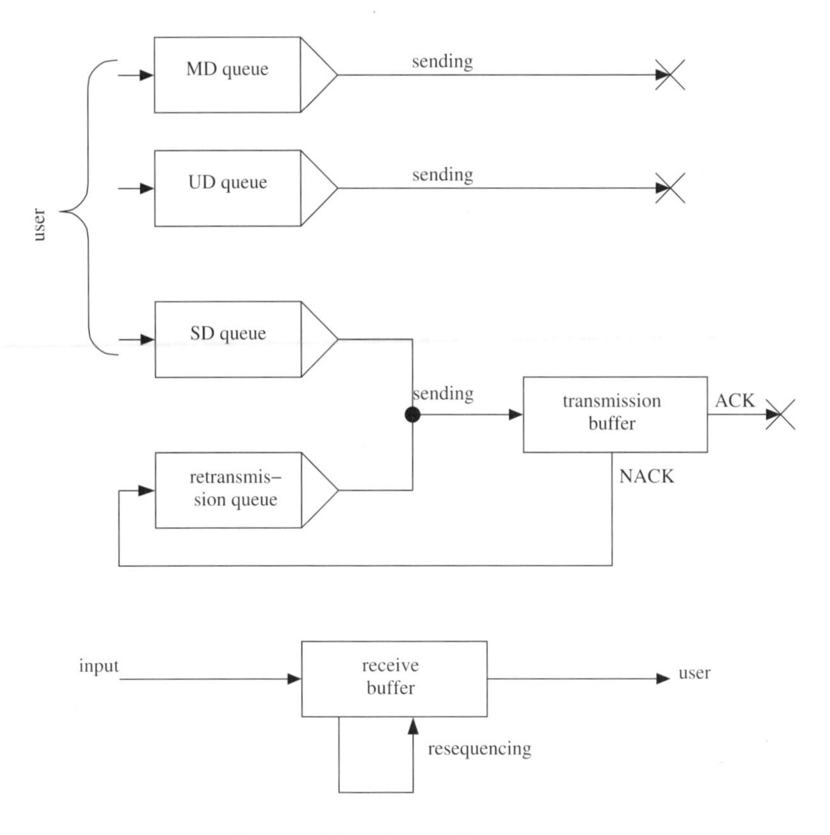

Figure 5.24: SSCOP buffers and queues

The SSCOP provides an additional mechanism called local message retrieval. As we have seen in the preceding sections, there are circumstances where messages can be lost, namely if a connection is released, but there are still messages in tranmission queues and buffers, in the case of protocol errors and during resynchronisation. For some applications this is not a problem, because the higher level protocols do not use mechanisms like resynchronisation and include additional mechanisms, like retransmissions, to ensure integrity. For some applications, however, the loss of messages is not allowable. We will give an example—the SSCF at the public NNI—later in this chapter. For these applications local message retrieval can be used to ensure that messages do not get lost.

Local message retrieval is controlled by a configuration parameter of SSCOP—the `Clear-Buffers` parameter. The signal `AA-RETRIEVE.request` is used to initiate retrieval, the `AA-RETRIEVE.indication` is used by the SSCOP to deliver the retrieved PDUs to the user, and `AA-RETRIEVE-COMPLETE.indication` signals to the user that all messages have been retrieved. The `Clear-Buffers` parameter controls the time at which the SSCOP destroys SD messages. If this parameter is set to `YES`, messages are destroyed in the following cases:

- When a new connection is established the transmission queue and buffer are cleared. This is done on the side requesting the connection as well as on the responding side. The transmitter is also cleared when a connection is re-established during release.

- After the SSCOP user has responded to a resynchronisation indication. Both the transmission queue and buffer are cleared.
- When resynchronisation is invoked, the transmission buffer and queue are cleared, as well as the receive buffer.
- If error recovery is acknowledged, the receiver buffer is cleared.
- When the connection is released, all buffers and queues are cleared.
- When messages are in-sequence or selectively acknowledged.

Additionally, requests to send data are ignored while error recovery is in progress if `Clear-Buffers` is `YES`.

To summarise: if `Clear-Buffers` is set to `YES` a message is deleted whenever it seems that the peer has got it, when there is no chance that the peer will get it (because the connection is released), when there is a protocol error or when the user invokes a procedure that interrupts normal SSCOP operation (resynchronisation or release). Even an acknowledged message may get lost if, a short time after the acknowledgement was received, the connection is released—if it was a selectively acknowledged message, it may still wait in the peer's receive buffer and will be destroyed without beeing delivered to the user.

For situations where loss of messages is not permitted, `Clear-Buffers` can be set to `NO`. This changes the behaviour of the SSCOP in the following way:

- A message is cleared from the transmitter only if: the peer has acknowledged it in sequence or the local user had a chance to retrieve it from the SSCOP.
- The local data retrieval is enabled to get messages back from the SSCOP that are not yet in-sequence acknowledged from the peer.

Local retrieval works by sending an `AA-RETRIEVE.request` to the SSCOP. This request carries a parameter that can have the following values:

N where N is an SD message sequence number. In this case all messages with a sequence number higher than N are retrieved from the transmission buffer and all messages from the transmission queue. Because now only in-sequence acknowledged messages are removed from the transmitter, this returns a contiguous set of messages from $N + 1$ up to the last message delivered to the SSCOP.

Unknown All messages from the transmission queue (SD queue) are retrieved.

Total All messages from the transmission buffer and the transmission queue are retrieved.

For each message that is retrieved, the SSCOP generates an `AA-RETRIEVE.indication` which contains the retrieved message and its message number. These messages are removed entirely from SSCOP, i.e. they will not be sent or cleared.

When all messages specified in the request have been retrieved, an `AA-RETRIEVE-COMPLETE` signal is sent to the user.

In Figure 5.25 a situation is shown with a number of messages waiting in each of the buffers and queues. Messages 24–27 are waiting in the SD queue (they may be blocked because of retransmissions or the window being closed), messages 23 and 20 are waiting

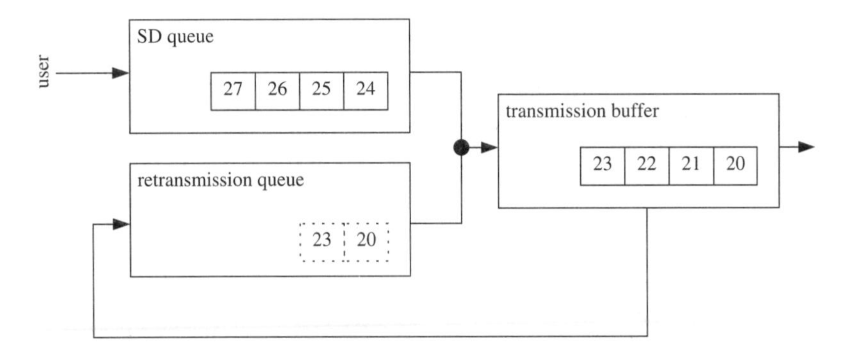

Figure 5.25: SSCOP retrieval example

for retransmission and messages 21 and 22 have been successfully acknowledged by the peer SSCOP. The last two messages are held in the transmission buffer because it is assumed that `Clear-Buffers` is set to `YES`. Because in this case only in-sequence acknowledged messages can be removed from the buffer and the acknowledgement for message 10 is still outstanding, the two messages are held in the transmission buffer. If in this situation the SSCOP starts a data retrieval, the following happens for different values of the parameter of the `AA-RETRIEVE.request`:

25 In this case messages 24, 25, 26 and 27 are retrieved. Messages 20–23 are left where they are.

21 Messages 22–27 are retrieved. Messages 21 and 20 remain in the transmission buffer (remember that the retransmission queue consists only of links into the transmission buffer; message 20 is also contained in the transmission buffer).

Unknown Messages 24–27 are retrieved. Messages 20–23 remain in the transmission buffer.

Total Messages 20–27 are retrieved. The transmission buffer and both queues are empty.

After all messages that where selected by the signal parameter are retrieved, an `AA-RETRIEVE-COMPLETE` signal is sent.

5.1.12 Interface to Layer Management

The interface between the SSCOP and layer management consists of one signal that signals SSCOP errors to the layer management (`MAA-ERROR.indication`) and the signal `MAA-DATA` (request and indication) which can be used to send unassured data between the management instances. As of this writing, no uses for `MAA-DATA` are defined.

The error signal contains two parameters: an error code in the form of a single character and, for error code “V”, a number which indicates the number of retransmitted SD PDUs. Table 5.2 gives all possible error codes and their meaning.

The last two are not really error conditions.

Table 5.2: SSCOP error codes

Error Type	Error Code	Description
Receipt of unexpected or inappropriate message	A	Receipt of SD message in wrong state.
	B	Receipt of BGN message in wrong state.
	C	Receipt of BGAK message in wrong state.
	D	Receipt of BGREJ message in wrong state.
	E	Receipt of END message in wrong state. This error never occurs.
	F	Receipt of ENDAK message in wrong state.
	G	Receipt of POLL message in wrong state.
	H	Receipt of STAT message in wrong state.
	I	Receipt of USTAT message in wrong state.
	J	Receipt of RS message in wrong state.
	K	Receipt of RSAK message in wrong state.
	L	Receipt of ER message in wrong state.
	M	Receipt of ERAK message in wrong state.
Unsuccessful retransmission	O	Too many retransmissions of connection control messages ($VT(CC) \geq MaxCC$).
	P	Timer NO-RESPONSE expiry, i.e. no answer to POLL message for a long period.
Sequence number and length errors	Q	An SD PDU is received that is already in the receive buffer or a POLL PDU indicates a next new SD PDU number that is lower than what the receiver has already seen.
	R	A STAT message indicates a POLL sequence number outside the window of expected numbers (either this number was already acknowledged by an earlier STAT or a POLL with this number was never sent).
	S	The in-sequence acknowledge number in a STAT PDU is outside the window of allowed values, i.e. it is either lower than what was already acknowledged or higher than the highest SD PDU sent up to now.
	T	One of the SD sequence numbers in a USTAT message is wrong.
	U	The size of a received PDU is either less than four (so it cannot contain a valid trailer), is not a multiple of four or has the wrong size for the given type of message.
SD loss	V	SD messages were retransmitted. This signal contains the number of retransmitted messages.
Credit condition	W	The window was closed by the receiver.
	X	The window was reopened by the receiver.

5.2 SSCF UNI: Service Specific Coordination Function at the UNI

The top layer of the SAAL is the so-called Service Specific Coordination Function (SSCF). For the support of the UNI this sublayer is defined in [Q.2130]. Although the standard contains over 50 pages, the SSCF at the UNI is not much more than a null-layer, which more or less provides a simple mapping between the signals at its lower and upper boundaries.

The signals exchanged at the upper boundary of the SSCF (and also the SAAL) are:

AAL-ESTABLISH
: This signal is used to establish a connection from the SAAL (in the form of a request and a confirmation) or to inform the upper layer that an incoming connection has been established (indication). All three forms can take user data as a parameter. This data may be delivered to the peer, or has been received from the peer.

AAL-RELEASE
: This is used to release a connection (request and confirmation) or to inform the SAAL user that the connection has been released (indication). As in the AAL-ESTABLISH case, these signals may carry user data.

AAL-DATA
: This is used by the SAAL user to send a data packet and by the SAAL to hand out a received packet to its user. The parameter is the user data. There is a request and an indication form.

AAL-UNITDATA
: This may be used by the SAAL user for unassured data transfer to the peer (request and indication).

The SSCF simplifies the upper layer interface of the SAAL with regard to the SSCOP. As an example the sequence of signals during connection establishment is show in Figure 5.26.

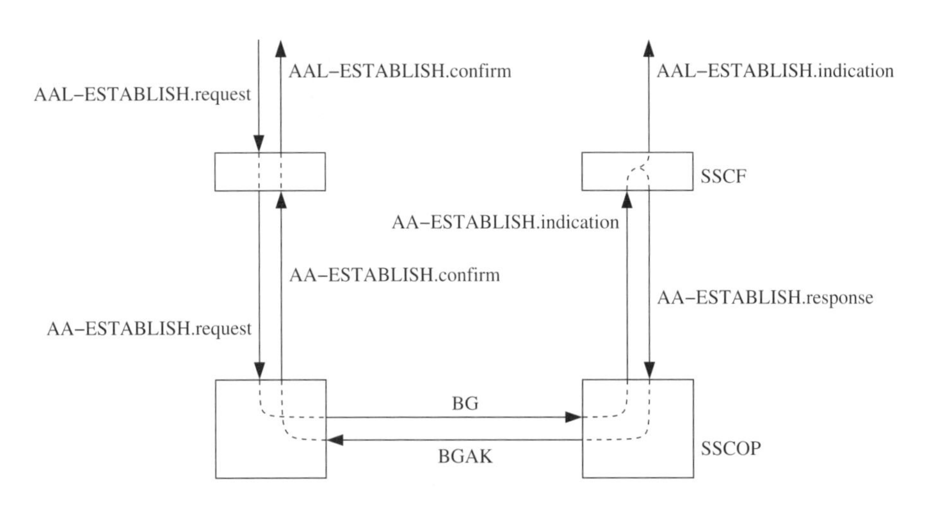

Figure 5.26: Establishment of an SAAL connection

On the side where the SAAL user requests the establishment of the connection, the `AAL-ESTABLISH.request` is directly mapped to an `AA-ESTABLISH.request`. The confirmation from SSCOP is mapped accordingly. On the incoming side, the SSCF answers directly with an `AA-ESTABLISH.confirm` to SSCOP—the user receives only an indication.

Another example is the mapping of `AAL-ESTABLISH.requests` to the resynchronisation procedure in the case when the connection is already in the READY state (see Figure 5.27). If the establishment of an AAL connection is requested when the connection is already established, this request is mapped to an `AA-RESYNC.request`. In the example both SAALs invoke the procedure at the same time and SSCOP executes a simultaneous resynchronisation. Both SAAL users receive an `AA-ESTABLISH.confirm` after successful resynchronisation.

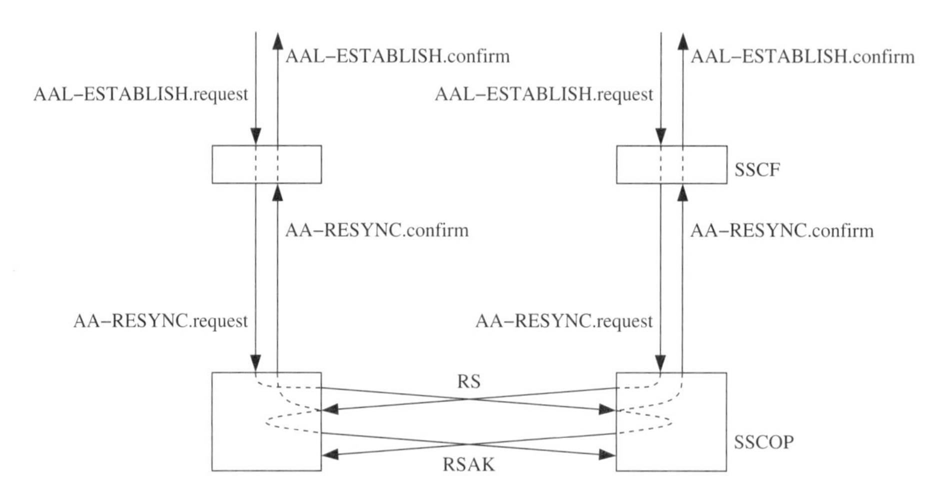

Figure 5.27: Re-establishment of an SAAL connection with collision

Error recovery and connection tear-down are handled in a similar manner. All cases of simultaneous invocation of the different SSCOP procedures are handled. In all the numerous cases the SAAL user receives sequences of establish and release indications. The action of the user should normally be: clean up all resources in the case of a SAAL release; try to find out the peer state in the case of an establish indication.

The SSCF standard also provides values for the parameters to the SSCOP (see Table 5.3). The parameters in this table are default values that can be changed according to the local operation environment (for example, for satellite links).

5.3 SSCF NNI: Service Specific Coordination Function at the NNI

In contrast to the SSCF at the UNI, the SSCF at the NNI is a complex protocol [Q.2140]. The place of this protocol in the public NNI stack is shown in Figure 5.28.

On top of the SAAL resides the Message Transfer Part 3 (broadband) (MTP-3b). The responsibility of the MTP is the reliable transfer of datagrams through a network. In other words, MTP-3b implements a connectionless network on top of SAAL connections. Users of the MTP-3b are: B-ISUP and application stacks that consist of SSCP, TCAP and

Table 5.3: SSCOP parameters for the UNI

SSCOP parameter	Value	Meaning
MaxCC	4	number of retransmissions of control messages
Timer CC	1 s	retransmission timeout for control messages
Timer KEEP-ALIVE	2 s	switch to the idle phase
Timer NO-RESPONSE	7 s	timeout to declare connection dead
Timer POLL	750 ms	maximum interval between POLL messages in the active phase
Timer IDLE	15 s	maximum interval between POLL messages in the idle phase
k	4096	maximum number of user bytes in an SD message
j	4096	maximum number of user bytes in control messages (not used by the UNI)
MaxPD	25	maximum number of SD messages between POLL messages
Clear-Buffers	YES	

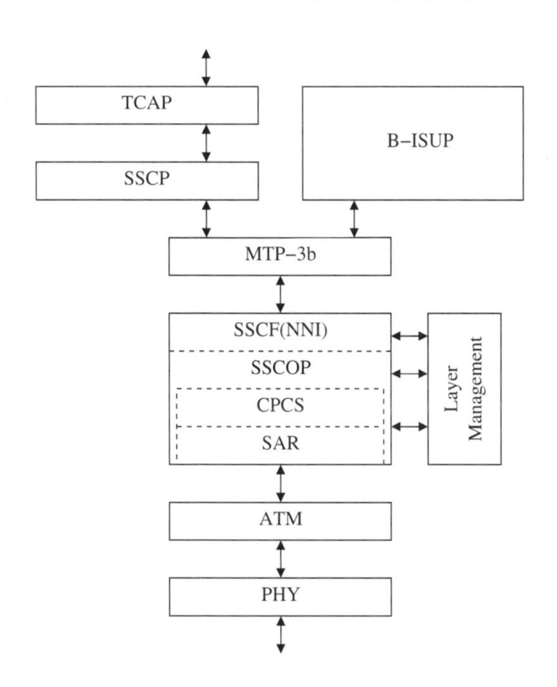

Figure 5.28: NNI protocol stack

application layers. B-ISUP is responsible for the establishment of ATM user connections through the network, while the application stacks can be used to implement intelligent network services [VH98].

Because of the very high performance requirements that are placed on public networks, the SAAL layer tries to ensure that messages never get lost. There are features built into the SSCF at the NNI that provide enhanced reliability in the SAAL, when compared with the UNI:

- Before declaring an SAAL connection available, the connection is checked for reliability and performance for some time.
- Timers are set to values that ensure very fast reaction to link failures and congestion situations.
- While the SAAL connection is in use its performance characteristics are monitored and, if necessary, rechecking is initiated.
- Message loss is prevented by using the local retrieval feature of SSCOP and handing over untransmitted messages to SSCOPs on alternative links.

Before an SAAL connection is declared as usable to the SAAL user, the link is proved (see the state diagram in Figure 5.29).

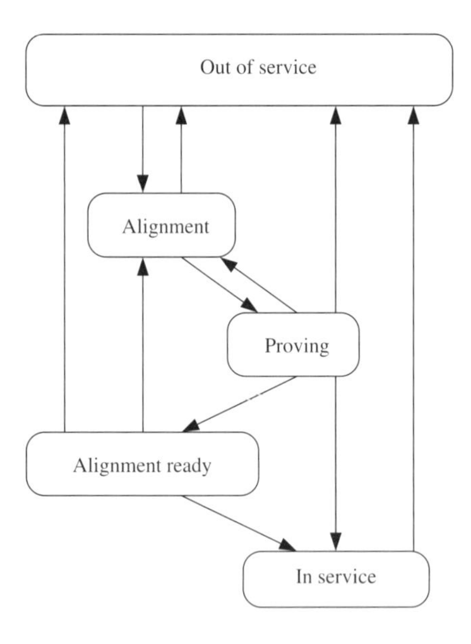

Figure 5.29: SSCF states at the NNI

The Proving procedure is initiated by a start request from the SAAL user. This start request moves the SSCF into the Alignment state and generates an `AA-ESTABLISH.request` to the SSCOP. This request contains an SSCF PDU in the user-to-user field. When the SSCOP connection is successfully established (the SSCF receives the `AA-ESTABLISH.confirm`), the SSCF moves to the Proving state. In the Proving state traffic is generated by the SSCF at approximately 50% of the link bandwidth and the operation of the SSCOP is monitored (retransmission rate and credit). If these parameters seem useful, the SSCF stops proving, emits an SSCF PDU that informs the peer about successful proving and moves to the Alignment Ready state. If the peer also enters the Alignment Ready state, both SAALs finally declare the link In Service. In the In Service state, information can be exchanged by the SSCF users.

The SSCF at the NNI is a real protocol in that it uses PDUs for communication between the peer entities. SSCF PDUs can be carried either as normal sequenced SSCOP data or in the user-to-user fields of SSCOP connection control messages. All SSCF PDUs have a size of four bytes and all SAAL user PDUs are required to have a minimum size of five bytes. This makes it easy to distinguish SSCF and SSCF user information: messages less than four bytes are silently discarded, messages of four bytes are handled by the SSCF and messages greater than four bytes are handed out to the user. Figure 5.30 shows the format of an SSCF PDU.

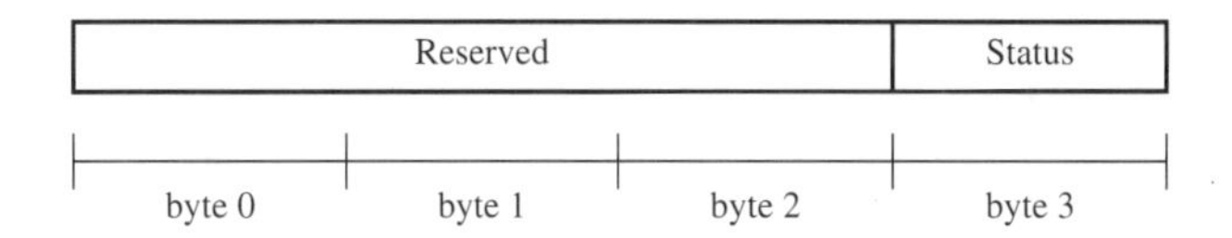

Figure 5.30: SSCF at the NNI PDU

Three bytes of the PDU are reserved and should contain zeros. The status field in the SSCF PDUs is used to communicate the current status to the peer in case it changes. The possible values are shown in Table 5.4.

Table 5.4: Status field of NNI SSCF PDUs

Value	Status
0x01	out of service
0x02	processor outage
0x03	in service
0x04	normal
0x05	emergency
0x07	alignment not successful
0x08	management initiated
0x09	protocol error
0x0a	proving not successful

To minimise loss of signalling messages in the network, the MTP-3b layer uses the local retrieval feature of the SSCOP in the case of failures of a signalling link. NNI links are usually configured with backup links and alternativ routes. If it is not possible to send a signalling message on one link due to failures or congestion, the message can be sent on another link. The procedure of switching to an alternativ link in the case of a failure is called changeover [Q.704]. Figure 5.31 shows a configuration with two signalling links to a peer.

In this configuration there are two links between two MTP-3 peer entities: link-1 and link-2.

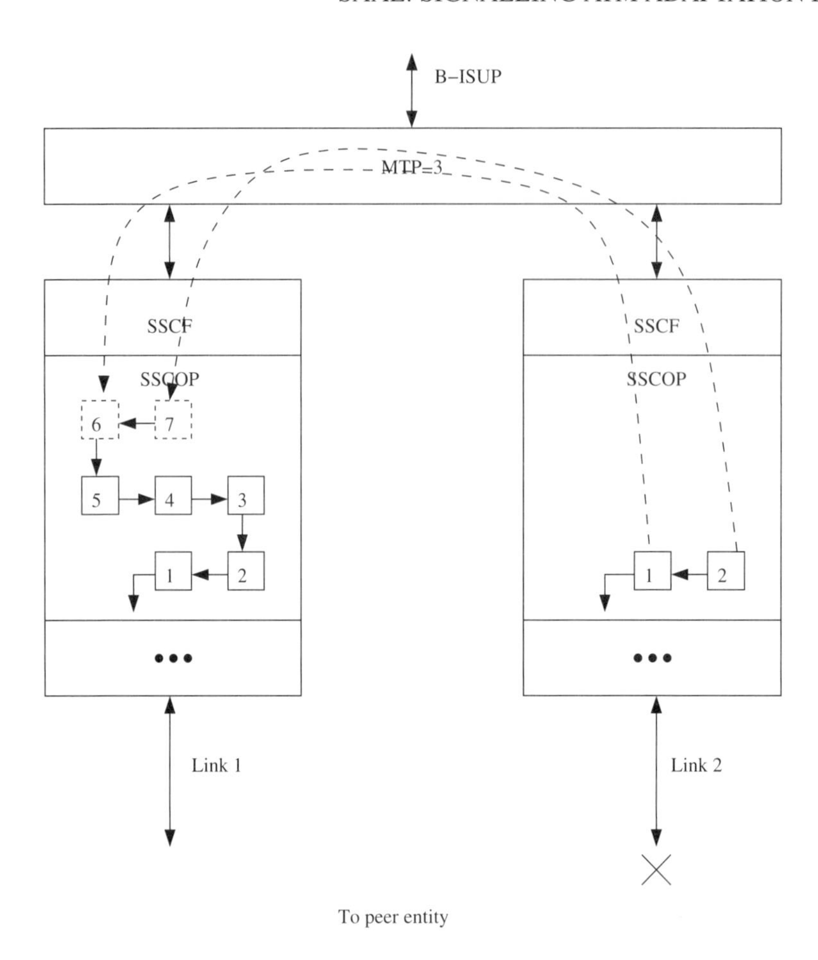

Figure 5.31: Changeover on the NNI

Each of the links has a SAAL stack and each of the SSCOPs has a message queue: the SSCOP on link-1 has five messages and the SSCOP on link-2 two messages. If in this situation link-2 breaks (suppose an excavator were to cut the cable), then the MTP-3, as soon as it is informed about the failed link, uses the SSCOP retrieval feature to get all messages out of the SSCOP which have not be fully received by the peer SSCOP, and hands these messages over to the working link. This feature prevents the loss of messages under almost all circumstances (of course, if the excavator cuts all your cables, then you are lost).

Just like the SSCF at the UNI, the SSCF at the NNI specifies default values for the SSCOP. As one can see from Table 5.5, timer settings are quite different from the UNI. The small timer values enable a faster reaction to link failures and congestion situations. Of course, these parameters can be adjusted for operation in unusual environments, for example on satellite links.

Table 5.5: SSCOP parameters for the NNI

SSCOP parameter	Value	Meaning
MaxCC	4	number of retransmissions of control messages
Timer CC	200 ms	retransmission timeout for control messages
Timer KEEP-ALIVE	100 ms	switch to the idle phase
Timer NO-RESPONSE	1.5 s	timeout to declare connection dead
Timer POLL	100 ms	maximum interval between POLL messages in the active phase
Timer IDLE	100 ms	maximum interval between POLL messages in the idle phase
k	4096	maximum number of user bytes in SD messages
j	4	maximum number of user bytes in control messages
MaxPD	500	maximum number of SD messages between POLL messages
Clear-Buffers	NO	

5.4 Summary

In this chapter we have looked at the layer beneath the UNI and the PNNI signalling, namely the Signalling ATM Adaptation Layer (SAAL). We have seen that the SAAL consists of a complicated transport protocol, the Service Specific Connection Oriented Protocol (SSCOP), which provides assured data transfer, keep-alive and flow control. This transport protocol sits on top of a standard AAL5 sublayer. The upper sublayer of the SAAL is the Service Specific Coordination Function (SSCF) which comes in two types: the SSCF at the UNI and the SSCF at the NNI. Whereas the SSCF at the UNI provides a simple mapping of interface signals, the SSCF at the NNI is a real protocol which is used below MTP-3b in public networks. Both SSCFs provide default parameters for the parameterisation of the SSCOP. We have seen the operation of the SSCOP and the SSCF at the UNI for many situations.

6

PNNI: Private Network Node Interface

6.1 Introduction

This chapter describes the PNNI protocol family that is used in private ATM networks. The abbreviation PNNI stands for either Private Network Node Interface or Private Network-to-Network Interface, reflecting two possible applications. The first application is the connection of private ATM switches, the second is the connection of groups of private ATM switches.

The PNNI protocols were defined by the ATM-Forum in [PNNI]. The public NNI protocol was considered too complicated, needing too many resources and beeing too static to be implemented in cheap private ATM switches. Therefore, a set of new protocols was developed.

The Private Network Node Interface (PNNI) protocol family consists of two protocols. The first protocol, called the PNNI routing protocol, is used to distribute topology information between switches and groups of switches. This information will be needed for later routing of connections.

The PNNI signalling protocol, which is the second in the family, is responsible for the establishment of point-to-point and point-to-multipoint connections across the network. This protocol is based on the UNI protocol described in Chapter 3.

6.1.1 Introduction to the PNNI Routing Protocol

The PNNI routing protocol is used to distribute topology information between switches and groups of switches. It is part of the ATM control plane and runs on top of the AAL5. A hierarchy mechanism ensures that this protocol scales well for large ATM networks.

A PNNI network like other networks consists of ATM switches and physical ATM links. Data is passed through these nodes and links between end systems. End systems are the originating and terminating points of connections. For routing purposes the 19 most significant bytes of the 20 byte ATM end system addresses are used (see Chapter 4). The last byte of the address is only interpreted inside the end system and is ignored by PNNI routing.

If a PNNI network were organised in a non-hierarchical way, then each node would have to maintain the entire topology of the network, including information for every physical link and every node in the network. This works well for small networks, but in large networks the exchange of routing information creates an enormous overhead. Therefore, the PNNI supports a hierarchy that can reduce the overhead while providing efficient routing. It is important to recognise that the hierarchy is only used for distribution and storage of topology information and for finding a route from the origin to the destination of a connection. The final user data transport uses a flat network and does not produce any bottlenecks at higher levels of the hierarchy.

In this book we focus on flat, i.e. non-hierarchical, PNNI networks. We do not explore the PNNI hierarchy in detail because the authors of this book could not perform any experiments on such a hierarchy and therefore could not verify the behaviour described in the standard documents. However, a small survey is needed, because for the PNNI the flat network is only a special case of the hierarchical network. The terminology of hierarchical networks is used everywhere.

The nodes of the lowest level of the PNNI hierarchy (the switches) are organised into groups. Such a node in the context of the lowest hierarchy level is also called a "logical node" or "node". A node is an ordinary ATM switch that talks the PNNI protocols. A logical node is identified by a logical node ID.

Nodes are grouped into so-called "peer groups". Each node of such a peer group exchanges all information with other members of the same peer group, such that all members of the peer group maintain an identical topology database, i.e. they all have the same view of the network. All nodes of a peer group have the same 13-byte peer group ID which is configurable by management means. Neighbouring nodes exchange their own peer group IDs in "Hello packets", as described in Section 6.2.5. If they have the same peer group ID then they know that they belong to the same peer group. If the IDs are different, then they belong to different peer groups.

In each peer group one node has an exposed position. This node is called the "peer group leader" (PGL). The PGL is determined by a peer group leader election process. The criterion for the election is the leadership priority parameter that is configured for every node. The peer group leader election process is running all the time to avoid problems if the current PGL fails. But what is the PGL good for? A peer group is represented in the next hierarchy level by a single node called a "logical group node" (LGN). This is a kind of virtual node. But the functionality of this LGN must be implemented by some real node. This is done by the PGL of the peer group.

Let us summarise. We have nodes, which are grouped into peer groups. Each peer group will be represented by a (logical group) node at the next hierarchy level. At this next hierarchy level the nodes are grouped into peer groups of a higher level. Each higher level peer group will be represented by a node at a higher lever—and so on. Finally, we have a hierarchy that is constructed by the recursive mechanism just described. The top of the hierarchy is a peer group without a PGL.

If we have a flat PNNI network we only have one peer group and this peer group has no PGL.

On the basis of the structure of peer groups, topology information is collected in every peer group. This collected information is distributed to other peer groups to allow global routing. To enable scalability the information is aggregated and passed to the next higher hierarchy level. This higher level is responsible for further distribution inside its peer group and further higher levels. The aggregation comprises link aggregation, nodal aggregation and address summarisation.

Link aggregation means that several links between two peer groups are represented by one logical link at the next highest level of the hierarchy. In this case link parameters are aggregated.

Nodal aggregation means that a peer group is represented by one LGN at the next highest hierarchy level. If such a group is not simply represented by one point this results in increased complexity. In this case the LGN can have a very complicated internal structure, for example, a star topology or even a more complex structure, with many parameters. In most cases we have a simple nodal aggregation where a peer group will be represented by one point with simple parameters.

Address summarisation is a very interesting feature of the PNNI which is also important in non-hierarchical PNNI networks. It simply means that a node, which has several end systems attached to it, can advertise the reachability of one or more address prefixes, where each address prefix can comprise many end system addresses. This means that each end system address need not be distributed individually.

We have already mentioned the distribution of topology information. But what does this mean? Topology information is encoded in so-called "PNNI Topology State Elements" (PTSEs). These PTSEs are distributed by the PNNI routing protocol.

We can distinguish between two kinds of distribution. If the network (or parts of it) starts operation an initial database exchange is performed using so-called "Database Summary Packets". If the network is up and running and the initial database exchange is finished, then a flooding mechanism with "PNNI Topology State Packets" (PTSPs) is used. PTSPs contain one or more PTSE. Received PTSPs are acknowledged.

6.1.2 Introduction to the PNNI Signalling Protocol

The PNNI signalling protocol is responsible for the establishment of point-to-point and point-to-multipoint connections across the network. The protocol is based on the UNI protocol described in Chapter 3 ([UNI4.0]).

The main differences to the UNI are:

- The PNNI does not support some of the UNI features like proxy signalling, leaf-initiated join or some supplementary services.
- The PNNI is symmetric.
- The PNNI contains additional information elements to support dynamic connection setup.
- Associated signalling is supported.

The PNNI signalling protocol usually runs in VCI 5 of VPI 0 in non-associated mode. Other VPIs can contain their own signalling connections in VCI 5, but they operate in associated mode. The signalling in VPI 0 controls all VPIs that are not controlled by other signalling channels.

6.2 Routing Protocol

6.2.1 Addressing

6.2.1.1 Introduction

The addresses of ATM systems and subsystems play a very important role for PNNI routing. They need to be configured properly in a working PNNI environment (see Figure 6.1).

The addressing and identification of components of the PNNI routing hierarchy are based on Network Service Access Point (NSAP) addresses used in private networks. These ATM end system addresses are 20 bytes long. Embedded E.164 NSAP addresses can be used to cover public network addresses.

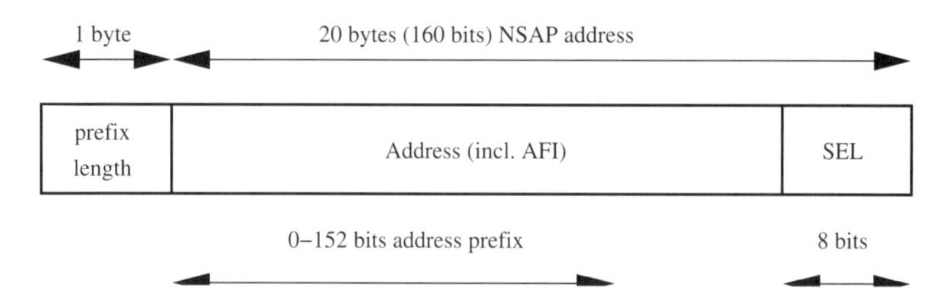

Figure 6.1: PNNI address prefix

A specific characteristic of PNNI routing is that it only operates on the first (leftmost) 19 bytes of the address. The last (rightmost) byte (called the Selector/ SEL) has only local significance to the end system and is ignored by the PNNI. It distinguishes destinations reachable at the same ATM end system interface and therefore does not influence the routing. The first byte of the address, the Authority and Format Identifier (AFI), is used by the PNNI to distinguish between individual addresses and group addresses. Valid AFI values are specified in the UNI 4.0 standard.

To support scalability PNNI uses the concept of address prefixes. A prefix of an ATM end system address is the first (leftmost) p bits of the address. The value of p may vary from zero to 152. Prefixes are used to summarise some portion of the addressing domain. Shorter prefixes summarise greater portions than do longer prefixes, i.e. longer prefixes are more specific.

6.2.1.2 Node Addresses and End System Addresses

Nodes (switches) involved in the PNNI are addressed using ATM End System Addresses. A single switching system may contain multiple nodes, e.g. if the switching system implements the functionality of different LGNs at different levels in the hierarchy. Each of these nodes requires a unique ATM end system address. The switching system can generate these different addresses by using the same 19-byte prefix plus different selector values, i.e. by varying the last byte of the ATM address.

PNNI routing uses the mechanism whereby a node advertises which end system it can reach. To allow scalability, a node will not advertise each reachable end system address, but will advertise the reachability of a group of end systems by using prefixes of ATM end system addresses to form summaries. If an end system attached to a node does not fit into one of the node's configured summaries,[1] then it will be necessary for the node to make an explicit advertisement for that end system which will necessarily carry a full 152-bit (all 19 bytes) prefix length. True summaries of end system reachability will have prefix lengths less than 152 bits.

Let us consider an example. We assume that we have a node with a configured prefix

0x39.0000.1111.2222.3333.44 of length 80 bits

Three end systems are attached to this node with addresses:

0x39.0000.1111.2222.3333.4444.0001.ff1a4ce80001.00

---

[1] Both ATM end system addresses and the node summaries can be configured. Therefore they may fit or may not.

0x39.0000.1111.2222.3333.4444.0022.ff1a4ce70001.00
0x39.0000.9999.2222.3333.4444.0333.ff1a4ce90001.00

Then this node will advertise the reachability of the two address prefixes:

0x39.0000.1111.2222.3333.44 with length 80 bits
0x39.0000.9999.2222.3333.4444.0333.ff1a4ce90001 with length 152 bits

Large networks use hierarchies. Where the addressing hierarchy follows the topology hierarchy, it is possible to advertise reachability of a large number of end systems using a single prefix, which is good for scalability. Shorter prefixes (i.e. lower prefix length values) summarise greater numbers of addresses and vice versa.

It is possible that there is more than one summary which matches a given destination. In such a case the PNNI routing mechanism has to select one summary (which corresponds to one node) to direct the call. PNNI route computation will always direct calls to a node that is advertising the best match for a given destination. The best match is defined to be the matching (summary) advertisement with the longest prefix. The PNNI route computation will also consider the scope of the advertised summary. A scope is used to mark summaries that are only valid in a part of the routing domain.

6.2.2 *Logical Links*

Logical links are an abstract representation of the connectivity between two logical nodes. Different implementations of these logical links are possible, e.g. physical links, VPCs, uplinks (only used in a hierarchical PNNI network) and aggregation of logical links. Logical links are mainly used to carry user data. There may be parallel logical links between two nodes.

If VPCs are used to connect two lowest level nodes, then they must be configured by network management.

6.2.3 *PNNI Routing Control Channels*

A PNNI Routing Control Channel (RCC) is a Virtual Channel Connection (VCC) used for the exchange of PNNI routing protocol messages, such as Hellos packets, PNNI Topology State Packets (PTSPs) and PNNI Topology State Element (PTSE) acknowledgement packets. These RCCs are used between logical nodes that are logically or physically adjacent. In contrast to logical links, RCCs are only used to carry PNNI routing control data.

Three different implementations of a VCC exist:

- *Over a physical link.* A reserved VCC with VPI=0 and VCI=18 will be used.
- *Over a VPC with VPI=X.* A VCC with VPI=X and VCI=18 will be used.
- *Over a Switched Virtual Channel Connection (SVCC).* The VPI and VCI are assigned by PNNI signalling in the normal way

Different service categories (nrt-VBR, rt-VBR, CBR, ABR, UBR) and different traffic parameters can be used for an RCC. The preferred way is to use of the nrt-VBR service category.

A PNNI RCC uses the AAL Type 5 (AAL5) without an AAL Service Specific Convergence Sublayer (SSCS). This allows an unassured information transfer and a mechanism to detect corrupted AAL-SDUs. Error correction is handled by the PNNI above the AAL. One complete PNNI packet is encapsulated in exactly one AAL-SDU.

SVCCs are only important in hierarchical structured PNNI systems. They are used to connect a higher level LGN with another Logical Group Node (LGN) and to connect border nodes with LGNs of the neighbouring group. The problem with SVCCs is that the topology data bases of the involved nodes is incomplete when RCC-SVCCs are established. However, the algorithms solve this problem.

At the lowest level of the hierarchy there may be multiple RCCs between two nodes. However, only one RCC is used for the purpose of database synchronisation and flooding. The other RCCs are only used as backup links in the case of a link failure.

6.2.4 Identifiers and Indicators

6.2.4.1 Level Indicators

The PNNI can be used in a hierarchy consisting of several levels. So the PNNI entities (nodes, links and peer groups) occur at various hierarchy levels. The indicator level is an integer number that ranges from zero to 104. It specifies the number of significant bits used for the peer group ID which will be used to address such a peer group. This concept is similar to the address prefix concept we saw in the previous section.

If we have two peer groups, where one is an ancestor (parent, grandparent, etc.) of the other in the hierarchy, the ancestor is a higher level peer group and will have a smaller level indicator than the other. A peer group with an ID of n bits length may have a parent peer group whose identifier ranges anywhere from zero to $n - 1$ bits in length. Similarly, a peer group with an ID of m bits length may have a child peer group whose identifier ranges anywhere from $m + 1$ to 104 bits in length. Not all levels need to be be used in a specific topology, i.e. a hierarchy can consist, for example, of levels 80, 32 and 16.

6.2.4.2 Peer Group and Node Identifiers

A peer group is a group of connected nodes of the same hierarchy level. They have the same peer group identifier configured. A peer group leader identifier is a string of bits between zero and 104 bits (13 bytes) in length. Peer group identifiers are encoded using 14 bytes: one byte level indicator followed by 13 bytes of identifier information. The identifier information field must be encoded with the $104 - n$ rightmost bits set to zero, where n is the level of the peer group (Figure 6.2).

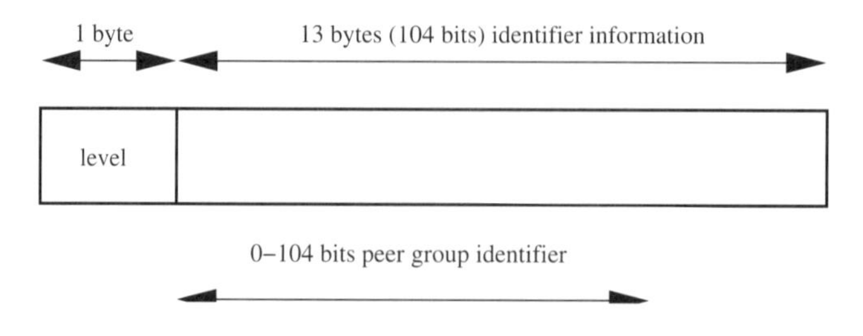

Figure 6.2: PNNI peer group identifiers

Each node is identified by its node identifier (node ID). The node identifier is 22 bytes in length and consists of one byte level indicator followed by a 21 opaque value, i.e. these 21 bytes have no internal structure, and no internal structure may be assumed when analysing node identifiers. This opaque value must be unique within the entire routing domain. The level of a node is the same as the level of its containing peer group. Usually, the node opaque value will be automatically generated and is based on the ATM end system addresses of the node and level indicators related to the node. The node ID is not allowed to change while the node is operational.

6.2.4.3 Scope of Addresses

Addresses and address prefixes used in PNNI can have scopes associated with them. The scopes are used to determine the scope of the advertisement of reachable addresses. The scope can also be used to force the user to use group addresses (e.g. an ATM Anycast address) instead of individual address if the individual addresses have a smaller scope and can therefore not be reached directly.

For the PNNI, the scope of reachable addresses is specified by a level indicator and therefore ranges from zero to 104. At the UNI/Integrated Local Management Interface (ILMI), the scope is indicated by a scope identifier between 1 (very local) and 15 (global). The mapping between these 15 scopes and the PNNI level indicator can be configured by the administrator. The default mapping is defined as in Table 6.1.

Table 6.1: Default mapping for UNI/ILMI scope and PNNI reachable addresses level indicator

UNI/ILMI scope	PNNI reachable addresses level indicator
1–3	96
4–5	80
6–7	72
8–10	64
11–12	48
13–14	32
15 (global)	0 (global)

6.2.4.4 Port IDs and Logical Links

Nodes are connected by logical links. These links must be distinguished when handling topology and routing information. Therefore, logical links and the ports they are attached to have identifiers.

A port ID is a 32-bit number assigned by a node to unambiguously identify a point of attachment of a logical link to that node. Port IDs are only meaningful in the context of the assigning node, identified by its node ID. The values zero and 0xFFFFFFFF are reserved and not used to identify physical ports.

A logical link is identified by the node ID of either node at the end of that logical link and the port ID assigned by that node, i.e. each link has two identifiers because it is attached to two nodes. For logical links each end node advertises the port ID and outbound link characteristics

(characteristics of one direction) inside a peer group. This is important because ATM links are bi-directional and potentially have different characteristics (e.g. different QoS parameters) in each direction. Therefore, the characteristics of both directions should be advertised. Two nodes of a logical link exchange their port IDs with each other using PNNI Hello packets.

6.2.4.5 Aggregation Tokens

Aggregation tokens are only used in hierarchical PNNI networks. An aggregation token, along with the remote node ID, identifies uplinks which are to be aggregated at the next highest level of a hierarchy. Aggregation tokens are four-byte identifiers. All links between a pair of logical group nodes with the same value of the aggregation token must be advertised as one logical link. The aggregation token is included in a PNNI Topology State Element (PTSE) which describes a link. The scope of significance of an aggregation token is limited to pair-wise logical group nodes. However, the large space allows globally unique token values for ease of administration.

6.2.5 Hello Protocol

6.2.5.1 Introduction

The PNNI Hello protocol is used in order to discover[2] and verify the identity of neighbouring nodes and to determine the status of the links to those nodes.

The Hello protocol runs on all RCCs all the time by exchanging PNNI Hello packets at regular intervals.

Another task of the PNNI Hello protocol is the negotiation of the PNNI version to be used between neighbouring nodes. A PNNI implementation generally supports a range of protocol versions in protocol messages. Each Hello packet contains version fields that are unsigned integers. Every node advertises the oldest and the newest supported version in the Hello packet. The newest version that is supported by both nodes is finally used. If no common version can be found, then the communication cannot be continued and an error will be reported to the network management.

Figure 6.3 shows a PNNI Hello packet. The first six fields are included in any PNNI packet. The value 0x0001 of the packet type field denotes that this is a PNNI Hello packet. The next field is the length of the whole packet in bytes. The version fields are used to negotiate the versions of neighbours, as just explained. The node IDs in the packet are used to identify the node that is originating the packet and the node that is expected to receive it, i.e. the neighbour of the originating node on the link. The neighbour's node ID is known from the last Hello packet in the opposite direction. If it is not yet known, then its node ID is set to zero. The ATM end system address is used if somebody wants to contact the originating node by a signalled ATM connection (only in a hierarchical network). The peer group ID of the originating node is used by the neighbour to identify whether the originating node is in the same peer group or not. The port ID identifies the port of the originating node to which the current link is attached. The remote port ID is doing the same for the remote node. The Hello

---

[2] The discovery of neighbouring nodes is only needed if a physical link or a Permanent Virtual Connection (PVC) is used as an RCC. If an SVC is used (in a hierarchical system), then the identity must be clear before connection setup.

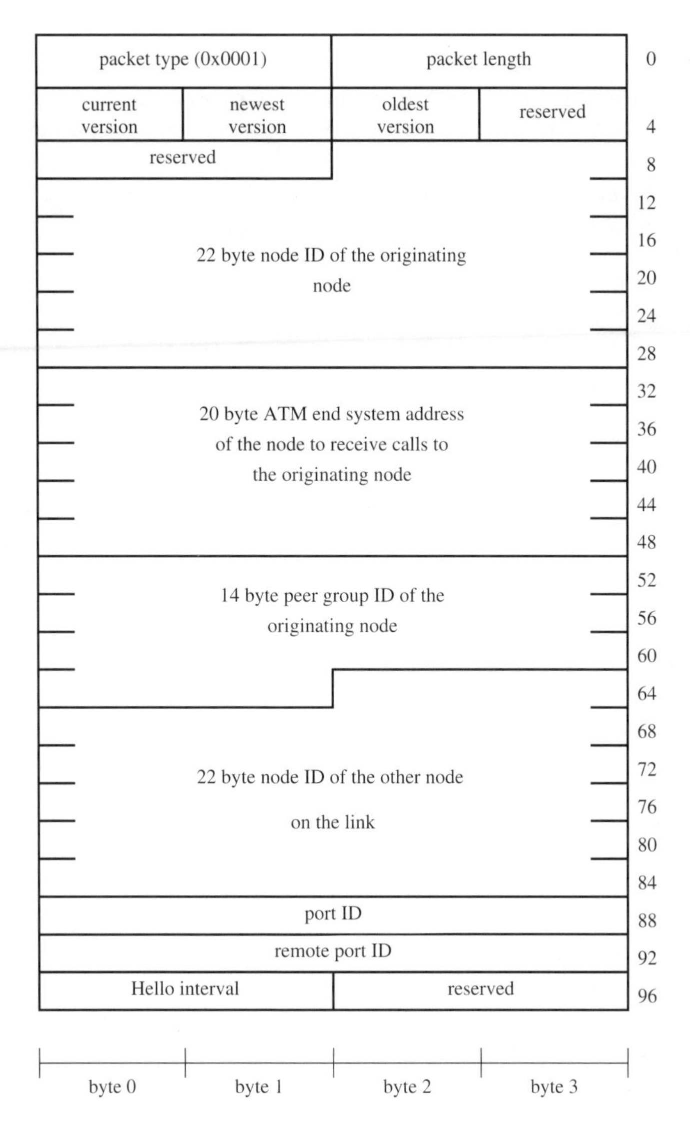

Figure 6.3: PNNI Hello packet. In addition, the aggregation token, nodal hierarchy list, uplink information attribute and LGN horizontal link extension information groups can be included

interval is the number of seconds between two subsequent Hello packets. The remote node assumes a link failure if a couple of these packets are lost. The default interval is 1 second.

If a PNNI hierarchy is used, then the PNNI Hello packet is longer and contains also an aggregation token, a nodal hierarchy list, an uplink information attribute and an LGN horizontal link extension.

A small trace of two ATM switches talking Hello is shown in Section 6.2.11.1.

6.2.5.2 Hello State Machine

Between two neighbouring nodes each RCC has its own instance of the Hello protocol. An instance of the Hello protocol is made up of its Hello state machine. To increase the fault tolerance, it is possible to have more than one RCC between two lowest level neighbour nodes, i.e. two or more parallel physical links and/or VPCs between the nodes are established by network management. In such a case each RCC will run its own Hello protocol with its own Hello state machines on both ends. However, for the purpose of data base synchronisation and other PNNI protocols, only one RCC is used.

Each side of an RCC has its own Hello protocol instance and thus its own protocol state. This protocol state will be analysed by other parts of the PNNI. The following states of a Hello state machine are possible:

- *Down.* Unusable (physical) link. No PNNI routing packets can be sent or received over such a link.
- *Attempt.* In this state attempts are made to contact the neighbour by periodically sending Hello packets with period determined by a configuration parameter $HelloInterval$ (default: *HelloInterval* = 1 second). If a link remains in this state, there is probably something wrong.
- *1-WayInside.* Hello packets have been received from the neighbour. Both nodes are in the same peer group. But in the neighbour's Hello packet the remote node ID and the remote port ID were set to zero, i.e. the neighbour did not yet recognise me as its neighbour.
- *1-WayOutside.* Hello packets have been received from the neighbour. Both nodes are in different peer groups. But in the neighbour's Hello the remote node ID and the remote port ID were set to zero, i.e. the neighbour did not recognise me as its neighbour yet.
- *2-WayInside.* Hello packets have been exchanged with the neighbour. Both nodes are in the same peer group. This state indicates that a bi-directional communication over this link between the two nodes has been achieved. Routing information (PNNI Topology State Packet (PTSP)s, PTSE acknowledgement packets, etc.) can be exchanged over this link.
- *2-WayOutside.* Hello packets have been exchanged with the neighbour. Both nodes are in different peer groups, but the nodal hierarchy list advertised by the neighbour does not include a common higher level peer group. The node still searches for a common higher level peer group that contains both this node and the neighbour node.
- *CommonOutside.* Hello packets have been exchanged with the neighbour. Both nodes are in different peer groups. A common higher level peer group has be found. This link can be used as an uplink to exchange routing information (PTSPs, PTSE acknowledgement packets, etc.).

The states 2-WayInside (if the neighbour node is in the same peer group) and CommonOutside (if the neighbour is in a different peer group) indicate a full operational connection to a neighbour. The states 1-WayOutside, 2-WayOutside and CommonOutside are unimportant for non-hierarchical PNNI systems.

6.2.5.3 Sending Hello Packets

In all states other than Down, Hello packets are transmitted periodically, by default every 1 second. This value is configurable. In the configuration we used for our experiments the value was 15 seconds. In addition, Hello packets are sent if the state of a link has changed. But the period between successive transmissions of Hello packets is limited by a lower limit. This prevents a node from sending Hello packets at unacceptably high rates if several state changes occur, and is important to avoid a fast sequence number overflow.

If the instance does not know the remote node ID and the remote port ID (i.e. the instance is in state Attempt) then these fields are set to zero in the transmitted Hello packets.

In hierarchical systems the nodal hierarchy list is included in the Hello packet on links across the border to another peer group, i.e. when the state is 1-WayOutside, 2-WayOutside or CommonOutside. The nodal hierarchy list describes all the node's higher level node IDs, peer group IDs and LGN ATM end system addresses as received from the PGL in its higher level binding information. If no higher levels of the hierarchy are known, then an empty nodal hierarchy list must be included in the Hello packet. Whenever a change occurs in the number or content of known higher levels, as expressed in the nodal hierarchy list, in the node ID, peer group ID or ATM address at the lowest level, then the sequence number of the nodal hierarchy list must be incremented and a new Hello packet must be sent. The sequence number of the first nodal hierarchy list sent to any neighbour must be greater than zero. Note that when the hierarchy is still coming up, the number of levels included in the nodal hierarchy list may increase with each successive Hello packet.

If the link is in state 1-WayOutside, 2-WayOutside or CommonOutside, i.e. it is a link across the border, then an Uplink Information Attribute Information Group is included in the Hello packet. This ULIA IG passes aggregated information about the peer group internal link structure to the neighbour in the other peer group.

6.2.5.4 Receiving Hello Packets

PNNI Hello packets can be received in any state. If the protocol version of a Hello packet is not supported, that packet will be discarded.

The node ID found in a received packet is important to the receiving node because it identifies the neighbour node.

If the received Hello packet contains a new instance of the nodal hierarchy list, as indicated by a new sequence number, then the nodal hierarchy list must be searched for the lowest level peer group that both nodes have in common. Even though it does not explicitly appear in the list, the neighbour must be considered to be the lowest level component of the nodal hierarchy list.

6.2.5.5 LGN Hello Protocol

An LGN is a node on a higher level of the hierarchy of nodes. If two of these LGNs are neighbours then they also use the Hello protocol to exchange information.

Two LGN neighbours are usually not connected by a direct link. Therefore, a Switched Virtual Channel Connection (SVCC) based RCC must be used to connect these nodes. The Hello protocol runs on this SVCC. As opposed to lowest level neighbour nodes, LGNs will never have more than one RCC (SVCC) between them. Once the SVCC is established by signalling, a Hello protocol instance is initiated. This protocol is essentially the same as the protocol that runs between lowest level neighbours, with a few modifications. A port value of 0xFFFFFFFF is always used as the port ID filled in the Hello messages. This works because there is always only one RCC between two LGNs. SVCC based RCCs are always inside one peer group. Therefore, the Hello protocol instance may only be in the state Down, Attempt,

1-Way(Inside) or 2-Way(Inside). For the LGN, which is the calling party for the SVCC based RCC, the uplink PNNI Topology State Element must necessarily be received before the SVCC is established. Otherwise the path to establish the SVCC is unknown and the call cannot be routed. If a called party LGN receives a SETUP from a node which it still has to recognise as a neighbour, the called party LGN must accept the call, but ignore Hello packets until an uplink PTSE is received indicating that node as a neighbour.

The Hello protocol between the LGNs is used to monitor the status of the SVCC used as an RCC between the two LGNs to increase robustness. If the monitoring function detects a failure, then it will cause the SVCC to be re-established. In this situation the states of the state machines on both sides of the logical link will not change until a specific timer has expired.

6.2.6 Database Synchronisation

6.2.6.1 Introduction

The reason for the existence of the PNNI routing protocol is the need for a distributed knowledge of topology information that can be used to route calls in the network. The database synchronisation is a key feature to distribute the knowledge.

When a node learns about the existence of a neighbouring peer node that resides in the same peer group from the Hello protocol, it initiates a database exchange process in oder to synchronise the topology databases. Both nodes send their own knowledge to the neighbour so that each node can complete its topology database.

The process of database synchronisation involves the exchange of a sequence of Database Summary packets, which contain identifying information of all PNNI Topology State Elements in a node's topology database, i.e. each node sends a list of indices of the entries contained in its database to the neighbour. One side sends a Database Summary packet and the other side responds (and implicitly acknowledges the received packet) with its own Database Summary packet.

After a node has received a Database Summary packet from a neighbouring peer, it examines its topology database for the presence of each PTSE indexed in the packet. If the PTSE is not found in the database, or the database entry is older than the PTSE listed in the Database Summary of the neighbour, the node requests the PTSE from its peer.

6.2.6.2 Sending Database Summary Packets

Database Summary packets, as shown in Figure 6.4, are exchanged between neighbouring nodes. To perform this a master–slave relationship is used. One node is the master and the other node is the slave. The master sends the first Database Summary packet. The slave can only respond to the master's Database Summary packet. At any given time only one Database Summary packet may be outstanding. The master is responsible for retransmission.

In the Negotiating state, the node tries to find out whether it is the master or the slave of an RCC. Each node sends empty Database Summary packets, with the Initialise, More and Master flag bits set to 1. The node with the largest node ID will be the master. After the Negotiating state the nodes are in the Exchanging state. In this state it is clear which node is the master and which node is the slave.

In the Exchanging state the Database Summary packets contain summaries of topology state information contained in the node's database. The PNNI Topology State Packet and the

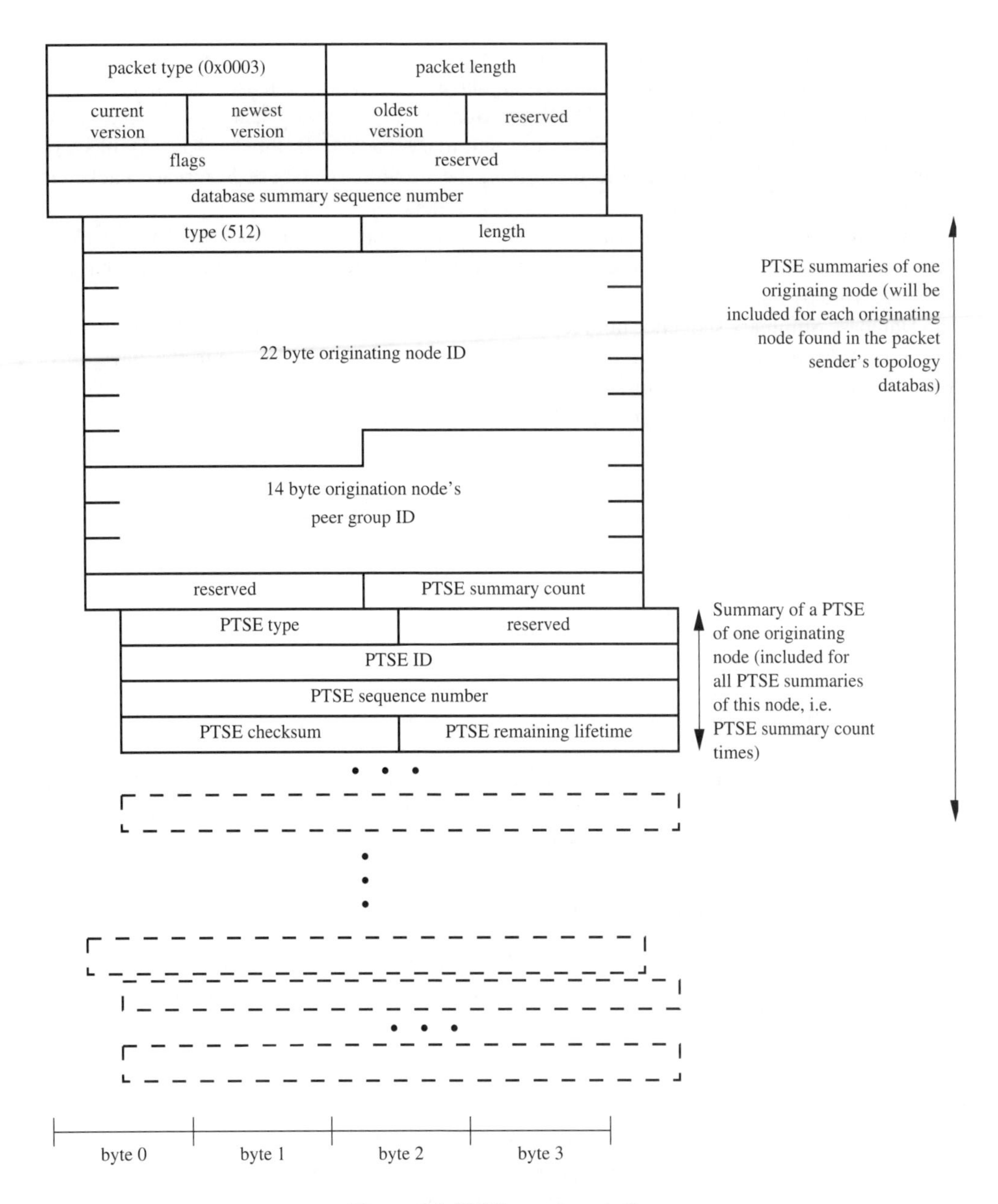

Figure 6.4: PNNI summary packet

PNNI Topology State Element header information of each such PTSE is listed in one of the node's Database Summary packets. PTSEs for which new instances are received after the Exchanging state has been entered are not needed to be included in any Database Summary packet, since they will be handled by the normal flooding procedure which is responsible for database updates.

In the Exchange state, the determination when to send a Database summary packets depends on whether the node is master or slave. The first Database Summary packet is sent by the master. Any new Database Summary packet is sent by the master if the slave acknowledged the last Database Summary packet. If the last packet has not been acknowledged after a specific time, then the previous Database Summary packet is retransmitted.

The slave sends Database Summary packets only in response to Database Summary packets received from the master. If the received packet from the master is new, i.e. this packet has not been received by the slave before, then the slave acknowledges the received packet by echoing the Database Summary sequence number as part of the new Data Summary packet sent by the slave. If the received packet from the master is old, i.e. it is already known to the slave, then the slave retransmits the previous Database Summary packet sent by the slave. If the slave has no more data to send, it sends an empty Database Summary packet with the More flag bit set to zero.

6.2.6.3 Receiving Database Summary Packets

The incoming Database Summary packet is associated with a neighbouring peer by the interface over which it was received.

If a node is in the Negotiating state it still has to find out whether it will be the master or the slave. If it receives an empty Database Summary packet where the Initialise, More and Master flag bits are 1, then it determines who is the master. The node with the larger node ID is the master. In any case the Initialise bit is now set to zero. The Master bit is only set to one if the originator of the packet is the master.

If the flags and the sequence number of the received packet are consistent, then the packet is processed. If a listed PTSE is new or newer compared with the own topology database, the corresponding PTSE will be requested soon by a PTSE Request packet.

6.2.6.4 Sending PTSE Request Packets

After a Database Summary packet is received and processed the outstanding PTSEs must be requested. To request the needed PTSEs, the node sends a PTSE Request packet containing a list of one or more PTSEs. Figure 6.5 shows a packet.

The PTSE Request packet is transmitted to a particular neighbouring peer. The node waits for one or more responding PNNI Topology State Packets containing all requested PTSEs. If all requested PTSEs are not received within a specific time, then a new PTSE Request packet including the missing PTSEs and/or any other needed PTSEs is transmitted. There is at most one PTSE Request packet outstanding at any time for each neighbouring peer.

The process continues until all needed PTSEs from any neighbouring node, whose Database Summary indicates new PTSEs, are known to the node.

6.2.6.5 Receiving PTSE Request Packets

What happens when a node receives a PTSE Request packet? For each PTSE specified in the PTSE Request packet, the PTSE must be looked up in the receiver's node topology database. The requested PTSEs are then bundled into one ore more PTSPs and are transmitted to the neighbouring node that was sending the PTSE Request packet.

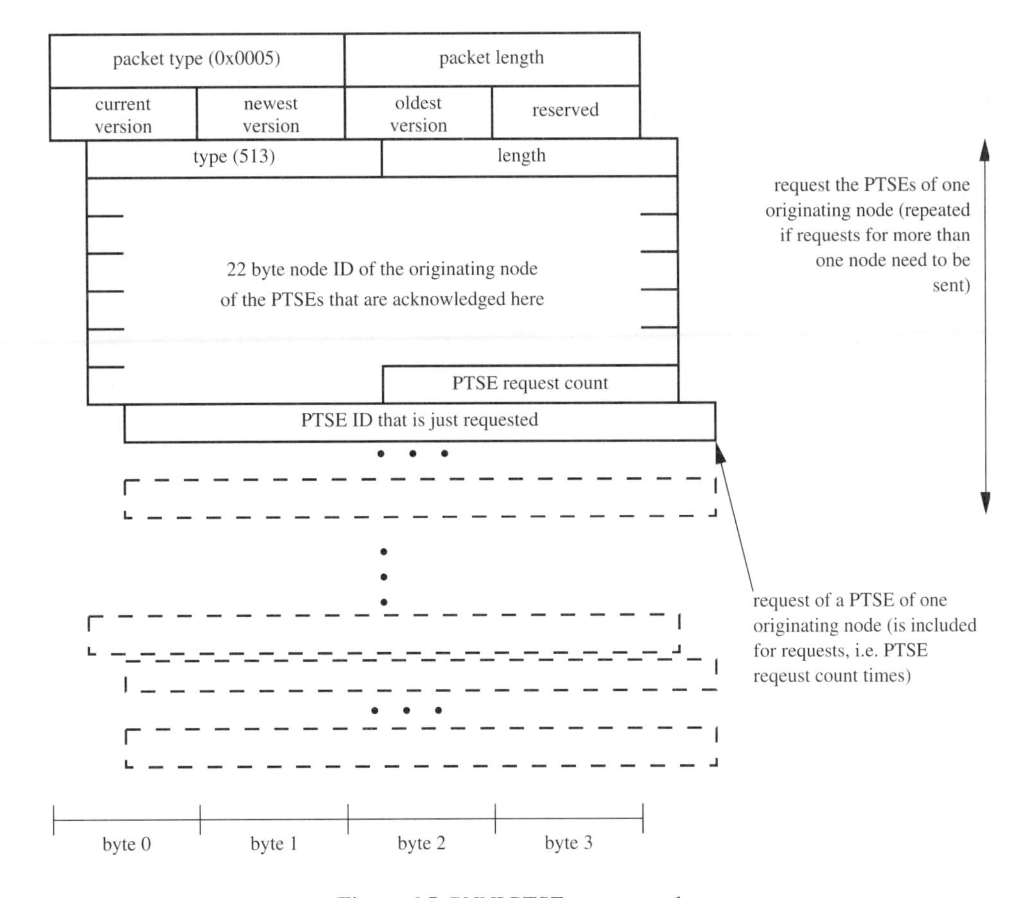

Figure 6.5: PNNI PTSE request packet

6.2.7 Topology Description and Distribution

Topology information includes topology state parameters and nodal information. Topology information is encoded in a flexible Type/Length/Value (TLV) format to allow future extensions.

The topology state parameter is a generic term that refers to either a link state parameter or a nodal state parameter. Two kinds of topology state parameters are known: topology metric and topology attribute.

A topology metric combines the parameters of all links and nodes along a given path to determine whether the path is acceptable for carrying a given connection, whereas a topology attribute describes a single link or node.

A link state parameter provides information that captures an aspect or property of a link. This kind of parameter is individual for both directions of a bi-directional ATM link.

A nodal state parameter provides information that captures an aspect or property of a node. Nodal state parameters are used to construct PNNI complex node representations. These complex node representations are only needed in complex and hierarchical systems and not investigated further in this book.

The Resource Availability Information Group (RAIG) contains information that is used to attach values of the topology state parameters to nodes, links, and reachable addresses. Different parameters can be carried in a RAIG:

- Cell Delay Variation (CDV).
- maximum Cell Transfer Delay (maxCTD). This is the sum of the fixed delay components across the link/node and CDV.
- Administrative Weight (AW). This is a value set by the network operator. A lower value indicates a more desirable link or node.
- Cell Loss Ratio (CLR) for CLP $= 0$ (CLR_0).
- Cell Loss Ratio (CLR) for CLP $= 0 + 1$ (CLR_{0+1}).
- maximum Cell Rate (maxCR).
- Available Cell Rate (AvCR). This is a measure for the effective available capacity.
- Cell Rate Margin (CRM). A very specific and optional Variable Bit Rate (VBR) parameter.
- Variance Factor (VR). A very specific and optional VBR parameter.

6.2.8 Advertising and Summarising Reachable Addresses

6.2.8.1 Scope of Advertisement of Addresses

Each address in a PNNI network has a scope of advertisement. This scope is used to determine how far an address should be advertised in the network and only makes sense in networks with more than one hierarchy level.

The advertisement scope of a reachable address is specified by a level indicator, which means that the address will be advertised up to this level, but not into any higher level of PNNI routing. Nodes outside the peer group of the destination node will not know the existence of the advertised address. If the level indicator is set to zero, then the advertisement scope is unlimited, which means that the address may be advertised throughout the entire PNNI routing domain. Usually, the advertisement scope is unlimited. See Section 6.2.4.3 for further information.

6.2.8.2 Summary Address and Suppressed Summary Address

To allow routing of calls each node must know how to route calls to any of the end systems attached to the network. This means that each node must store routing information for any possible destination address. This information must not just be stored—it must also be distributed all around the PNNI network.

It is clear that this would be very inefficient in large networks. Therefore, the concept of summary addresses is used. A summary address is an address that includes a range of individual addresses. See also Section 6.2.1.2.

By default, a lowest level node has one default internal summary address. This is the 13-byte prefix of the node's address. Only if the node is at level 104 does it have no default summary address. By default an LGN has one internal summary address which is identical to the lower level Peer Group ID it represents.

When overlapping addresses are present, the longest match summary address will be used to determine whether an address is to be advertised explicitly or indirectly through the use of a summary address.

A suppressed summary address is used to suppress the advertisement of addresses which match a specified prefix, regardless of scope. By default a node has no suppressed summary address

6.2.9 *Flooding*

Flooding is the mechanism used in the PNNI to distribute topology information (PTSEs) hop-by-hop inside a peer group. PNNI flooding is a reliable process, because packets are acknowledged and, if necessary, retransmitted. The flooding of topology information starts at the highest hierarchy level. To allow routing of a call the topology information collected by the highest level peer group must be available to lower level nodes. This enables every node in the network to directly calculate appropriate routes. But it requires that all PNNI nodes have information not only of their own peer group, but also of all of their ancestor peer groups.

How does it work? Higher level PTSEs are flooded to all nodes of their peer group, and in addition to all descendant peer groups, i.e. to all lower level nodes contained in the lower level peer groups represented by the nodes in this peer group, and so on. When a higher level node floods a new or updated PTSE then this higher level node floods the PTSE to the Peer Group Leader (PGL) of the lower level which this higher level node represents, as well as to all neighbouring peers at its level (if it was received from another node then the PTSE would not be sent back to this node). The PGL of the lower level peer group will in turn flood the PTSE in the lower level peer group.

PTSEs generated in a given peer group never get flooded to a higher level peer group. Instead, the PGL summarises the topology of the peer group (based on the PTSEs generated within the peer group) and floods the new PTSEs originated by the LGN at the parent peer group's level. This mechanism allows for scalability.

Flooding in a non-hierarchical network is quite simple. Every new PTSE is passed to all other nodes in the peer group. Summarisation is not needed.

How is flooding implemented in the PNNI? If a PTSE needs to be distributed, then it is encapsulated in a PNNI Topology State Packet (PTSP) (see Figure 6.6). This PTSP is then distributed. Received packets are acknowledged by PTSE acknowledgement packets (Figure 6.7). Non-acknowledged packets are retransmitted.

The process of flooding new topology information is running all the time. The information exchanged and stored is then used by the Generic Call Admission Control (GCAC) when a call must be routed.

6.2.10 *Hierarchy*

The PNNI is designed to be scalable to large networks. To support this it can be hierarchically structured. In this book we do not explore the PNNI hierarchy in detail because the authors could not perform any experiments on such a hierarchy and they could therefore not verify the behaviour that is described in the standards.

6.2.11 *Communication Examples*

This section shows seven typical PNNI routing protocol communication examples. All these examples show the communication between the six ATM interfaces attached to the three switches as presented in Figure 6.8. The PNNI is running on all the switches. All switches are in the same peer group and constitute a non-hierarchical PNNI network.

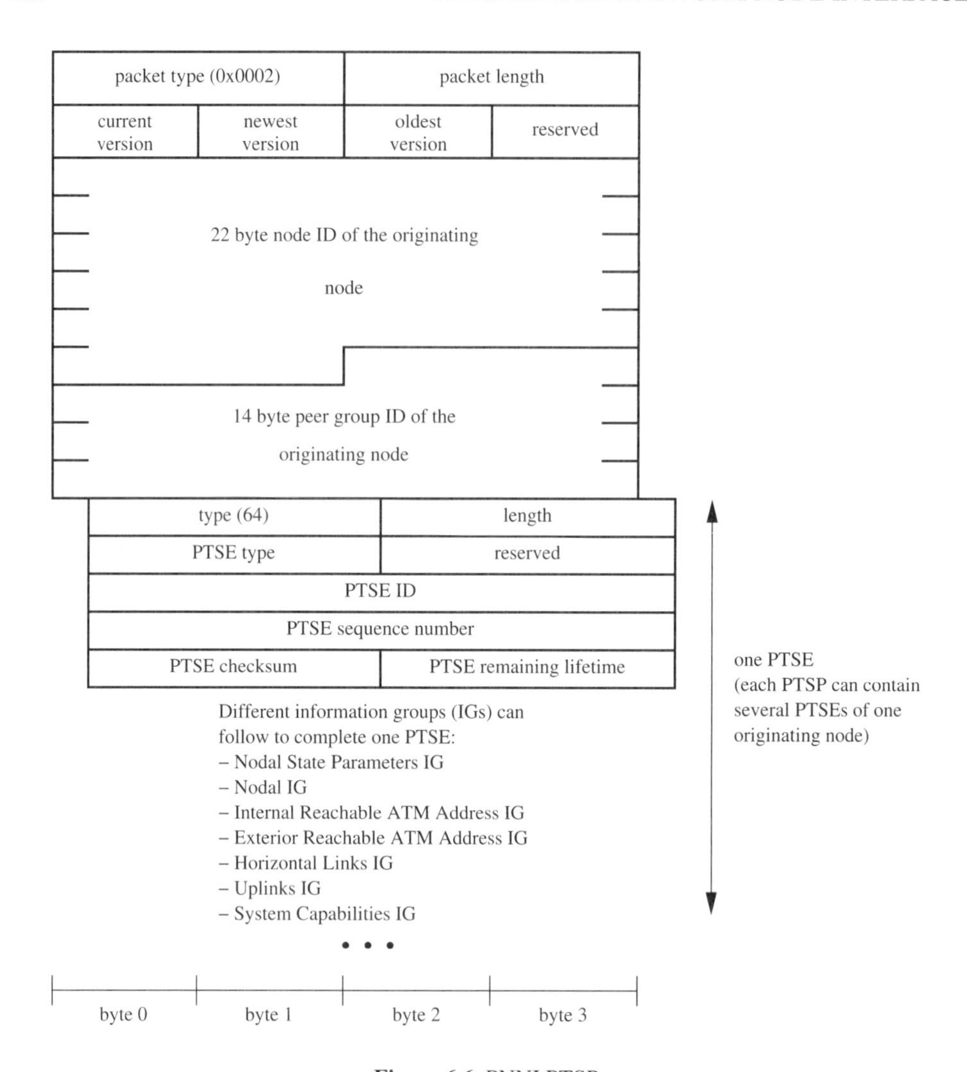

Figure 6.6: PNNI PTSP

The communication examples were recorded in the GMD Fokus laboratory. Optical splitting boxes were inserted into the three physical links between the switches to allow recording of both communication directions. The six output links of the splitting boxes (two communication streams per switch-to-switch connection) were multiplexed to two output links (on different VPIs) by an additional crossconnect switch. These two output links of the crossconnect were recorded by two additional ATM interface cards (in our case devices "`/dev/ty0`" and "`/dev/ty1`") inserted into an extra end system (lovina). The recorded ATM communication was decoded by the `a5r` tool and the `pnnidump` tool of the Tina tool set. To decode the messages of one communication stream (PNNI routing protocol messages use VPI=0 and VCI=18, but in some cases the VPI is different because of the multiplexing switch) the commands

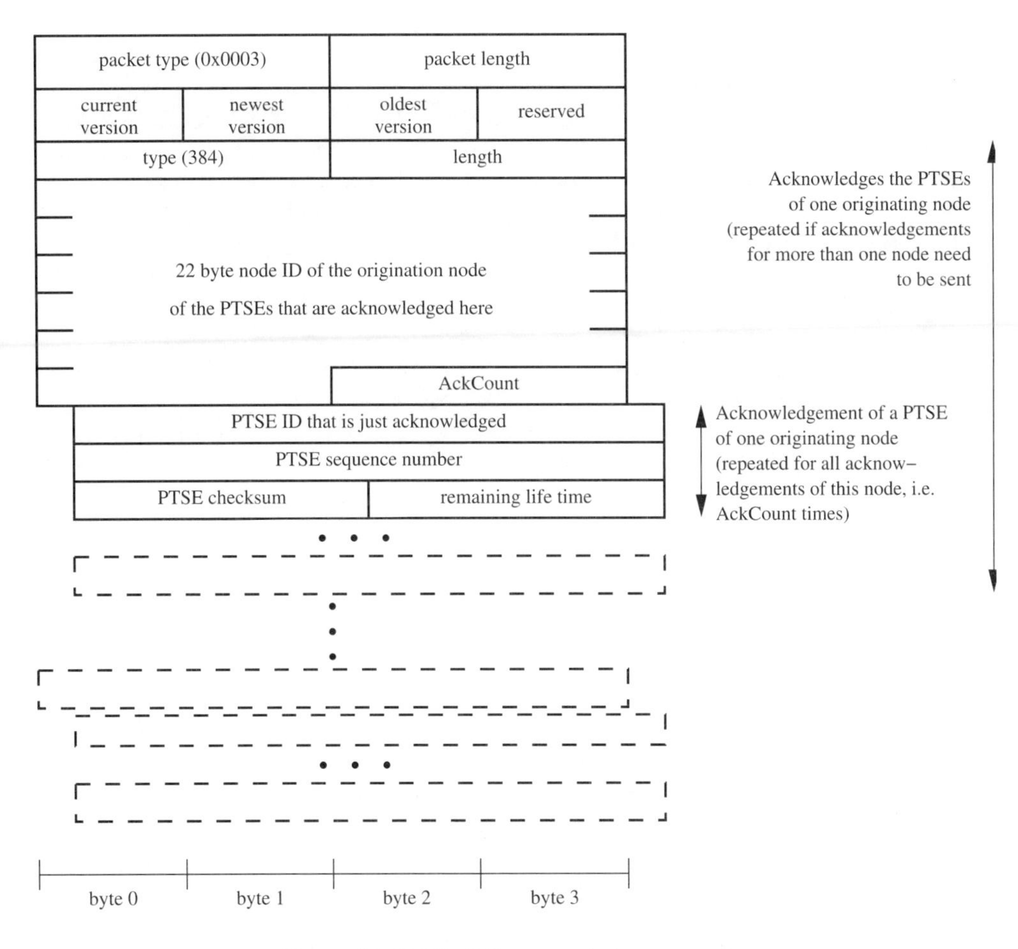

Figure 6.7: PNNI PTSE acknowledgement packet

```
a5r -c 18 < /dev/ty0 | pnnidump -Fhcp
```

were executed at lovina under the assumption that the stream was received on VPI=0. See Section 1.3 for details about capturing and decoding of ATM communications.

The following sections show the output of the Tina tools. The messages are marked with 1 to 6 depending on their direction (see Figure 6.8 for the six directions).

6.2.11.1 A Standalone ATM Switch

This first example shows the switch forelle that is already up and waiting for other switches. The other switches are still down. The following messages are transmitted every 15 s over forelle's outgoing links to forest (2) and to foreplay (5):

```
1  R  2⇒   0.000
2           hello
```

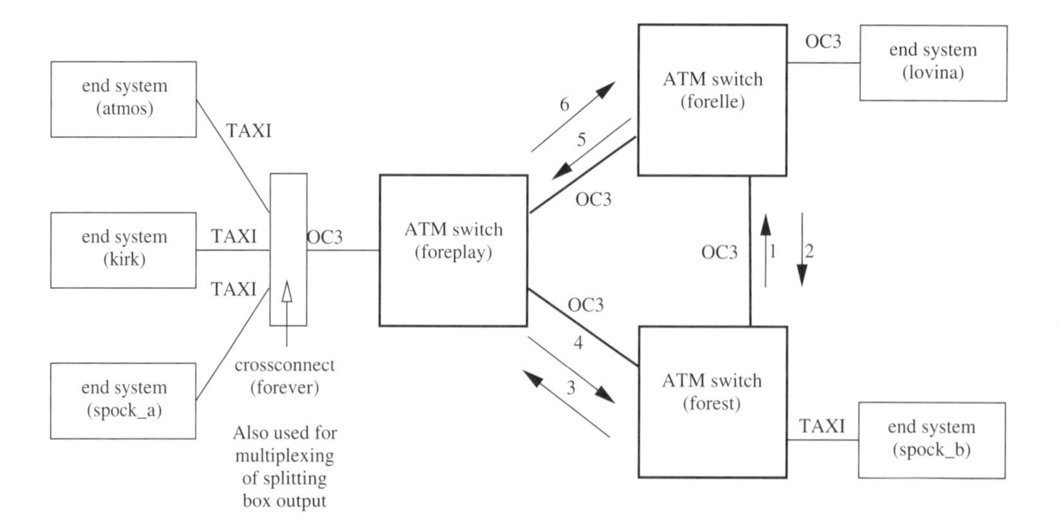

Figure 6.8: PNNI example scenario

```
3              node={level=80,node=0x39.0000.11.222222.0000.0000.0001.ff1a4ce70001.00}
4              addr=0x39.0000.11.222222.0000.0000.0001.ff1a4ce70001.00
5              pgid={level=80,id=39000011222222000000000000} rnode={level=0,
6              peergroup={level=0,id=00000000000000000000000000},esi=00:00:00:00:00:00}
7              port=0x10000010 rport=0x0 hinter=15
8    R  3⇐    0.157
9              hello
10             node={level=80,node=0x39.0000.11.222222.0000.0000.0001.ff1a4ce70001.00}
11             addr=0x39.0000.11.222222.0000.0000.0001.ff1a4ce70001.00
12             pgid={level=80,id=39000011222222000000000000} rnode={level=0,
13             peergroup={level=0,id=00000000000000000000000000},esi=00:00:00:00:00:00}
14             port=0x1000001b rport=0x0 hinter=15
```

6.2.11.2 Two Nodes of a Peer Group are Going Up

This example shows the PNNI messages when the two nodes (switches), foreplay and forelle, are going up. They exchange messages to get synchronised databases that store the peer group information.

Note that in this example both nodes advertise the same address prefix. This is an unsuitable configuration for production networks, but was used to demonstrate easily crankback in later sections.

First the switch foreplay goes up. As shown in Section 6.2.11.1 it transmits Hello messages to its neighbours forest (4) and forelle (6), but they are still down:

```
1    R  4⇒    0.988
2              hello
3              node={level=80,node=0x39.0000.11.222222.0000.0000.0002.ff0f0dc60001.00}
4              addr=0x39.0000.11.222222.0000.0000.0001.ff0f0dc60001.00
```

```
 5              pgid={level=80,id=39000011222222000000000000} rnode={level=0,
 6              peergroup={level=0,id=00000000000000000000000000},esi=00:00:00:00:00:00}
 7              port=0x10000012 rport=0x0 hinter=15
 8   R  6⇒     3.716
 9              hello
10              node={level=80,node=0x39.0000.11.222222.0000.0000.0002.ff0f0dc60001.00}
11              addr=0x39.0000.11.222222.0000.0000.0001.ff0f0dc60001.00
12              pgid={level=80,id=39000011222222000000000000} rnode={level=0,
13              peergroup={level=0,id=00000000000000000000000000},esi=00:00:00:00:00:00}
14              port=0x10000013 rport=0x0 hinter=15
```

Then switch forelle goes up. It starts with sending its own Hello messages as shown in Section 6.2.11.1:

```
 1   R  5⇐     17.700
 2              hello
 3              node={level=80,node=0x39.0000.11.222222.0000.0000.0001.ff1a4ce70001.00}
 4              addr=0x39.0000.11.222222.0000.0000.0001.ff1a4ce70001.00
 5              pgid={level=80,id=39000011222222000000000000} rnode={level=0,
 6              peergroup={level=0,id=00000000000000000000000000},esi=00:00:00:00:00:00}
 7              port=0x10000010 rport=0x0 hinter=15
 8   R  2⇒     17.857
 9              hello
10              node={level=80,node=0x39.0000.11.222222.0000.0000.0001.ff1a4ce70001.00}
11              addr=0x39.0000.11.222222.0000.0000.0001.ff1a4ce70001.00
12              pgid={level=80,id=39000011222222000000000000} rnode={level=0,
13              peergroup={level=0,id=00000000000000000000000000},esi=00:00:00:00:00:00}
14              port=0x1000001b rport=0x0 hinter=15
```

Now the switches forelle and foreplay have received the first Hello messages from each other (switch forest could not receive anything because it is still down). The next Hello messages show that they now know each other (see the changed remote node [`rnode`] field):

```
 1   R  6⇒     17.630
 2              hello
 3              node={level=80,node=0x39.0000.11.222222.0000.0000.0002.ff0f0dc60001.00}
 4              addr=0x39.0000.11.222222.0000.0000.0001.ff0f0dc60001.00
 5              pgid={level=80,id=39000011222222000000000000} rnode={level=80,
 6              node=0x39.0000.11.222222.0000.0000.0001.ff1a4ce70001.00}
 7              port=0x10000013 rport=0x10000010 hinter=15
 8   R  5⇐     17.703
 9              hello
10              node={level=80,node=0x39.0000.11.222222.0000.0000.0001.ff1a4ce70001.00}
11              addr=0x39.0000.11.222222.0000.0000.0001.ff1a4ce70001.00
12              pgid={level=80,id=39000011222222000000000000} rnode={level=80,
13              node=0x39.0000.11.222222.0000.0000.0002.ff0f0dc60001.00}
14              port=0x10000010 rport=0x10000013 hinter=15
```

At the same time they start exchanging topology information:

```
1   R  6⇒   17.632
2            dbsum flags=<init,more,master> seq=925413206
3   R  6⇒   17.633
4            dbsum flags=<,,master> seq=925413207
5            ptsesum node={level=80,
6            node=0x39.0000.11.222222.0000.0000.0002.ff0f0dc60001.00}
7            pgid={level=80,id=39000011222222000000000000} scount=2
8            ptype=internal id=1342177285 seq=2 csum=59306 life=3575
9            ptype=nodal id=1 seq=1 csum=32489 life=3574
10  R  6⇒   17.669
11           ptsereqp ! illegal IG 5 within ptsereq
12  R  6⇒   17.670
13           ptsp
14           node={level=80,node=0x39.0000.11.222222.0000.0000.0002.ff0f0dc60001.00}
15           pgid={level=80,id=39000011222222000000000000} ptse ptype=internal
16           id=1342177285 seq=2 csum=59306 life=3574  internal vpcap=0 port=0x0
17           scope=0 ail=20 aic=1 prefix={104,39000011222222000000000001}
18           ptse ptype=nodal id=1 seq=1 csum=32489 life=3573  nodal
19           nsap=0x39.0000.11.222222.0000.0000.0001.ff0f0dc60001.00 lprio=0
20           flags=<,transit,simple,nobranch,normal>
21           leader={level=0,peergroup={level=0,id=00000000000000000000000000},
22           esi=00:00:00:00:00:00}
23  R  6⇒   17.672
24           ptsp
25           node={level=80,node=0x39.0000.11.222222.0000.0000.0002.ff0f0dc60001.00}
26           pgid={level=80,id=39000011222222000000000000} ptse ptype=hlinks
27           id=268435475 seq=1 csum=28202 life=3599  hlinks vpcap=0
28           rnode={level=80,node=0x39.0000.11.222222.0000.0000.0001.ff1a4ce70001.00}
29           rport=0x10000010 lport=0x10000013 atoken=0  ora flags=<cbr,,,,,>
30           aweight=5040 mcr=353207 acr=352367 ctd=745 cdv=725 clr0=8 clr01=8  ora
31           flags=<,rt-vbr,,,,> aweight=5040 mcr=353207 acr=352367 ctd=745 cdv=725
32           clr0=8 clr01=8  ora flags=<,,nrt-vbr,,,> aweight=5040 mcr=353207
33           acr=352367 ctd=745 cdv=725 clr0=8 clr01=8  ora flags=<,,,,ubr,>
34           aweight=5040 mcr=353207 acr=352367 ctd=745 cdv=725 clr0=8 clr01=8
35  R  6⇒   17.674
36           ptseackp ptseack
37           node={level=80,node=0x39.0000.11.222222.0000.0000.0001.ff1a4ce70001.00}
38           acount=3  id=1342177283 seq=2 csum=48237 life=3597  id=268435472 seq=1
39           csum=28205 life=3599  id=1 seq=1 csum=10091 life=3596
40           ! unexpected 56 padding bytes ...
41  R  5⇐   17.701
42           dbsum flags=<init,more,master> seq=925416771
43  R  5⇐   17.704
44           dbsum flags=<,more,> seq=925413206
45           ptsesum node={level=80,
46           node=0x39.0000.11.222222.0000.0000.0001.ff1a4ce70001.00}
47           pgid={level=80,id=39000011222222000000000000} scount=2  ptype=internal
48           id=1342177283 seq=2 csum=48237 life=3599  ptype=nodal id=1 seq=1
49           csum=10091 life=3598
50  R  5⇐   17.708
51           dbsum flags=<,,> seq=925413207
52  R  5⇐   17.709
53           ptsereqp ! illegal IG 5 within ptsereq
```

```
54  R  5⇐   17.710
             ptsp node={level=80,
             node=0x39.0000.11.222222.0000.0000.0001.ff1a4ce70001.00}
             pgid={level=80,id=39000011222222000000000000} ptse ptype=hlinks
             id=268435472 seq=1 csum=28205 life=3599  hlinks vpcap=0
             rnode={level=80,node=0x39.0000.11.222222.0000.0000.0002.ff0f0dc60001.00}
60           rport=0x10000013 lport=0x10000010 atoken=0  ora flags=<cbr,,,,,>
             aweight=5040 mcr=353207 acr=352367 ctd=745 cdv=725 clr0=8 clr01=8
             ora flags=<,rt-vbr,,,,> aweight=5040 mcr=353207 acr=352367 ctd=745
             cdv=725 clr0=8 clr01=8  ora flags=<,,nrt-vbr,,,> aweight=5040
             mcr=353207 acr=352367 ctd=745 cdv=725 clr0=8 clr01=8
             ora flags=<,,,,ubr,> aweight=5040 mcr=353207 acr=352367
             ctd=745 cdv=725 clr0=8 clr01=8
    R  5⇐   17.712
             ptsp node={level=80,
             node=0x39.0000.11.222222.0000.0000.0001.ff1a4ce70001.00}
70           pgid={level=80,id=39000011222222000000000000} ptse ptype=internal
             id=1342177283 seq=2 csum=48237 life=3597  internal vpcap=0 port=0x0
             scope=0 ail=20 aic=1 prefix={104,39000011222222000000000001}
             ptse ptype=nodal id=1 seq=1 csum=10091 life=3596  nodal
             nsap=0x39.0000.11.222222.0000.0000.0001.ff1a4ce70001.00 lprio=0
             flags=<,transit,simple,nobranch,normal>
             leader={level=0,peergroup={level=0,id=00000000000000000000000000},
             esi=00:00:00:00:00:00}
    R  5⇐   17.713
             ptseackp ptseack node={level=80,
80           node=0x39.0000.11.222222.0000.0000.0002.ff0f0dc60001.00} acount=3
             id=1342177285 seq=2 csum=59306 life=3574  id=268435475 seq=1
             csum=28202 life=3599  id=1 seq=1 csum=32489 life=3573
             ! unexpected 56 padding bytes ...
```

6.2.11.3 A Third Node of a Peer Group is Going Up

Now the switches foreplay and forelle know each other and switch forest goes up. It starts with sending its Hello messages to its neighbours:

```
1   R  3⇐   0.647
             hello
             node={level=80,node=0x39.0000.11.222222.0000.0000.0001.ff1a36900001.00}
             addr=0x39.0000.11.222222.0000.0000.0001.ff1a36900001.00
5            pgid={level=80,id=39000011222222000000000000} rnode={level=0,
             peergroup={level=0,id=00000000000000000000000000},esi=00:00:00:00:00:00}
             port=0x10000012 rport=0x0 hinter=15
    R  1⇐   1.609
             hello
10           node={level=80,node=0x39.0000.11.222222.0000.0000.0001.ff1a36900001.00}
             addr=0x39.0000.11.222222.0000.0000.0001.ff1a36900001.00
             pgid={level=80,id=39000011222222000000000000} rnode={level=0,
             peergroup={level=0,id=00000000000000000000000000},esi=00:00:00:00:00:00}
             port=0x1000001b rport=0x0 hinter=15
```

So the nodes forest and forelle get to know each other and exchange corresponding Hello messages:

```
1   R  2⇒   1.063
2            hello
3            node={level=80,node=0x39.0000.11.222222.0000.0000.0001.ff1a4ce70001.00}
4            addr=0x39.0000.11.222222.0000.0000.0001.ff1a4ce70001.00
5            pgid={level=80,id=39000011222222000000000000} rnode={level=80,
6            node=0x39.0000.11.222222.0000.0000.0001.ff1a36900001.00}
7            port=0x1000001b rport=0x1000001b hinter=15
8   R  1⇐   1.611
9            hello
10           node={level=80,node=0x39.0000.11.222222.0000.0000.0001.ff1a36900001.00}
11           addr=0x39.0000.11.222222.0000.0000.0001.ff1a36900001.00
12           pgid={level=80,id=39000011222222000000000000} rnode={level=80,
13           node=0x39.0000.11.222222.0000.0000.0001.ff1a4ce70001.00}
14           port=0x1000001b rport=0x1000001b hinter=15
```

The switches forest and foreplay also know each other and therefore exchange corresponding Hello messages (not shown here).

Now all switches know all their neighbours, because every switch in the peer group has a direct connection to every other switch in the peer group.

Parallel to the exchange of Hello messages the switches start to exchange topology information (regular Hello messages are also exchanged but they are not shown here):

```
1   R  3⇐   0.648
2            dbsum flags=<init,more,master> seq=925416945
3   R  3⇐   0.650
4            dbsum flags=<,more,> seq=925413256 ptsesum
5            node={level=80,node=0x39.0000.11.222222.0000.0000.0001.ff1a36900001.00}
6            pgid={level=80,id=39000011222222000000000000} scount=2
7            ptype=internal id=1342177284 seq=2 csum=4995 life=3599
8            ptype=nodal id=1 seq=1 csum=54679 life=3598
9   R  3⇐   0.652
10           ptsp
11           node={level=80,node=0x39.0000.11.222222.0000.0000.0001.ff1a4ce70001.00}
12           pgid={level=80,id=39000011222222000000000000} ptse
13           ptype=internal id=1342177283 seq=2 csum=48237 life=3548  internal
14           vpcap=0 port=0x0 scope=0 ail=20 aic=1
15           prefix={104,39000011222222000000000001}
16           ptse ptype=hlinks id=268435472 seq=1 csum=28205 life=3550
17           hlinks vpcap=0 rnode={level=80,
18           node=0x39.0000.11.222222.0000.0000.0002.ff0f0dc60001.00}
19           rport=0x10000013 lport=0x10000010 atoken=0
20           ora flags=<cbr,,,,,> aweight=5040 mcr=353207 acr=352367
21           ctd=745 cdv=725 clr0=8 clr01=8
22           ora flags=<,rt-vbr,,,,> aweight=5040 mcr=353207 acr=352367
23           ctd=745 cdv=725 clr0=8 clr01=8  ora flags=<,,nrt-vbr,,,> aweight=5040
24           mcr=353207 acr=352367 ctd=745 cdv=725 clr0=8 clr01=8
25           ora flags=<,,,,ubr,> aweight=5040 mcr=353207 acr=352367 ctd=745
26           cdv=725 clr0=8 clr01=8
27           ptse ptype=nodal id=1 seq=1 csum=10091 life=3547
28           nodal nsap=0x39.0000.11.222222.0000.0000.0001.ff1a4ce70001.00 lprio=0
29           flags=<,transit,simple,nobranch,normal> leader={level=0,
30           peergroup={level=0,id=00000000000000000000000000},esi=00:00:00:00:00:00}
31  R  3⇐   0.793
```

```
32          ptsp
33          node={level=80,node=0x39.0000.11.222222.0000.0000.0001.ff1a36900001.00}
34          pgid={level=80,id=39000011222222000000000000}
35          ptse ptype=hlinks id=268435483 seq=1 csum=39502 life=3599
36          hlinks vpcap=0 rnode={level=80,
37          node=0x39.0000.11.222222.0000.0000.0001.ff1a4ce70001.00}
38          rport=0x1000001b lport=0x1000001b atoken=0
39          ora flags=<cbr,,,,,> aweight=5040 mcr=353207 acr=352341 ctd=745
40          cdv=725 clr0=8 clr01=8
41          ora flags=<,rt-vbr,,,,> aweight=5040 mcr=353207 acr=352341 ctd=745
42          cdv=725 clr0=8 clr01=8
43          ora flags=<,,nrt-vbr,,,> aweight=5040 mcr=353207 acr=352341 ctd=745
44          cdv=725 clr0=8 clr01=8  ora flags=<,,,,ubr,> aweight=5040 mcr=353207
45          acr=352341 ctd=745 cdv=725 clr0=8 clr01=8
46  R  3⇐   0.795
47          ptsp node={level=80,
48          node=0x39.0000.11.222222.0000.0000.0002.ff0f0dc60001.00}
49          pgid={level=80,id=39000011222222000000000000}
50          ptse ptype=internal id=1342177285 seq=2 csum=59306 life=3524
51          internal vpcap=0 port=0x0 scope=0 ail=20 aic=1
52          prefix={104,39000011222222000000000001}  ptse ptype=hlinks
53          id=268435475 seq=1 csum=28202 life=3549  hlinks vpcap=0
54          rnode={level=80,node=0x39.0000.11.222222.0000.0000.0001.ff1a4ce70001.00}
55          rport=0x10000010 lport=0x10000013 atoken=0
56          ora flags=<cbr,,,,,> aweight=5040 mcr=353207 acr=352367 ctd=745 cdv=725
57          clr0=8 clr01=8  ora flags=<,rt-vbr,,,,> aweight=5040 mcr=353207 acr=352367
58          ctd=745 cdv=725 clr0=8 clr01=8
59          ora flags=<,,nrt-vbr,,,> aweight=5040 mcr=353207 acr=352367 ctd=745 cdv=725
60          clr0=8 clr01=8  ora flags=<,,,,ubr,> aweight=5040 mcr=353207 acr=352367
61          ctd=745 cdv=725 clr0=8 clr01=8  ptse ptype=nodal id=1 seq=1
62          csum=32489 life=3523
63          nodal nsap=0x39.0000.11.222222.0000.0000.0001.ff0f0dc60001.00
64          lprio=0 flags=<,transit,simple,nobranch,normal> leader={level=0,
65          peergroup={level=0,id=00000000000000000000000000},
66          esi=00:00:00:00:00:00}
67  R  3⇐   0.796
68          dbsum flags=<,,> seq=925413257
69  R  3⇐   0.797
70          ptsp node={level=80,
71          node=0x39.0000.11.222222.0000.0000.0001.ff1a36900001.00}
72          pgid={level=80,id=39000011222222000000000000}
73          ptse ptype=hlinks id=268435474 seq=1 csum=54971 life=3599
74          hlinks vpcap=0 rnode={level=80,
75          node=0x39.0000.11.222222.0000.0000.0002.ff0f0dc60001.00} rport=0x10000012
76          lport=0x10000012 atoken=0  ora flags=<cbr,,,,,> aweight=5040
77          mcr=353207 acr=352365 ctd=745 cdv=725 clr0=8 clr01=8
78          ora flags=<,rt-vbr,,,,> aweight=5040 mcr=353207 acr=352365 ctd=745
79          cdv=725 clr0=8 clr01=8  ora flags=<,,nrt-vbr,,,> aweight=5040
80          mcr=353207 acr=352365 ctd=745 cdv=725 clr0=8 clr01=8
81          ora flags=<,,,abr,,> aweight=5040 mcr=353207 acr=352365 ctd=745
82          cdv=725 clr0=8 clr01=8  ora flags=<,,,,ubr,> aweight=5040 mcr=353207
83          acr=352365 ctd=745 cdv=725 clr0=8 clr01=8
84  R  3⇐   0.798
85          ptsp node={level=80,
86          node=0x39.0000.11.222222.0000.0000.0001.ff1a4ce70001.00}
87          pgid={level=80,id=39000011222222000000000000}
```

```
88              ptse ptype=hlinks id=268435483 seq=1 csum=39502 life=3598
89              hlinks vpcap=0 rnode={level=80,
90              node=0x39.0000.11.222222.0000.0000.0001.ff1a36900001.00}
91              rport=0x1000001b lport=0x1000001b atoken=0
92              ora flags=<cbr,,,,,> aweight=5040 mcr=353207 acr=352341 ctd=745
93              cdv=725 clr0=8 clr01=8
94              ora flags=<,rt-vbr,,,,> aweight=5040 mcr=353207 acr=352341 ctd=745
95              cdv=725 clr0=8 clr01=8  ora flags=<,,nrt-vbr,,,> aweight=5040
96              mcr=353207 acr=352341 ctd=745 cdv=725 clr0=8 clr01=8
97              ora flags=<,,,,ubr,> aweight=5040 mcr=353207 acr=352341 ctd=745
98              cdv=725 clr0=8 clr01=8
99   R  3⇐     0.799
100             ptsp node={level=80,
101             node=0x39.0000.11.222222.0000.0000.0001.ff1a36900001.00}
102             pgid={level=80,id=39000011222222000000000000}
103             ptse ptype=internal id=1342177284 seq=2 csum=4995 life=3597
104             internal vpcap=0 port=0x0 scope=0 ail=20 aic=1
105             prefix={104,39000011222222000000000001}
106             ptse ptype=nodal id=1 seq=1 csum=54679 life=3596
107             nodal nsap=0x39.0000.11.222222.0000.0000.0001.ff1a36900001.00
108             lprio=0 flags=<,transit,simple,nobranch,normal> leader={level=0,
109             peergroup={level=0,id=00000000000000000000000000},
110             esi=00:00:00:00:00:00}
111  R  3⇐     0.800
112             ptseackp ptseack node={level=80,
113             node=0x39.0000.11.222222.0000.0000.0002.ff0f0dc60001.00} acount=1
114             id=268435474 seq=1 csum=50496 life=3599
115             ! unexpected 56 padding bytes ...
116  R  2⇒     1.064
117             dbsum flags=<init,more,master> seq=925416821
118  R  2⇒     1.065
119             dbsum flags=<,,master> seq=925416822 ptsesum
120             node={level=80,node=0x39.0000.11.222222.0000.0000.0002.ff0f0dc60001.00}
121             pgid={level=80,id=39000011222222000000000000} scount=3
122             ptype=internal id=1342177285 seq=2 csum=59306 life=3526
123             ptype=hlinks id=268435475 seq=1 csum=28202 life=3551
124             ptype=nodal id=1 seq=1 csum=32489 life=3525
125             ptsesum
126             node={level=80,node=0x39.0000.11.222222.0000.0000.0001.ff1a4ce70001.00}
127             pgid={level=80,id=39000011222222000000000000} scount=3
128             ptype=internal id=1342177283 seq=2 csum=48237 life=3550
129             ptype=hlinks id=268435472 seq=1 csum=28205 life=3552
130             ptype=nodal id=1 seq=1 csum=10091 life=3549
131  R  2⇒     1.066
132             ptsereqp ! illegal IG 5 within ptsereq
133  R  2⇒     1.067
134             ptsp
135             node={level=80,node=0x39.0000.11.222222.0000.0000.0001.ff1a4ce70001.00}
136             pgid={level=80,id=39000011222222000000000000}
137             ptse ptype=internal id=1342177283 seq=2 csum=48237 life=3549
138             internal vpcap=0 port=0x0 scope=0 ail=20 aic=1
139             prefix={104,39000011222222000000000001}
140             ptse ptype=hlinks id=268435472 seq=1 csum=28205 life=3551
141             hlinks vpcap=0 rnode={level=80,
142             node=0x39.0000.11.222222.0000.0000.0002.ff0f0dc60001.00}
143             rport=0x10000013 lport=0x10000010 atoken=0
```

```
144            ora flags=<cbr,,,,,> aweight=5040 mcr=353207 acr=352367 ctd=745
145            cdv=725 clr0=8 clr01=8
146            ora flags=<,rt-vbr,,,,> aweight=5040 mcr=353207 acr=352367 ctd=745
147            cdv=725 clr0=8 clr01=8
148            ora flags=<,,nrt-vbr,,,> aweight=5040 mcr=353207 acr=352367 ctd=745
149            cdv=725 clr0=8 clr01=8  ora flags=<,,,,ubr,> aweight=5040 mcr=353207
150            acr=352367 ctd=745 cdv=725 clr0=8 clr01=8
151            ptse ptype=nodal id=1 seq=1 csum=10091 life=3548
152            nodal nsap=0x39.0000.11.222222.0000.0000.0001.ff1a4ce70001.00
153            lprio=0 flags=<,transit,simple,nobranch,normal> leader={level=0,
154            peergroup={level=0,id=00000000000000000000000000},
155            esi=00:00:00:00:00:00}
156  R  2⇒    1.180
157            ptsp
158            node={level=80,node=0x39.0000.11.222222.0000.0000.0002.ff0f0dc60001.00}
159            pgid={level=80,id=390000112222220000000000000} ptse ptype=internal
160            id=1342177285 seq=2 csum=59306 life=3525
161            internal vpcap=0 port=0x0 scope=0 ail=20 aic=1
162            prefix={104,39000011222222000000000001}
163            ptse ptype=hlinks id=268435475 seq=1 csum=28202 life=3550
164            hlinks vpcap=0 rnode={level=80,
165            node=0x39.0000.11.222222.0000.0000.0001.ff1a4ce70001.00}
166            rport=0x10000010 lport=0x10000013 atoken=0
167            ora flags=<cbr,,,,,> aweight=5040 mcr=353207 acr=352367 ctd=745
168            cdv=725 clr0=8 clr01=8
169            ora flags=<,rt-vbr,,,,> aweight=5040 mcr=353207 acr=352367 ctd=745
170            cdv=725 clr0=8 clr01=8  ora flags=<,,nrt-vbr,,,> aweight=5040
171            mcr=353207 acr=352367 ctd=745 cdv=725 clr0=8 clr01=8
172            ora flags=<,,,,ubr,> aweight=5040 mcr=353207 acr=352367 ctd=745
173            cdv=725 clr0=8 clr01=8  ptse ptype=nodal id=1 seq=1 csum=32489
174            life=3524
175            nodal nsap=0x39.0000.11.222222.0000.0000.0001.ff0f0dc60001.00
176            lprio=0 flags=<,transit,simple,nobranch,normal> leader={level=0,
177            peergroup={level=0,id=00000000000000000000000000},
178            esi=00:00:00:00:00:00}
179  R  2⇒    1.181
180            ptsp
181            node={level=80,node=0x39.0000.11.222222.0000.0000.0001.ff1a4ce70001.00}
182            pgid={level=80,id=390000112222220000000000000}
183            ptse ptype=hlinks id=268435483 seq=1 csum=39502 life=3599
184            hlinks vpcap=0 rnode={level=80,
185            node=0x39.0000.11.222222.0000.0000.0001.ff1a36900001.00}
186            rport=0x1000001b lport=0x1000001b atoken=0
187            ora flags=<cbr,,,,,> aweight=5040 mcr=353207 acr=352341 ctd=745
188            cdv=725 clr0=8 clr01=8
189            ora flags=<,rt-vbr,,,,> aweight=5040 mcr=353207 acr=352341 ctd=745
190            cdv=725 clr0=8 clr01=8  ora flags=<,,nrt-vbr,,,> aweight=5040
191            mcr=353207 acr=352341 ctd=745 cdv=725 clr0=8 clr01=8
192            ora flags=<,,,,ubr,> aweight=5040 mcr=353207 acr=352341 ctd=745
193            cdv=725 clr0=8 clr01=8
194  R  2⇒    1.182
195            ptsp
196            node={level=80,node=0x39.0000.11.222222.0000.0000.0002.ff0f0dc60001.00}
197            pgid={level=80,id=390000112222220000000000000}
198            ptse ptype=hlinks id=268435474 seq=1 csum=50496 life=3598
199            hlinks vpcap=0 rnode={level=80,
```

```
200             node=0x39.0000.11.222222.0000.0000.0001.ff1a36900001.00}
201             rport=0x10000012 lport=0x10000012 atoken=0
202             ora flags=<cbr,,,,,> aweight=5040 mcr=353207 acr=352367 ctd=745
203             cdv=725 clr0=8 clr01=8
204             ora flags=<,rt-vbr,,,,> aweight=5040 mcr=353207 acr=352367 ctd=745
205             cdv=725 clr0=8 clr01=8  ora flags=<,,nrt-vbr,,,> aweight=5040
206             mcr=353207 acr=352367 ctd=745 cdv=725 clr0=8 clr01=8
207             ora flags=<,,,,ubr,> aweight=5040 mcr=353207 acr=352367 ctd=745
208             cdv=725 clr0=8 clr01=8
209   R  2⇒    1.183
210             ptseackp ptseack
211             node={level=80,node=0x39.0000.11.222222.0000.0000.0001.ff1a36900001.00}
212             acount=4  id=1342177284 seq=2 csum=4995 life=3598
213             id=268435483 seq=1 csum=39502 life=3599
214             id=268435474 seq=1 csum=54971 life=3599
215             id=1 seq=1 csum=54679 life=3597
216             ! unexpected 56 padding bytes ...
217   R  1⇐    1.612
218             dbsum flags=<init,more,master> seq=925416946
219   R  1⇐    1.613
220             dbsum flags=<,more,> seq=925416821
221             ptsesum node={level=80,
222             node=0x39.0000.11.222222.0000.0000.0001.ff1a36900001.00}
223             pgid={level=80,id=39000011222222000000000000} scount=2
224             ptype=internal id=1342177284 seq=2 csum=4995 life=3600
225             ptype=nodal id=1 seq=1 csum=54679 life=3599
226   R  1⇐    1.614
227             dbsum flags=<,,> seq=925416822
228   R  1⇐    1.615
229             ptsereqp ! illegal IG 5 within ptsereq
230   R  4⇒    1.629
231             dbsum flags=<init,more,master> seq=925413256
232   R  4⇒    1.630
233             dbsum flags=<,,master> seq=925413257
234             ptsesum node={level=80,
235             node=0x39.0000.11.222222.0000.0000.0002.ff0f0dc60001.00}
236             pgid={level=80,id=39000011222222000000000000} scount=3
237             ptype=internal id=1342177285 seq=2 csum=59306 life=3526
238             ptype=hlinks id=268435475 seq=1 csum=28202 life=3552
239             ptype=nodal id=1 seq=1 csum=32489 life=3525
240             ptsesum node={level=80,
241             node=0x39.0000.11.222222.0000.0000.0001.ff1a4ce70001.00}
242             pgid={level=80,id=39000011222222000000000000} scount=3
243             ptype=internal id=1342177283 seq=2 csum=48237 life=3549
244             ptype=hlinks id=268435472 seq=1 csum=28205 life=3551
245             ptype=nodal id=1 seq=1 csum=10091 life=3548
246   R  4⇒    1.632
247             ptseackp ptseack node={level=80,
248             node=0x39.0000.11.222222.0000.0000.0001.ff1a4ce70001.00} acount=3
249             id=1342177283 seq=2 csum=48237 life=3548
250             id=268435472 seq=1 csum=28205 life=3550
251             id=1 seq=1 csum=10091 life=3547
252             ! unexpected 56 padding bytes ...
253   R  4⇒    1.633
254             ptseackp ptseack node={level=80,
255             node=0x39.0000.11.222222.0000.0000.0002.ff0f0dc60001.00} acount=3
```

```
256              id=1342177285 seq=2 csum=59306 life=3524
257              id=268435475 seq=1 csum=28202 life=3549
258              id=1 seq=1 csum=32489 life=3523
259              ! unexpected 56 padding bytes ...
260   R   4⇒    1.634
261              ptsereqp ! illegal IG 5 within ptsereq
262   R   4⇒    1.635
263              ptsp node={level=80,
264              node=0x39.0000.11.222222.0000.0000.0001.ff1a4ce70001.00}
265              pgid={level=80,id=39000011222222000000000000}
266              ptse ptype=hlinks id=268435483 seq=1 csum=39502 life=3598
267              hlinks vpcap=0 rnode={level=80,
268              node=0x39.0000.11.222222.0000.0000.0001.ff1a36900001.00}
269              rport=0x1000001b lport=0x1000001b atoken=0
270              ora flags=<cbr,,,,,> aweight=5040 mcr=353207 acr=352341 ctd=745
271              cdv=725 clr0=8 clr01=8
272              ora flags=<,rt-vbr,,,,> aweight=5040 mcr=353207 acr=352341 ctd=745
273              cdv=725 clr0=8 clr01=8
274              ora flags=<,,nrt-vbr,,,> aweight=5040 mcr=353207 acr=352341 ctd=745
275              cdv=725 clr0=8 clr01=8
276              ora flags=<,,,,ubr,> aweight=5040 mcr=353207 acr=352341 ctd=745
277              cdv=725 clr0=8 clr01=8
278   R   4⇒    1.705
279              ptsp node={level=80,
280              node=0x39.0000.11.222222.0000.0000.0002.ff0f0dc60001.00}
281              pgid={level=80,id=39000011222222000000000000}
282              ptse ptype=hlinks id=268435474 seq=1 csum=50496 life=3599
283              hlinks vpcap=0 rnode={level=80,
284              node=0x39.0000.11.222222.0000.0000.0001.ff1a36900001.00}
285              rport=0x10000012 lport=0x10000012 atoken=0
286              ora flags=<cbr,,,,,> aweight=5040 mcr=353207 acr=352367 ctd=745
287              cdv=725 clr0=8 clr01=8
288              ora flags=<,rt-vbr,,,,> aweight=5040 mcr=353207 acr=352367 ctd=745
289              cdv=725 clr0=8 clr01=8
290              ora flags=<,,nrt-vbr,,,> aweight=5040 mcr=353207 acr=352367 ctd=745
291              cdv=725 clr0=8 clr01=8
292              ora flags=<,,,,ubr,> aweight=5040 mcr=353207 acr=352367 ctd=745
293              cdv=725 clr0=8 clr01=8
294   R   1⇐    1.743
295              ptsp node={level=80,
296              node=0x39.0000.11.222222.0000.0000.0001.ff1a36900001.00}
297              pgid={level=80,id=39000011222222000000000000}
298              ptse ptype=hlinks id=268435483 seq=1 csum=39502 life=3599
299              hlinks vpcap=0 rnode={level=80,
300              node=0x39.0000.11.222222.0000.0000.0001.ff1a4ce70001.00}
301              rport=0x1000001b lport=0x1000001b atoken=0
302              ora flags=<cbr,,,,,> aweight=5040 mcr=353207 acr=352341 ctd=745
303              cdv=725 clr0=8 clr01=8  ora flags=<,rt-vbr,,,,> aweight=5040
304              mcr=353207 acr=352341 ctd=745 cdv=725 clr0=8 clr01=8
305              ora flags=<,,nrt-vbr,,,> aweight=5040 mcr=353207 acr=352341 ctd=745
306              cdv=725 clr0=8 clr01=8  ora flags=<,,,,ubr,> aweight=5040
307              mcr=353207 acr=352341 ctd=745 cdv=725 clr0=8 clr01=8
308   R   1⇐    1.744
309              ptsp node={level=80,
310              node=0x39.0000.11.222222.0000.0000.0001.ff1a36900001.00}
311              pgid={level=80,id=39000011222222000000000000}
```

```
312           ptse ptype=internal id=1342177284 seq=2 csum=4995 life=3598
313           internal vpcap=0 port=0x0 scope=0 ail=20 aic=1
314           prefix={104,39000011222222000000000001}
315           ptse ptype=nodal id=1 seq=1 csum=54679 life=3597
316           nodal nsap=0x39.0000.11.222222.0000.0000.0001.ff1a36900001.00
317           lprio=0 flags=<,transit,simple,nobranch,normal> leader={level=0,
318           peergroup={level=0,id=00000000000000000000000000},
319           esi=00:00:00:00:00:00}
320  R  1⇐   1.746
321           ptsp node={level=80,
322           node=0x39.0000.11.222222.0000.0000.0001.ff1a36900001.00}
323           pgid={level=80,id=39000011222222000000000000}
324           ptse ptype=hlinks id=268435474 seq=1 csum=54971 life=3599
325           hlinks vpcap=0 rnode={level=80,
326           node=0x39.0000.11.222222.0000.0000.0002.ff0f0dc60001.00}
327           rport=0x10000012 lport=0x10000012 atoken=0
328           ora flags=<cbr,,,,,> aweight=5040 mcr=353207 acr=352365 ctd=745
329           cdv=725 clr0=8 clr01=8  ora flags=<,rt-vbr,,,,> aweight=5040
330           mcr=353207 acr=352365 ctd=745 cdv=725 clr0=8 clr01=8
331           ora flags=<,,nrt-vbr,,,> aweight=5040 mcr=353207 acr=352365 ctd=745
332           cdv=725 clr0=8 clr01=8
333           ora flags=<,,,abr,,> aweight=5040 mcr=353207 acr=352365 ctd=745
334           cdv=725 clr0=8 clr01=8
335           ora flags=<,,,,ubr,> aweight=5040 mcr=353207 acr=352365 ctd=745
336           cdv=725 clr0=8 clr01=8
337  R  4⇒   3.465
338           ptseackp ptseack node={level=80,
339           node=0x39.0000.11.222222.0000.0000.0001.ff1a36900001.00}
340           acount=3  id=1342177284 seq=2 csum=4995 life=3597
341           id=268435474 seq=1 csum=54971 life=3599
342           id=1 seq=1 csum=54679 life=3596
343           ! unexpected 56 padding bytes ...
344  R  1⇐   3.505
345           ptsp node={level=80,
346           node=0x39.0000.11.222222.0000.0000.0002.ff0f0dc60001.00}
347           pgid={level=80,id=39000011222222000000000000}
348           ptse ptype=hlinks id=268435474 seq=1 csum=50496 life=3598
349           hlinks vpcap=0 rnode={level=80,
350           node=0x39.0000.11.222222.0000.0000.0001.ff1a36900001.00}
351           rport=0x10000012 lport=0x10000012 atoken=0
352           ora flags=<cbr,,,,,> aweight=5040 mcr=353207 acr=352367 ctd=745
353           cdv=725 clr0=8 clr01=8
354           ora flags=<,rt-vbr,,,,> aweight=5040 mcr=353207 acr=352367 ctd=745
355           cdv=725 clr0=8 clr01=8
356           ora flags=<,,nrt-vbr,,,> aweight=5040 mcr=353207 acr=352367 ctd=745
357           cdv=725 clr0=8 clr01=8
358           ora flags=<,,,,ubr,> aweight=5040 mcr=353207 acr=352367 ctd=745
359           cdv=725 clr0=8 clr01=8
360  R  1⇐   3.506
361           ptseackp ptseack node={level=80,
362           node=0x39.0000.11.222222.0000.0000.0002.ff0f0dc60001.00} acount=3
363           id=1342177285 seq=2 csum=59306 life=3525
364           id=268435475 seq=1 csum=28202 life=3550  id=1 seq=1 csum=32489
365           life=3524
366           ! unexpected 56 padding bytes ...
```

6.2.11.4 Two or More Nodes are Up and Running

This example shows the PNNI messages that are exchanged between two switches (here forelle and forest) that are up and running.

The following messages are transmitted approximately every 15 s on the link (1/2) between forelle and forest:

```
1   R  1⇐   0.000
2            hello node={level=80,
3            node=0x39.0000.11.222222.0000.0000.0001.ff1a36900001.00}
4            addr=0x39.0000.11.222222.0000.0000.0001.ff1a36900001.00
5            pgid={level=80,id=39000011222222000000000000}
6            rnode={level=80,node=0x39.0000.11.222222.0000.0000.0001.ff1a4ce70001.00}
7            port=0x1000001b rport=0x1000001b hinter=15
8   R  2⇒   0.021
9            hello
10           node={level=80,node=0x39.0000.11.222222.0000.0000.0001.ff1a4ce70001.00}
11           addr=0x39.0000.11.222222.0000.0000.0001.ff1a4ce70001.00
12           pgid={level=80,id=39000011222222000000000000}
13           rnode={level=80,node=0x39.0000.11.222222.0000.0000.0001.ff1a36900001.00}
14           port=0x1000001b rport=0x1000001b hinter=15
```

If we have a situation as at the end of Section 6.2.11.3, then these kinds of messages are exchanged on all links (1/2), (3/4) and (5/6).

6.2.11.5 A Link is Going Down

All three switches are up and running as at the end of Section 6.2.11.3 and all links are ready. Then the link (1/2) between forest and forelle fails (at time 0.000).

Because of the outstanding Hello messages over the link (1/2) the switches forest and forelle recognise the link failure after about 20 s. Then the topology databases of all switches of the peer group need to be updated. Therefore over the remaining links some packets are exchanged.

First forest and foreplay exchange messages over their common link (regular Hello messages were also exchanged but they are not shown here):

```
1   R  3⇐   21.234
2            ptsp node={level=80,
3            node=0x39.0000.11.222222.0000.0000.7001.ff1a36900001.00}
4            pgid={level=80,id=39000011222222000000000000}
5            ptse ptype=hlinks id=268435483 seq=6 csum=39241 life=0
6   R  4⇒   21.334
7            ptsp node={level=80,
8            node=0x39.0000.11.222222.0000.0000.8001.ff1a4ce70001.00}
9            pgid={level=80,id=39000011222222000000000000}
10           ptse ptype=hlinks id=268435483 seq=6 csum=39241 life=0
11  R  3⇐   21.235
12           ptseackp ptseack node={level=80,
13           node=0x39.0000.11.222222.0000.0000.8001.ff1a4ce70001.00} acount=1
14           id=268435483 seq=6 csum=39241 life=0
15           ! unexpected 56 padding bytes ...
```

```
16  R  4⇒   21.337
              ptseackp ptseack node={level=80,
              node=0x39.0000.11.222222.0000.0000.7001.ff1a36900001.00} acount=1
              id=268435483 seq=6 csum=39241 life=0
20            ! unexpected 56 padding bytes ...
```

Then forelle and foreplay exchange some messages over their common link (regular Hello messages were also exchanged but they are not shown here):

```
1   R  6⇒   21.404
2             ptsp node={level=80,
3             node=0x39.0000.11.222222.0000.0000.7001.ff1a36900001.00}
4             pgid={level=80,id=39000011222222000000000000}
5             ptse ptype=hlinks id=268435483 seq=6 csum=39241 life=0
6   R  5⇐   21.407
7             ptsp node={level=80,
8             node=0x39.0000.11.222222.0000.0000.8001.ff1a4ce70001.00}
9             pgid={level=80,id=39000011222222000000000000}
10            ptse ptype=hlinks id=268435483 seq=6 csum=39241 life=0
11  R  6⇒   21.405
12            ptseackp ptseack node={level=80,
13            node=0x39.0000.11.222222.0000.0000.8001.ff1a4ce70001.00} acount=1
14            id=268435483 seq=6 csum=39241 life=0
15            ! unexpected 56 padding bytes ...
16  R  5⇐   21.408
17            ptseackp ptseack node={level=80,
18            node=0x39.0000.11.222222.0000.0000.7001.ff1a36900001.00} acount=1
19            id=268435483 seq=6 csum=39241 life=0
20            ! unexpected 56 padding bytes ...
```

Now all nodes of the peer group know the new topology.

6.2.11.6 A Link is Going Up

Next the link (1/2) is going up again and messages are exchanged to handle the new situation:

```
1   R  1⇐   24.243
2             hello node={level=80,
3             node=0x39.0000.11.222222.0000.0000.7001.ff1a36900001.00}
4             addr=0x39.0000.11.222222.0000.0000.7001.ff1a36900001.00
5             pgid={level=80,id=39000011222222000000000000} rnode={level=0,
6             peergroup={level=0,id=00000000000000000000000000},
7             esi=00:00:00:00:00:00} port=0x1000001b rport=0x0 hinter=15
8   R  2⇒   24.302
9             hello node={level=80,
10            node=0x39.0000.11.222222.0000.0000.8001.ff1a4ce70001.00}
11            addr=0x39.0000.11.222222.0000.0000.8001.ff1a4ce70001.00
12            pgid={level=80,id=39000011222222000000000000}
13            rnode={level=0,peergroup={level=0,id=00000000000000000000000000},
14            esi=00:00:00:00:00:00} port=0x1000001b rport=0x0 hinter=15
15  R  1⇐   24.306
16            hello node={level=80,
```

```
17                node=0x39.0000.11.222222.0000.0000.7001.ff1a36900001.00}
18                addr=0x39.0000.11.222222.0000.0000.7001.ff1a36900001.00
19                pgid={level=80,id=3900001122222200000000000} rnode={level=80,
20                node=0x39.0000.11.222222.0000.0000.8001.ff1a4ce70001.00}
21                port=0x1000001b rport=0x1000001b hinter=15
22    R   1⇐    24.307
23                dbsum flags=<init,more,master> seq=925482970
24    R   1⇐    24.748
25                dbsum flags=<init,more,master> seq=925482970
26    R   2⇒    24.823
27                dbsum flags=<init,more,master> seq=925482845
28    R   2⇒    24.825
29                hello node={level=80,
30                node=0x39.0000.11.222222.0000.0000.8001.ff1a4ce70001.00}
31                addr=0x39.0000.11.222222.0000.0000.8001.ff1a4ce70001.00
32                pgid={level=80,id=3900001122222200000000000} rnode={level=80,
33                node=0x39.0000.11.222222.0000.0000.7001.ff1a36900001.00}
34                port=0x1000001b rport=0x1000001b hinter=15
```

Because of the successful transmission of Hello messages over the link (1/2) the switches forest and forelle recognise that the link is up (at time 24 s). Then the topology databases of all nodes of the peer group need to be updated (Hello messages are not shown here):

```
1     R   3⇐    39.070
2                 ptsp node={level=80,
3                 node=0x39.0000.11.222222.0000.0000.7001.ff1a36900001.00}
4                 pgid={level=80,id=3900001122222200000000000}
5                 ptse ptype=hlinks id=268435483 seq=1 csum=39166 life=3599
6                 hlinks vpcap=0 rnode={level=80,
7                 node=0x39.0000.11.222222.0000.0000.8001.ff1a4ce70001.00}
8                 rport=0x1000001b lport=0x1000001b atoken=0
9                 ora flags=<cbr,,,,,> aweight=5040 mcr=353207 acr=352365 ctd=745
10                cdv=725 clr0=8 clr01=8
11                ora flags=<,rt-vbr,,,,> aweight=5040 mcr=353207 acr=352365 ctd=745
12                cdv=725 clr0=8 clr01=8
13                ora flags=<,,nrt-vbr,,,> aweight=5040 mcr=353207 acr=352365 ctd=745
14                cdv=725 clr0=8 clr01=8
15                ora flags=<,,,,ubr,> aweight=5040 mcr=353207 acr=352365 ctd=745
16                cdv=725 clr0=8 clr01=8
17    R   4⇒    39.202
18                ptsp node={level=80,
19                node=0x39.0000.11.222222.0000.0000.8001.ff1a4ce70001.00}
20                pgid={level=80,id=3900001122222200000000000}
21                ptse ptype=hlinks id=268435483 seq=1 csum=39166 life=3598
22                hlinks vpcap=0 rnode={level=80,
23                node=0x39.0000.11.222222.0000.0000.7001.ff1a36900001.00}
24                rport=0x1000001b lport=0x1000001b atoken=0
25                ora flags=<cbr,,,,,> aweight=5040 mcr=353207 acr=352365 ctd=745
26                cdv=725 clr0=8 clr01=8
27                ora flags=<,rt-vbr,,,,> aweight=5040 mcr=353207 acr=352365 ctd=745
28                cdv=725 clr0=8 clr01=8
29                ora flags=<,,nrt-vbr,,,> aweight=5040 mcr=353207 acr=352365 ctd=745
30                cdv=725 clr0=8 clr01=8
31                ora flags=<,,,,ubr,> aweight=5040 mcr=353207 acr=352365 ctd=745
```

```
32            cdv=725 clr0=8 clr01=8
33  R  4⇒    39.203
34            ptseackp ptseack node={level=80,
35            node=0x39.0000.11.222222.0000.0000.7001.ff1a36900001.00} acount=1
36            id=268435483 seq=1 csum=39166 life=3599
37            ! unexpected 56 padding bytes ...
38  R  5⇐    39.278
39            ptsp node={level=80,
40            node=0x39.0000.11.222222.0000.0000.7001.ff1a36900001.00}
41            pgid={level=80,id=39000011222222000000000000}
42            ptse ptype=hlinks id=268435483 seq=1 csum=39166 life=3598
43            hlinks vpcap=0 rnode={level=80,
44            node=0x39.0000.11.222222.0000.0000.8001.ff1a4ce70001.00}
45            rport=0x1000001b lport=0x1000001b atoken=0
46            ora flags=<cbr,,,,,> aweight=5040 mcr=353207 acr=352365 ctd=745
47            cdv=725 clr0=8 clr01=8
48            ora flags=<,rt-vbr,,,,> aweight=5040 mcr=353207 acr=352365 ctd=745
49            cdv=725 clr0=8 clr01=8
50            ora flags=<,,nrt-vbr,,,> aweight=5040 mcr=353207 acr=352365 ctd=745
51            cdv=725 clr0=8 clr01=8
52            ora flags=<,,,,ubr,> aweight=5040 mcr=353207 acr=352365 ctd=745
53            cdv=725 clr0=8 clr01=8
54  R  5⇐    39.279
55            ptsp node={level=80,
56            node=0x39.0000.11.222222.0000.0000.8001.ff1a4ce70001.00}
57            pgid={level=80,id=39000011222222000000000000}
58            ptse ptype=hlinks id=268435483 seq=1 csum=39166 life=3599
59            hlinks vpcap=0 rnode={level=80,
60            node=0x39.0000.11.222222.0000.0000.7001.ff1a36900001.00}
61            rport=0x1000001b lport=0x1000001b atoken=0
62            ora flags=<cbr,,,,,> aweight=5040 mcr=353207 acr=352365 ctd=745
63            cdv=725 clr0=8 clr01=8
64            ora flags=<,rt-vbr,,,,> aweight=5040 mcr=353207 acr=352365 ctd=745
65            cdv=725 clr0=8 clr01=8
66            ora flags=<,,nrt-vbr,,,> aweight=5040 mcr=353207 acr=352365 ctd=745
67            cdv=725 clr0=8 clr01=8
68            ora flags=<,,,,ubr,> aweight=5040 mcr=353207 acr=352365 ctd=745
69            cdv=725 clr0=8 clr01=8
70  R  6⇒    39.315
71            ptsp node={level=80,
72            node=0x39.0000.11.222222.0000.0000.7001.ff1a36900001.00}
73            pgid={level=80,id=39000011222222000000000000}
74            ptse ptype=hlinks id=268435483 seq=1 csum=39166 life=3598
75            hlinks vpcap=0 rnode={level=80,
76            node=0x39.0000.11.222222.0000.0000.8001.ff1a4ce70001.00}
77            rport=0x1000001b lport=0x1000001b atoken=0
78            ora flags=<cbr,,,,,> aweight=5040 mcr=353207 acr=352365 ctd=745
79            cdv=725 clr0=8 clr01=8
80            ora flags=<,rt-vbr,,,,> aweight=5040 mcr=353207 acr=352365 ctd=745
81            cdv=725 clr0=8 clr01=8
82            ora flags=<,,nrt-vbr,,,> aweight=5040 mcr=353207 acr=352365 ctd=745
83            cdv=725 clr0=8 clr01=8
84            ora flags=<,,,,ubr,> aweight=5040 mcr=353207 acr=352365 ctd=745
85            cdv=725 clr0=8 clr01=8
86  R  1⇐    39.409
87            dbsum flags=<,more,> seq=925482845
```

```
88               ptsesum node={level=80,
89               node=0x39.0000.11.222222.0000.0000.9001.ff0f0dc60001.00}
90               pgid={level=80,id=390000112222220000000000000} scount=4
91               ptype=internal id=1342177281 seq=8 csum=41639 life=2490
92               ptype=hlinks id=268435475 seq=7 csum=28180 life=2492
93               ptype=hlinks id=268435474 seq=7 csum=50490 life=2943
94               ptype=nodal id=1 seq=7 csum=32451 life=3386
95               ptsesum node={level=80,
96               node=0x39.0000.11.222222.0000.0000.8001.ff1a4ce70001.00}
97               pgid={level=80,id=390000112222220000000000000} scount=3
98               ptype=internal id=1342177282 seq=7 csum=15337 life=2026
99               ptype=hlinks id=268435472 seq=7 csum=28183 life=2935
100              ptype=nodal id=1 seq=6 csum=9830 life=2025
101              ptsesum node={level=80,
102              node=0x39.0000.11.222222.0000.0000.7001.ff1a36900001.00}
103              pgid={level=80,id=390000112222220000000000000} scount=3
104              ptype=internal id=1342177283 seq=5 csum=41744 life=1998
105              ptype=hlinks id=268435474 seq=7 csum=54965 life=3394
106              ptype=nodal id=1 seq=6 csum=54450 life=2036
107   R  1⇐   39.409
108              dbsum flags=<,more,> seq=925482845
109   R  1⇐   39.410
110              dbsum flags=<,,> seq=925482846
111   R  1⇐   39.411
112              ptsp node={level=80,
113              node=0x39.0000.11.222222.0000.0000.7001.ff1a36900001.00}
114              pgid={level=80,id=390000112222220000000000000}
115              ptse ptype=hlinks id=268435483 seq=1 csum=39166 life=3599
116              hlinks vpcap=0 rnode={level=80,
117              node=0x39.0000.11.222222.0000.0000.8001.ff1a4ce70001.00}
118              rport=0x1000001b lport=0x1000001b atoken=0
119              ora flags=<cbr,,,,,> aweight=5040 mcr=353207 acr=352365 ctd=745
120              cdv=725 clr0=8 clr01=8
121              ora flags=<,rt-vbr,,,,> aweight=5040 mcr=353207 acr=352365 ctd=745
122              cdv=725 clr0=8 clr01=8
123              ora flags=<,,nrt-vbr,,,> aweight=5040 mcr=353207 acr=352365 ctd=745
124              cdv=725 clr0=8 clr01=8
125              ora flags=<,,,,ubr,> aweight=5040 mcr=353207 acr=352365 ctd=745
126              cdv=725 clr0=8 clr01=8
127   R  1⇐   39.412
128              ptseackp ptseack
129              node={level=80,
130              node=0x39.0000.11.222222.0000.0000.8001.ff1a4ce70001.00} acount=1
131              id=268435483 seq=1 csum=39166 life=3599
132              ! unexpected 56 padding bytes ...
133              ptseack node={level=80,
134              node=0x39.0000.11.222222.0000.0000.7001.ff1a36900001.00} acount=3
135              id=1342177283 seq=5 csum=41744 life=1997  id=268435474 seq=7
136              csum=54965 life=3393  id=1 seq=6 csum=54450 life=2036
137              ! unexpected 56 padding bytes ...
138   R  2⇒   39.495
139              dbsum flags=<init,more,master> seq=925482845
140   R  2⇒   39.497
141              dbsum flags=<,,master> seq=925482846
142              ptsesum node={level=80,
143              node=0x39.0000.11.222222.0000.0000.9001.ff0f0dc60001.00}
```

```
144            pgid={level=80,id=39000011222222000000000000} scount=4
145            ptype=internal id=1342177281 seq=8 csum=41639 life=2491
146            ptype=hlinks id=268435475 seq=7 csum=28180 life=2492
147            ptype=hlinks id=268435474 seq=7 csum=50490 life=2944
148            ptype=nodal id=1 seq=7 csum=32451 life=3386
149            ptsesum node={level=80,
150            node=0x39.0000.11.222222.0000.0000.8001.ff1a4ce70001.00}
151            pgid={level=80,id=39000011222222000000000000} scount=3
152            ptype=internal id=1342177282 seq=7 csum=15337 life=2027
153            ptype=hlinks id=268435472 seq=7 csum=28183 life=2937
154            ptype=nodal id=1 seq=6 csum=9830 life=2026
155            ptsesum node={level=80,
156            node=0x39.0000.11.222222.0000.0000.7001.ff1a36900001.00}
157            pgid={level=80,id=39000011222222000000000000} scount=3
158            ptype=internal id=1342177283 seq=5 csum=41744 life=1997
159            ptype=hlinks id=268435474 seq=7 csum=54965 life=3393
160            ptype=nodal id=1 seq=6 csum=54450 life=2036
161   R  2⇒   39.498
162            ptseackp ptseack node={level=80,
163            node=0x39.0000.11.222222.0000.0000.8001.ff1a4ce70001.00} acount=3
164            id=1342177282 seq=7 csum=15337 life=2026
165            id=268435472 seq=7 csum=28183 life=2935  id=1 seq=6 csum=9830
166            life=2025
167            ! unexpected 56 padding bytes ...
168            ptseack node={level=80,
169            node=0x39.0000.11.222222.0000.0000.7001.ff1a36900001.00} acount=1
170            id=268435483 seq=1 csum=39166 life=3599
171            ! unexpected 56 padding bytes ...
172   R  2⇒   39.501
173            ptsp node={level=80,
174            node=0x39.0000.11.222222.0000.0000.8001.ff1a4ce70001.00}
175            pgid={level=80,id=39000011222222000000000000}
176            ptse ptype=hlinks id=268435483 seq=1 csum=39166 life=3599
177            hlinks vpcap=0 rnode={level=80,
178            node=0x39.0000.11.222222.0000.0000.7001.ff1a36900001.00}
179            rport=0x1000001b lport=0x1000001b atoken=0
180            ora flags=<cbr,,,,,> aweight=5040 mcr=353207 acr=352365 ctd=745
181            cdv=725 clr0=8 clr01=8
182            ora flags=<,rt-vbr,,,,> aweight=5040 mcr=353207 acr=352365 ctd=745
183            cdv=725 clr0=8 clr01=8
184            ora flags=<,,nrt-vbr,,,> aweight=5040 mcr=353207 acr=352365 ctd=745
185            cdv=725 clr0=8 clr01=8
186            ora flags=<,,,,ubr,> aweight=5040 mcr=353207 acr=352365 ctd=745
187            cdv=725 clr0=8 clr01=8
188   R  3⇐   39.579
189            ptsp node={level=80,
190            node=0x39.0000.11.222222.0000.0000.8001.ff1a4ce70001.00}
191            pgid={level=80,id=39000011222222000000000000}
192            ptse ptype=hlinks id=268435483 seq=1 csum=39166 life=3598
193            hlinks vpcap=0 rnode={level=80,
194            node=0x39.0000.11.222222.0000.0000.7001.ff1a36900001.00}
195            rport=0x1000001b lport=0x1000001b atoken=0
196            ora flags=<cbr,,,,,> aweight=5040 mcr=353207 acr=352365 ctd=745
197            cdv=725 clr0=8 clr01=8
198            ora flags=<,rt-vbr,,,,> aweight=5040 mcr=353207 acr=352365 ctd=745
199            cdv=725 clr0=8 clr01=8
```

```
200          ora flags=<,,nrt-vbr,,,> aweight=5040 mcr=353207 acr=352365 ctd=745
201          cdv=725 clr0=8 clr01=8
202          ora flags=<,,,,ubr,> aweight=5040 mcr=353207 acr=352365 ctd=745
203          cdv=725 clr0=8 clr01=8
204   R  6⇒  39.738
205          ptseackp ptseack node={level=80,
206          node=0x39.0000.11.222222.0000.0000.8001.ff1a4ce70001.00} acount=1
207          id=268435483 seq=1 csum=39166 life=3599
208          ! unexpected 56 padding bytes ...
```

6.2.11.7 A New End System Appears

All three switches are up and running. Now the end system spock_b (39...7001.xxx) is plugged into the switch forest. The switch recognises the new situation because of ILMI. This is the only end system that is connected to this switch. Forest announces the reachability of end systems that start with forest's prefix (39...7001:104).

First forelle will be informed by forest and forelle acknowledges the information:

```
1   R  1⇐  0.322
           ptsp node={level=80,
3          node=0x39.0000.11.222222.0000.0000.7001.ff1a36900001.00}
4          pgid={level=80,id=39000011222222000000000000}
5          ptse ptype=internal id=1342177282 seq=2 csum=41748 life=3599
6          internal vpcap=0 port=0x0 scope=0 ail=20 aic=1
7          prefix={104,390000112222222000000007001}
8   R  2⇒  0.448
9          ptseackp ptseack node={level=80,
10         node=0x39.0000.11.222222.0000.0000.7001.ff1a36900001.00} acount=1
11         id=1342177282 seq=2 csum=41748 life=3599
12         ! unexpected 56 padding bytes ...
```

Then foreplay will be informed by forest. It also sends an acknowledgement:

```
1   R  3⇐  0.527
2          ptsp node={level=80,
3          node=0x39.0000.11.222222.0000.0000.7001.ff1a36900001.00}
4          pgid={level=80,id=39000011222222000000000000}
5          ptse ptype=internal id=1342177282 seq=2 csum=41748 life=3599
6          internal vpcap=0 port=0x0 scope=0 ail=20 aic=1
7          prefix={104,390000112222222000000007001}
8   R  4⇒  0.830
9          ptseackp ptseack node={level=80,
10         node=0x39.0000.11.222222.0000.0000.7001.ff1a36900001.00} acount=1
11         id=1342177282 seq=2 csum=41748 life=3599
12         ! unexpected 56 padding bytes ...
```

Now the nodes foreplay and forelle send each other the new information. Because they already have this information they do not send acknowledgements:

```
1   R  5⇐   1.021
2            ptsp node={level=80,
3            node=0x39.0000.11.222222.0000.0000.7001.ff1a36900001.00}
4            pgid={level=80,id=39000011222222000000000000}
5            ptse ptype=internal id=1342177282 seq=2 csum=41748 life=3598
6            internal vpcap=0 port=0x0 scope=0 ail=20 aic=1
7            prefix={104,39000011222222000000007001}
8   R  6⇒   1.307
9            ptsp node={level=80,
10           node=0x39.0000.11.222222.0000.0000.7001.ff1a36900001.00}
11           pgid={level=80,id=39000011222222000000000000}
12           ptse ptype=internal id=1342177282 seq=2 csum=41748 life=3598
13           internal vpcap=0 port=0x0 scope=0 ail=20 aic=1
14           prefix={104,39000011222222000000007001}
```

6.3 Signalling Protocol

As was previously mentioned, PNNI signalling is based on UNI4.0. It uses the same messages and information elements; however, the protocol is symmetric and some additional information elements are used.

A PNNI signalling instance can be in one of the following states:

NN0 – NULL
: The protocol instance is non-existent.

NN1 – Call initiated
: A SETUP message has been sent to the peer.

NN3 – Call proceeding sent
: A CALL PROCEEDING message has been sent in response to an incoming SETUP message.

NN4 – Alerting delivered
: An ALERTING message has been sent to the initiator of a SETUP (i.e. in the backward direction of a call).

NN6 – Call present
: A SETUP message has been received.

NN7 – Alerting received
: An ALERTING message has been received in the backward direction.

NN9 – Call proceeding received
: A CALL PROCEEDING message has been received in the backward direction.

NN10 – Active
: The connection is in the active state—data can be exchanged.

NN11 – Release request
: A RELEASE message has been sent to the next node.

NN12 – Release indication
A RELEASE message has been received.

Usually a switch creates two finite state machine (FSM) instances for a call routed through the switch: an incoming FSM and an outgoing FSM. The incoming FSM goes through the following states:

1. receive SETUP ⇒ NN6
2. send CALL PROCEEDING ⇒ NN3
3. send ALERTING ⇒ NN4 (optional)
4. send CONNECT ⇒ NN10

The FSM on the outgoing side of the node goes through the following states while talking to the next switch:

1. send SETUP ⇒ NN1
2. receive CALL PROCEEDING ⇒ NN9
3. receive ALERTING ⇒ NN7 (optional)
4. receive CONNECT ⇒ NN10

Note that sending the CALL PROCEEDING is mandatory in the PNNI and that it does not use CONNECT ACKNOWLEDGE messages. As in UNI4.0, timers T303, T301 and T310 are used.

Call clearing is done by sending a RELEASE message, starting timer T308 and entering NN11. The peer should go to NN12, forward the RELEASE, release resources, respond with a RELEASE COMPLETE and enter NN0. On receipt of the RELEASE COMPLETE the releasing side enters NN0. Clear collision is handled by interpreting a RELEASE in NN11 as a RELEASE COMPLETE.

Restart and status enquiry procedures are similar to [UNI4.0].

Four additional information elements are used in the PNNI (some information elements are also slightly changed). Two of the information elements belong to the support of semi-permanent VCs and VPs and are not of interest for our discussion. The remaining two are: the Designated Transit List (DTL) information element and the crankback information element.

The DTL IE (Figure 6.9) is computed by the first switch receiving the SETUP from the user. It contains a list of all switches through which the call should be routed to reach the destination system. This list is computed from the current topology database of the switch. This means that in contrast to IP, the PNNI is *source routed.*

The triple of logical node/port indicator, logical node ID and logical port ID subfields describes one hop of the connection. There can be up to 20 hops in a DTL IE. The current transit pointer is a pointer into the list of hops and is updated at each node when the SETUP message is forwarded to the next node.

The crankback IE (Figure 6.10) is included in the RELEASE message when a SETUP is rejected at a node because the DTL is wrong. This may happen because the topology information in the originating node was not up to date. The crankback IE enables the originating node to compute an alternative path for the SETUP.

The crankback level is the PNNI hierarchy level at which the crankback occurred. The blocked transit type is either 2 (the call or party has been blocked at the succeeding end of this interface), 3 (blocked node) or 4 (blocked link). In the second case the 22-byte node identifier

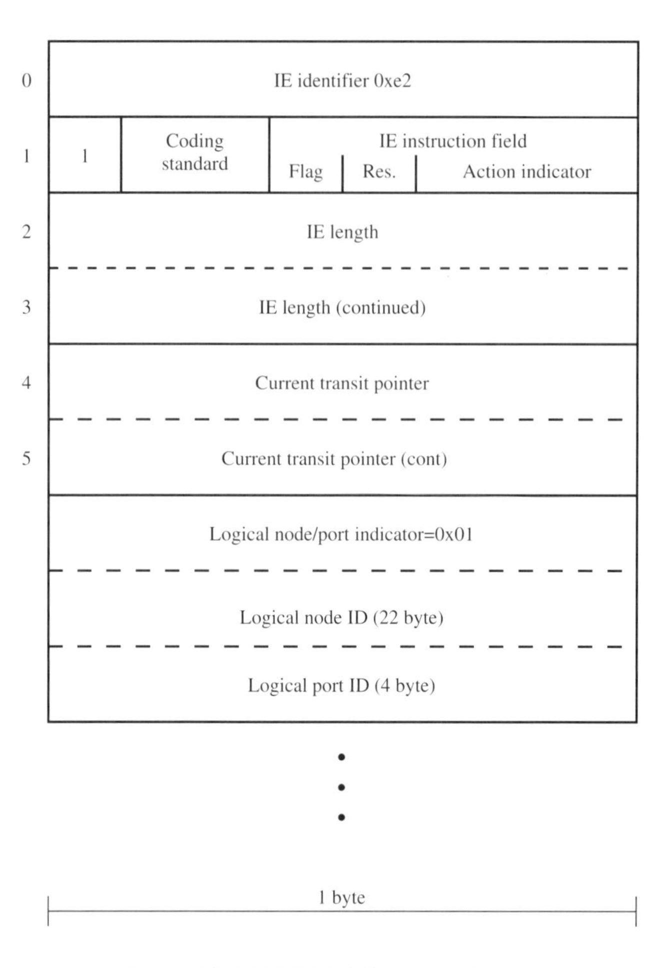

Figure 6.9: PNNI DTL information element

is included; in the third case the node IDs of both ends and the port ID (for a total of 48 byte) are included. The cause is one of the UNI cause codes or one of two additional codes: 0x80 ("Next node unreachable") or 0xa0 ("DTL transit not my node ID"). The diagnostics can optionally include topology state parameters for the generic connection admission control.

One extension to UNI4.0 which must be noted is the use of bit 4 in the second byte of the information element header. This bit is called the pass along request bit. If set, the information element is forwarded to the next hop, even if it is not recognised. This enables easy support of new information elements without changes to network nodes.

The protocol identifier for PNNI signalling is 0xf0.

6.3.1 Communication Examples

This section shows seven typical PNNI signalling protocol communication examples. All these examples show the communication between the same six ATM interfaces of the three

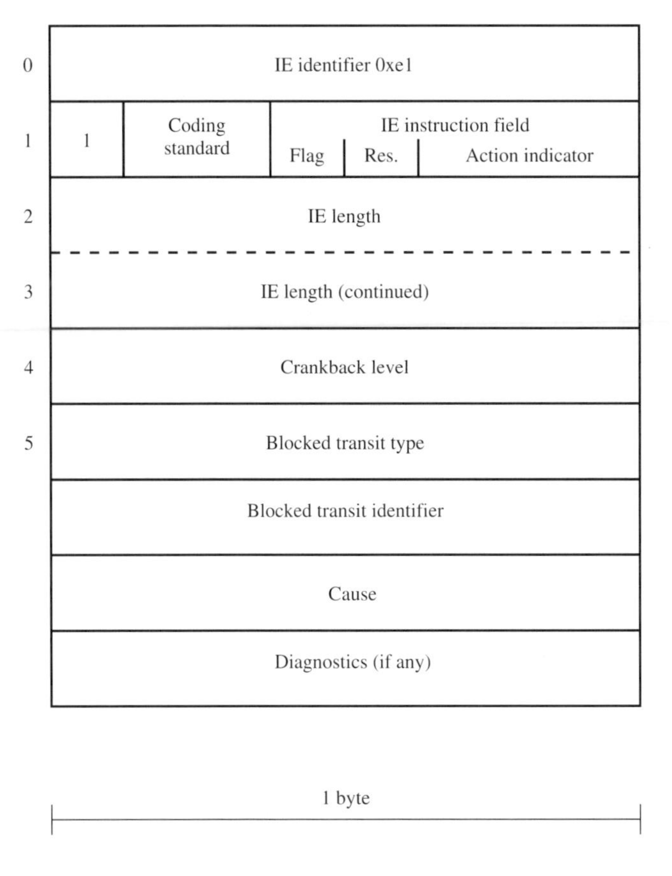

Figure 6.10: PNNI crankback information element

switches that were already presented earlier in Figure 6.8. The PNNI is running on all of the switches and, as in the routing protocol traces, all switches are in the same peer group.

The communication examples were recorded similar to Section 6.2.11. The recorded ATM communication was decoded by `a5r` (AAL decoder), by `sscopdump` (SSCOP decoder) and `sigdump` (UNI and PNNI signalling decoder). All these tools are part the Tina package described in Section 1.3. To decode the messages of one communication stream (PNNI signalling protocol messages used VPI=0 and VCI=5, but in some cases the VPI was different because of the crossconnect switch) the command

```
a5r -c 5 < /dev/ty0 | sscopdump -mo | sigdump -qp -Fhc
```

was executed at lovina for the VPI=0 case.

The following sections show the output of the Tina tools. The messages are marked with 1 to 6 depending on their direction.

6.3.1.1 A Connection Establishment

The first example shows the establishment of a simple point-to-point connection. End system spock_b makes a call to connect to lovina using link (1/2). Here are the setup, call proceed and connect messages:

```
1   S  1⇐   0.000
2             pnni cref={you,68} mtype=setup mlen=141
3             traffic={fpcr01=1200,bpcr01=1200,be}
4             bearer={class=bcob-x,atc=nrt-vbr_0,clip=not,user=p2p}
5             called={type=unknown,plan=aesa,addr=plan=aesa,
6             addr=0x39.0000.11.222222.0000.0000.8001.002048060070.00}
7             qos={forw=class0/unspecified,back=class0/unspecified}
8             repeat={indication=!illegal0xa}
9             dtl={cs=net,ptr=27,{node={level=80,
10            node=0x39.0000.11.222222.0000.0000.7001.ff1a36900001.00},
11            port=268435483},{node={level=80,
12            node=0x39.0000.11.222222.0000.0000.8001.ff1a4ce70001.00},
13            port=allports}}
14            calling={type=unknown,plan=aesa,
15            addr=0x39.276f.31.0001ef.0000.0401.7005.00204810078c.00}
16  S  2⇒   0.185
17            pnni cref={me,68} mtype=call_proc mlen=9
18             connid={vpass=explicit,pex=exclusive_vpci_vci,vpci=0,vci=67}
19  S  2⇒   0.188
20            pnni cref={me,68} mtype=connect mlen=0
```

6.3.1.2 A Connection Release

This example shows a connection release. We have an established connection between spock_b and lovina using link (1/2). The connection is released by lovina:

```
1   S  2⇒   0.000
2             pnni cref={me,68} mtype=release mlen=6
3             cause={loc=user,cvalue=normal,_unspecified,class=normal_event}
4   S  1⇐   0.244
5             pnni cref={you,68} mtype=rel_compl mlen=0
```

If spock_b were to release the connection we would see a release message on 1 and a release complete message on 2.

6.3.1.3 Setup with Crankback

This example shows a crankback. A crankback occurs if a call gets blocked on its way through the network. A called may be blocked, because the network topology may change while a SETUP travels, so the DTL that was computed at the originating switch does not match the changed topology. A crankback is triggered by rejecting the SETUP with a RELEASE COMPLETE that includes a crankback IE. This IE describes the reason for the rejection and allows the originating node to update its database. If there are alternative routes through the network the originating switch will try another route until the call succeeds or it cannot find new routes.

To get an example of a crankback we configured the same address prefix to all switches of our peer group (this is usually a bad idea because in normal operation we want to avoid crankbacks). This gives alternative routes to the same destination. If a switch wants to establish a connection to an end system that is not connected to its own port it cannot determine the destination switch by comparing the destination address and the switch address prefixes (all switches have the same address prefix). So the switch makes a guess of the correct destination switch. In our experiment we tried to establish connections from spock_b to lovina, from spock_b to atmos and from spock_b to kirk. In the last two cases the switch foreplay chooses the wrong way (to forelle) first. This triggers a crankback, after which the originating switch used the correct link.

When calling atmos from spock_b forest first tries to reach atmos at forelle:

```
1   S  1⇐   0.777
2            pnni cref={you,343} mtype=setup mlen=141
3            traffic={fpcr01=1200,bpcr01=1200,be}
4            bearer={class=bcob-x,atc=nrt-vbr_0,clip=not,user=p2p}
5            called={type=unknown,plan=aesa,addr=plan=aesa,
6            addr=0x39.0000.11.222222.0000.0000.0001.002048100898.00}
7            qos={forw=class0/unspecified,back=class0/unspecified}
8            repeat={indication=!illegal0xa}
9            dtl={cs=net,ptr=27,{node={level=80,
10           node=0x39.0000.11.222222.0000.0000.0001.ff1a36900001.00},
11           port=268435483},{node={level=80,
12           node=0x39.0000.11.222222.0000.0000.0001.ff1a4ce70001.00},
13           port=allports}}
14           calling={type=unknown,plan=aesa,
15           addr=0x39.276f.31.0001ef.0000.0401.7005.00204810078c.00}
```

But forelle has no end system atmos—it rejects the connection attempt and includes a crankback IE in the RELEASE COMPLETE:

```
1   S  2⇒   0.883
2            pnni cref={me,343} mtype=rel_compl mlen=62
3            cause={loc=private_network_serving_local_user,
4            cvalue=no_route_to_destination,class=normal_event,
5            reason=user_specific,cond=unknown}
6            crankback={cs=net,level=80,type=blocked_link,
7            precnode={level=80,
8            node=0x39.0000.11.222222.0000.0000.0001.ff1a4ce70001.00},
9            port=allports,succnode={level=0,peergroup={level=0,
10           id=00000000000000000000000000},esi=00:00:00:00:00:00},
11           cause=destination_unreachable}
```

Now forest tries an alternative switch: foreplay. The connection establishment via foreplay to atmos succeeds:

```
1   S  3⇐   0.908
2            pnni cref={you,344} mtype=setup mlen=141
```

```
3              traffic={fpcr01=1200,bpcr01=1200,be}
4              bearer={class=bcob-x,atc=nrt-vbr_0,clip=not,user=p2p}
5              called={type=unknown,plan=aesa,addr=plan=aesa,
6              addr=0x39.0000.11.222222.0000.0000.0001.002048100898.00}
7              qos={forw=class0/unspecified,back=class0/unspecified}
8              repeat={indication=!illegal0xa}
9              dtl={cs=net,ptr=27,{node={level=80,
10             node=0x39.0000.11.222222.0000.0000.0001.ff1a36900001.00},
11             port=268435474},{node={level=80,
12             node=0x39.0000.11.222222.0000.0000.0002.ff0f0dc60001.00},
13             port=allports}}
14             calling={type=unknown,plan=aesa,
15             addr=0x39.276f.31.0001ef.0000.0401.7005.00204810078c.00}
16   S  4⇒    0.993
17             pnni cref={me,344} mtype=call_proc mlen=9
18             connid={vpass=explicit,pex=exclusive_vpci_vci,vpci=0,vci=35}
19   S  4⇒    0.997
20             pnni cref={me,344} mtype=connect mlen=0
```

6.3.1.4 Routing over two Nodes

This example shows the establishment of a PNNI connection over the two PNNI links (3/4) and (5/6). Approximately at time 10.000 link (1/2) is removed. Because of the PNNI routing protocol all switches update their topology databases in about 30 seconds.

Now spock_b (connected to forest) calls lovina (connected to forelle) at time 60 s. Because the direct connection (1/2) is unavailable the call must use the links (3/4) and (5/6). We see the typical messages (SETUP, CALL PROCEEDING and CONNECT):

```
1    S  3⇐    60.508
2              pnni cref={you,116} mtype=setup mlen=168
3              traffic={fpcr01=1200,bpcr01=1200,be}
4              bearer={class=bcob-x,atc=nrt-vbr_0,clip=not,user=p2p}
5              called={type=unknown,plan=aesa,addr=plan=aesa,
6              addr=0x39.0000.11.222222.0000.0000.8001.002048060070.00}
7              qos={forw=class0/unspecified,back=class0/unspecified}
8              repeat={indication=!illegal0xa}
9              dtl={cs=net,ptr=27,{node={level=80,
10             node=0x39.0000.11.222222.0000.0000.7001.ff1a36900001.00},
11             port=268435474},{node={level=80,
12             node=0x39.0000.11.222222.0000.0000.9001.ff0f0dc60001.00},
13             port=268435475},{node={level=80,
14             node=0x39.0000.11.222222.0000.0000.8001.ff1a4ce70001.00},
15             port=allports}}
16             calling={type=unknown,plan=aesa,
17             addr=0x39.276f.31.0001ef.0000.0401.7005.00204810078c.00}
18   S  4⇒    60.615
19             pnni cref={me,116} mtype=call_proc mlen=9
20             connid={vpass=explicit,pex=exclusive_vpci_vci,vpci=0,vci=63}
21   S  6⇒    60.657
22             pnni cref={you,25} mtype=setup mlen=177 traffic=
23             {fpcr01=1200,bpcr01=1200,be}
24             bearer={class=bcob-x,atc=nrt-vbr_0,clip=not,user=p2p}
25             called={type=unknown,plan=aesa,addr=plan=aesa,
```

```
26            addr=0x39.0000.11.222222.0000.0000.8001.002048060070.00}
27            qos={forw=class0/unspecified,back=class0/unspecified}
28            connid={vpass=explicit,pex=exclusive_vpci_vci,vpci=0,vci=36}
29            repeat={indication=!illegal0xa}
30            dtl={cs=net,ptr=54,{node={level=80,
31            node=0x39.0000.11.222222.0000.0000.7001.ff1a36900001.00},
32            port=268435474},{node={level=80,
33            node=0x39.0000.11.222222.0000.0000.9001.ff0f0dc60001.00},
34            port=268435475},{node={level=80,
35            node=0x39.0000.11.222222.0000.0000.8001.ff1a4ce70001.00},
36            port=allports}}
37            calling={type=unknown,plan=aesa,
38            addr=0x39.276f.31.0001ef.0000.0401.7005.00204810078c.00}
39  S  5⇐  60.709
40            pnni cref={me,25} mtype=call_proc mlen=9
41            connid={vpass=explicit,pex=exclusive_vpci_vci,vpci=0,vci=36}
42  S  5⇐  60.797
43            pnni cref={me,25} mtype=connect mlen=0
44  S  4⇒  60.818
45            pnni cref={me,116} mtype=connect mlen=0
```

6.3.1.5 Hot Link Failure while Connection is Active

This example shows the establishment of a connection over the two PNNI links (3/4) and (5/6). The link (1/2) is down for this experiment. After the connection is established the link (5/6) is interrupted (hot link failure).

About 30 seconds after the connection establishment the link (5/6) is interrupted. The interruption is recognised by the switches forelle and forest because of the outstanding Hello packets of the PNNI routing protocol. After two outstanding Hello packets (about 30 seconds later) the switch foreplay disconnects the connection that uses the interrupted link and sends a release message to forest. Forest answers with a release complete message:

```
1   S  4⇒  125.312
2             pnni cref={me,116} mtype=release mlen=6
3             cause={loc=private_network_serving_local_user,
4             cvalue=destination_out_of_order,class=normal_event}
5   S  3⇐  125.134
6             pnni cref={you,116} mtype=rel_compl mlen=0
```

6.3.1.6 Point-to-multipoint: Setup of the First Leaf

This example shows the setup of a point-to-multipoint connection. spock_b connects the first leaf (kirk) using link (3/4). Here are the messages:

```
1   S  3⇐  0.000
2             pnni cref={you,92} mtype=setup mlen=148
3             traffic={fpcr01=1200,bpcr01=0,be}
4             bearer={class=bcob-x,atc=nrt-vbr_0,clip=not,user=p2mp}
5             called={type=unknown,plan=aesa,addr=plan=aesa,
6             addr=0x39.0000.11.222222.0000.0000.9001.0020481004cd.00}
```

```
7           qos={forw=class0/unspecified,back=class0/unspecified}
8           repeat={indication=!illegal0xa}
9           dtl={cs=net,ptr=27,{node={level=80,
10          node=0x39.0000.11.222222.0000.0000.7001.ff1a36900001.00},
11          port=268435474},{node={level=80,
12          node=0x39.0000.11.222222.0000.0000.9001.ff0f0dc60001.00},
13          port=allports}}
14          epref={type=local,flag=me,idval=0}
15          calling={type=unknown,plan=aesa,
16          addr=0x39.276f.31.0001ef.0000.0401.7005.00204810078c.00}
17  S  4⇒  0.434
18          pnni cref={me,92} mtype=call_proc mlen=16
19          connid={vpass=explicit,pex=exclusive_vpci_vci,vpci=0,vci=55}
20          epref={type=local,flag=notme,idval=0}
21  S  5⇐  0.437
22          pnni cref={me,92} mtype=connect mlen=7
23          epref={type=local,flag=notme,idval=0}
```

6.3.1.7 Point-to-multipoint: Adding a Second Leaf

We now add a second end system (lovina) to our point-to-multipoint tree:

```
1   S  1⇐  157.565
2           pnni cref={you,93} mtype=setup mlen=148
3           traffic={fpcr01=1200,bpcr01=0,be}
4           bearer={class=bcob-x,atc=nrt-vbr_0,clip=not,user=p2mp}
5           called={type=unknown,plan=aesa,addr=plan=aesa,
6           addr=0x39.0000.11.222222.0000.0000.8001.002048060070.00}
7           qos={forw=class0/unspecified,back=class0/unspecified}
8           repeat={indication=!illegal0xa}
9           dtl={cs=net,ptr=27,{node={level=80,
10          node=0x39.0000.11.222222.0000.0000.7001.ff1a36900001.00},
11          port=268435483},{node={level=80,
12          node=0x39.0000.11.222222.0000.0000.8001.ff1a4ce70001.00},
13          port=allports}}
14          epref={type=local,flag=me,idval=3***1}
15          calling={type=unknown,plan=aesa,
16          addr=0x39.276f.31.0001ef.0000.0401.7005.00204810078c.00}
17  S  2⇒  157.543
18          pnni cref={me,93} mtype=call_proc mlen=16
19          connid={vpass=explicit,pex=exclusive_vpci_vci,vpci=0,vci=73}
20          epref={type=local,flag=notme,idval=3***1}
21  S  2⇒  157.548
22          pnni cref={me,93} mtype=connect mlen=7
23          epref={type=local,flag=notme,idval=3***1}
```

6.3.1.8 Point-to-multipoint: Add a Second Leaf at the same End System

Now we add the second end system (lovina) a second time to our point-to-multipoint tree. This time we do not see a connection setup, but a real add party procedure. Here are the ADD PARTY and ADD PARTY ACKNOWLEDGE messages:

```
1  S  1⇐   251.282
             pnni cref={you,93} mtype=add_partymlen=122
             called={type=unknown,plan=aesa,addr=plan=aesa,
             addr=0x39.0000.11.222222.0000.0000.8001.002048060070.00}
             repeat={indication=!illegal0xa}
             dtl={cs=net,ptr=27,{node={level=80,
             node=0x39.0000.11.222222.0000.0000.7001.ff1a36900001.00},
             port=268435483},{node={level=80,
             node=0x39.0000.11.222222.0000.0000.8001.ff1a4ce70001.00},
             port=allports}}
             epref={type=local,flag=me,idval=4***2}
             calling={type=unknown,plan=aesa,
             addr=0x39.276f.31.0001ef.0000.0401.7005.00204810078c.00}
   S  2⇒   251.210
             pnni cref={me,93} mtype=add_party_ack
              mlen=7 epref={type=local,flag=notme,idval=4***2}
```

Further additions of the same end system look the same.

This situation would also be the same if a different end system was added but the new connection will be routed over the already existing connection of the same multicast call on link (1/2).

6.3.1.9 Point-to-multipoint: Drop a Leaf

Starting from the situation of the previous section spock_b drops one leaf of the end system lovina. Because there is still a second leaf of the current call on the same link (1/2) the PNNI connection will not be cleared, but a drop party procedure will be executed:

```
   S  1⇐   352.926
             pnni cref={you,93} mtype=drop_party mlen=13
             cause={loc=user,cvalue=normal,_unspecified,class=normal_event}
             epref={type=local,flag=me,idval=4***2}
   S  2⇒   352.939
             pnni cref={me,93} mtype=drop_party_ack mlen=13
             cause={loc=user,cvalue=normal,_unspecified,class=normal_event}
             epref={type=local,flag=notme,idval=4***2}
```

If there are no more leaves connected over link (1/2) the connection branch on this link will be released:

```
   S  1⇐   356.807
             pnni cref={you,93} mtype=release mlen=6
             cause={loc=user,cvalue=normal,_unspecified,class=normal_event}
   S  2⇒   357.912
             pnni cref={me,93} mtype=rel_compl mlen=6
             cause={loc=user,cvalue=normal,_unspecified,class=normal_event}
```

6.3.1.10 Point-to-multipoint: Release of all Leaves of a Call

Let us assume that we have a point-to-multipoint call with two branches rooted at spock_b. One branch with two leaves leads from forest to end systems on forelle (link 1/2) and one branch with three leaves to end systems on foreplay (link 3/4). The root node (spock_b) releases the entire call:

```
1   S  3⇐   452.251
2            pnni cref={you,92} mtype=drop_party mlen=13
3            cause={loc=user,cvalue=normal,_unspecified,class=normal_event}
4            epref={type=local,flag=me,idval=2}
5   S  3⇐   452.252
6            pnni cref={you,92} mtype=drop_party mlen=13
7            cause={loc=user,cvalue=normal,_unspecified,class=normal_event}
8            epref={type=local,flag=me,idval=1}
9   S  3⇐   452.253
10           pnni cref={you,92} mtype=release mlen=6
11           cause={loc=user,cvalue=normal,_unspecified,class=normal_event}
12  S  4⇒   452.685
13           pnni cref={me,92} mtype=drop_party_ack mlen=13
14           cause={loc=user,cvalue=normal,_unspecified,class=normal_event}
15           epref={type=local,flag=notme,idval=2}
16  S  4⇒   452.686
17           pnni cref={me,92} mtype=drop_party_ack mlen=13
18           cause={loc=user,cvalue=normal,_unspecified,class=normal_event}
19           epref={type=local,flag=notme,idval=1}
20  S  4⇒   452.687
21           pnni cref={me,92} mtype=rel_compl mlen=0
22  S  1⇐   452.800
23           pnni cref={you,93} mtype=drop_party mlen=13
24           cause={loc=user,cvalue=normal,_unspecified,class=normal_event}
25           epref={type=local,flag=me,idval=4}
26  S  1⇐   452.807
27           pnni cref={you,93} mtype=release mlen=6
28           cause={loc=user,cvalue=normal,_unspecified,class=normal_event}
29  S  2⇒   452.911
30           pnni cref={me,93} mtype=drop_party_ack mlen=13
31           cause={loc=user,cvalue=normal,_unspecified,class=normal_event}
32           epref={type=local,flag=notme,idval=4}
33  S  2⇒   452.912
34           pnni cref={me,93} mtype=rel_compl mlen=6
35           cause={loc=user,cvalue=normal,_unspecified,class=normal_event}
```

6.4 Summary

In this chapter we have looked at the protocol family that is used between network nodes in private networks—the PNNI. This family has two members: the PNNI routing protocol and the PNNI signalling protocol. The routing protocol is used to configure the network and distribute reachability information between the nodes of a network. The signalling protocol is used to establish and clear connections based on this information. We have seen PNNI routing and signalling operating in a non-hierarchical network.

7

ILMI: Integrated Local Management Interface

7.1 Introduction to ILMI

The ILMI[1] is used to provide ATM devices such as end systems or switches with status and configuration information. Such information is, for example, the registered ATM network prefixes, registered ATM addresses, VPCs, VCC and much more. ILMI can also be used on PNNI links to provide automatic configuration of the link parameters.

The typical application of ILMI is the exchange of ATM network prefixes and interface addresses to allow an automatic configuration of end systems that are attached to an ATM switch. See Section 7.4 for more details on the automatic configuration.

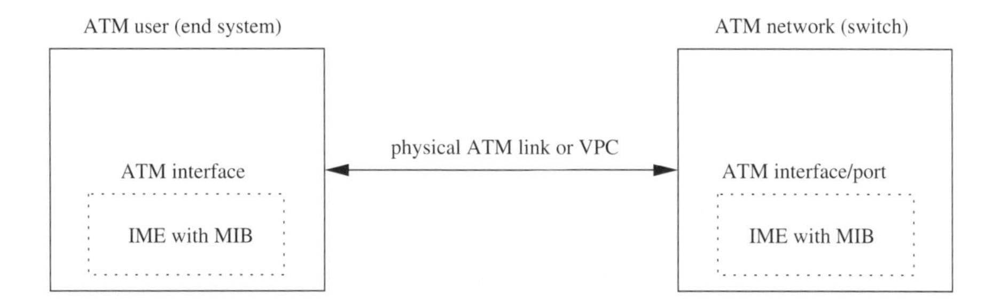

Figure 7.1: ILMI connection of an ATM end system and an ATM switch

Figure 7.1 shows a typical scenario where the ATM interface of an end system is attached via the UNI to an ATM port of a switch. In this context the term UNI only means that it is an interface between user and network and does not imply signalling. Different scenarios are possible.

The figure shows two ATM devices, the end system and the switch. Each of these two devices contains an ATM Interface Management Entity (IME). These IMEs are the functional units of the ATM devices that are responsible for the ILMI communication.

---

[1] ILMI is sometimes also called the Integrated Link Management Interface or the Interim Local Management Interface.

ILMI supports the bi-directional exchange of parameters between two IMEs. An ATM device with more than one ATM interface contains a separate IME for each ATM interface. For example, an ATM end system with one ATM interface contains one IME; an ATM switch contains one IME per ATM port. The IME is the functional unit that supports the ILMI function for the associated ATM interface. The ILMI communication takes place between two adjacent IMEs over a physical link or a Virtual Path Connection (VPC). Each IME contains a database, namely the ATM Management Information Base (MIB) that stores the information associated with the local ATM interface/IME.

To exchange the parameters the Simple Network Management Protocol (SNMP) is used on top of AAL5 and the usual SNMP terminology is applied. Every IME contains a server application (also called a management application) and a client application (also called an agent application). The server application of an IME manages the database that stores the local ATM Interface MIB. The client application of an IME can access the remote ATM Interface MIB via the remote server application. Because adjacent IMEs, which are connected by a link or VPC, contain server and client application, both IMEs can access the MIB of the adjacent IME. Details on the exchange of messages between IMEs can be found in Section 7.2.

Each IME contains an ATM Interface MIB. The structure of this MIB is the same for all IMEs. The MIB can be extended to support new or vendor-specific features. Some extensions, e.g. the auto configuration of a LAN Emulation Client (see Section 8.3 for details about LAN Emulation and [LECM2] for the definition of the MIB extension), are already defined. Vendor-specific extensions of the MIB are possible to allow the implementation of vendor-specific features. Details of the structure of the ATM Interface MIB can be found in Section 7.3.

The ILMI communication can be quite complex. Examples will help to understand the operation of ILMI. Section 7.5 shows and explains several ILMI communication examples of an end system and a switch.

7.2 The ILMI Protocol

The ILMI communication protocol is used on a physical link or VPC between two adjacent IMEs. The ILMI protocol is based on the SNMP version 1 as defined in [RFC1157]. A preconfigured Virtual Channel Connection (VCC) carries the protocol between the adjacent IMEs in the form of AAL5-encapsulated SNMP messages. The VPI/VCI of this VCC can be configured; the default values are VPI=0 and VCI=16.

The messages are encoded according to SNMP version 1. Four different operations to manipulate the MIB are defined: Get (to retrieve specific management information), GetNext (to retrieve, via traversal of the MIB, management information), Set (to modify management information), and Trap (to report extraordinary events, e.g. a coldStart trap signifies that the sending IME is reinitialising itself; the significance of the other parameters of the trap message is implementation specific). The implementation of these four operations uses five different PDU types: GetRequest-PDU, GetNextRequest-PDU, SetRequest-PDU, GetResponse-PDU (response to a Get, GetNext or Set operation), and Trap-PDU.

All SNMP messages used for ILMI use the community name “ILMI”. In all SNMP Traps the agent address field is set to the value 0.0.0.0. Addresses are not needed because there are only two communication IMEs.

According to the standard SNMP, messages may be up to 484 bytes long, the throughput of SNMP traffic should be no more than 1% of the link bandwidth and the burst length should be no more than 484 bytes.

The response time, i.e. the elapsed time from the submission on an SNMP message (e.g. GetRequest message), to the receipt of the corresponding SNMP messages (e.g. GetResponse message), is usually much less than 1 second.

7.3 The ATM Interface MIB

Each IME contains an ATM interface MIB. The structure of this MIB is the same for all IMEs. Every MIB is associated with exactly one ATM interface and stores the management information. This section provides a survey about the structure of the MIB. The reader will see what kind of information is stored in the MIB, but it is not a copy of the MIB definition. The complete MIB definition can be found in [ILMI4], [RFC1155], [RFC1213], vendor-specific documents and other related documents.

The managed information is stored in an MIB. The MIB is organised in a hierarchical fashion. The smallest piece of information stored in the MIB is a parameter. Each parameter is an instance of a type and has a unique name inside the MIB and a value. The unique name of a parameter is sometimes also called its variable name.

The name is a sequence of non-negative numbers, each separated by a dot (e.g. `1.3.6.1.2.1.1.1.0`). Usually, the last (least significant) number on the right-hand side (here 0) specifies the instance of a type (needed because sometimes more than one instance of a type exist; is always 0 if there is only one instance), the other (most significant) numbers on the left-hand side (here `1.3.6.1.2.1.1.1`) specify the name of the type of the parameter. The meaning of the parameters is specified in different documents, e.g. [ILMI4], [RFC1155] and [RFC1213]. All these documents together specify the variable names and possible values of all parameters. Not all parameters are always needed and implemented in an MIB.

To simplify the work with the parameters, symbolic names are associated with the sequences of non-negative numbers or iwth a most significant part of the sequence. For example, the sequence `1.3.6.1.2.1.1.1.0` corresponds to the symbolic name `iso.org.dod.internet.mgmt.mib-2.system.sysDescr.0`.

Most of the ILMI-relevant MIB parameters are placed in the `enterprise` subtree. The full name of this subtree is `iso.org.dod.internet.private.enterprises` or `1.3.6.1.4.1`. In the rest of this chapter we will always use the term `enterprises` as a shortcut for its full name. Inside the `enterprises` subtree we use the `atmForum` entry with the assigned number 353, i.e. `enterprises.atmForum` corresponds to the sequence `1.3.6.1.4.1.353`.

In the following sections we list the names of some parameter types of the MIB and provide a short description. The names are sometimes only the most significant and common part of a group of parameter types, i.e. they specify a subtree of the MIB. In such a case the group of parameters is described. For each parameter the sequences of non-negative numbers and the associated symbolic name are shown.

7.3.1 System Information MIB

The system information MIB is defined in [RFC1213]. The most important parameters are:

- `system`[2] `(1.3.6.1.2.1.1)`
  A subtree that stores information about the system (end system interface or switch port).

---

[2] The prefix `iso.org.dod.internet.mgmt.mib-2` is omitted.

- `sysDescr`[3] `(1.3.6.1.2.1.1.1)`
  Stores a description (human readable string) of the system supplied by the vendor.
- `sysUpTime`[3] `(1.3.6.1.2.1.1.3)`
  Stores the time (integer value that represents the time in hundredths of a second) since the system was going up. For example, a value of 8640000 means that the system is 1 day up.
- `sysName`[3] `(1.3.6.1.2.1.1.5)`
  Stores an administratively-assigned system name (human readable string).
- `sysLocation`[3] `(1.3.6.1.2.1.1.6)`
  Stores the physical location of the system (human readable string). This value can be set by the administrator.

7.3.2 Link Management MIB

The link management MIB is defined in [ILMI4]. The most important information is stored in two subtrees: the physical group and the ATM layer group.

The physical group `enterprises.atmForum.atmForumUni.atmfPhysicalGroup` `(1.3.6.1.4.1.353.2.1)` stores information associated with the physical interface. Important parameters are:

- `atmfPortTable`[4] `(1.3.6.1.4.1.353.2.1.1)`
  A subtree that stores status information associated with the physical layer of the ATM interface.
- `atmfMyIpNmAddress`[4] `(1.3.6.1.4.1.353.2.1.2)`
  Stores an IP address that can be used for external access to the MIB of this IME via IP/UDP. This parameter is optional and only present in the MIB if the IME supports this feature.
- `atmfMyOsiNmNsapAddress`[4] `(1.3.6.1.4.1.353.2.1.3)`
  Stores an ATM NSAP address that can be used to external access to the MIB of this IME via ATM/AAL5 without ILMI. This parameter is optional and only present in the MIB if the IME supports this feature.

The ATM layer group `enterprises.atmForum.atmForumUni.atmfAtmLayerGroup` `(1.3.6.1.4.1.353.2.2)` stores information associated with the ATM layer of the interface. This group contains a table `atmfAtmLayerTable (1.3.6.1.4.1.353.2.2.1.1)` which stores all possible properties of the ATM layer. Among the parameters in this group are:

- `atmfAtmLayerMaxVPCs`[5] `(1.3.6.1.4.1.353.2.2.1.1.2)`
  Stores the maximum number of VPCs supported by this interface (integer value between 0 and 4096).
- `atmfAtmLayerMaxVCCs`[5] `(1.3.6.1.4.1.353.2.2.1.1.3)`
  Stores the maximum number of VCCs supported by this interface (integer value between 0 and 268435456).
- `atmfAtmLayerUniVersion`[5] `(1.3.6.1.4.1.353.2.2.1.1.9)`
  Stores the latest version of UNI signalling that is supported on this interface (integer). Value

---

[3] The prefix `iso.org.dod.internet.mgmt.mib-2` is omitted.

[4] The prefix `enterprises.atmForum.atmForumUni.atmfPhysicalGroup` is omitted.

[5] The prefix `enterprises.atmForum.atmForumUni.atmfAtmLayerGroup.atmfAtmLayerTable.atmfAtm-LayerEntry` is omitted.

1 means UNI 2.0, value 2 means UNI 3.0, value 3 means UNI 3.1, value 4 means UNI 4.0 and value 5 means that UNI signalling is not supported.

- `atmfAtmLayerDeviceType`[6] (1.3.6.1.4.1.353.2.2.1.1.10)
  Stores the device type (integer). Value 1 means that the interface belongs to an ATM end system (user). Value 2 means that the interface belongs to an ATM switch (network node).

7.3.3 Address Registration MIB

The address registration MIB is defined in [ILMI4]. The most important parameters are:

- `atmfNetPrefixGroup.atmfNetPrefixTable`[7] (1.3.6.1.4.1.353.2.7.1)
  A subtree that stores all possible ATM network prefixes for the ATM interface. The table is stored in the ATM end system IME. The values are set by the ATM switch.
- `atmfNetPrefixGroup.atmfNetPrefixTable.atmfNetPrefixEntry.atmfNetPrefixStatus`[7] (1.3.6.1.4.1.353.2.7.1.1.3)
  Stores single ATM network prefixes. To create a new prefix this parameter with `.0.ATM-PREFIX` appended must be set to the integer value 1. For example, `...atmfNetPrefixStatus.0.13.1.1.2.2.3.3.4.4.5.5.6.6.7=1` creates a new table entry with the 13-byte network prefix 1.1.2.2.3.3.4.4.5.5.6.6.7. To remove an existing entry from the table the parameter plus `.0.ATM-PREFIX` must be set to the integer value 2. For example, `...atmfNetPrefixStatus.0.13.1.1.2.2.3.3.4.4.5.5.6.6.7=2` removes the table entry 1.1.2.2.3.3.4.4.5.5.6.6.7.
- `atmfAddressGroup.atmfAddressTable`[7] (1.3.6.1.4.1.353.2.6.1)
  A subtree that stores all ATM addresses for the ATM interface. The table is stored in the ATM switch IME. The values are set by the ATM end system.
- `atmfAddressGroup.atmfAddressTable.atmfAddressEntry.atmfAddressStatus`[7] (1.3.6.1.4.1.353.2.6.1.1.3)
  Stores single ATM addresses. Possible values are 1 and 2. The handling is analogous to `atmfNetPrefixStatus` above.

7.4 Automatic Configuration

This section describes the automatic configuration process of two connected ATM interfaces. This process consists of two subprocesses: the automatic link configuration and the automatic address registration. Both automatic configuration subprocesses are optional, and each can be replaced by manual configuration. If both subprocesses are executed, then the automatic link configuration is executed before the automatic address registration. Section 7.5 shows examples of a complete automatic configuration process.

7.4.1 Automatic Link Configuration

A private ATM user or node may automatically configure the interface type (Public UNI, Private UNI or PNNI) and the IME type (User-Side, Network-Side or Symmetric). If the

---

[6] The prefix `enterprises.atmForum.atmForumUni.atmfAtmLayerGroup.atmfAtmLayerTable.atmfAtmLayerEntry` is omitted.

[7] The prefix `enterprises.atmForum.atmForumUni` is omitted.

automatic configuration is used, then all other ATM activities are performed after the automatic link configuration is completed. This is because the type information, which is a result of the automatic link configuration, is used by other ATM protocols.

Each peers provides preconfigured information about the local device type (user/end system or network/switch) and the UNI type (private or public) as part of its MIB. This information is exchanged by using ILMI. Then, based on this information, the peers decide the interface type and the IME type of both peers. [ILMI4] contains a list of all possible combinations and the corresponding decisions of the peers. Some combinations (e.g. a user–user connection) are not defined.

7.4.2 Automatic Address Registration

This section describes the automatic address registration of a private ATM end system interface that is connected via a private UNI to a private ATM switch port. This registration can only be performed if the link is configured correctly, i.e. usually after the automatic link configuration. That is because it has to be clear who is the network side (switch), and who is the user side (end system). Symmetric links cannot use automatic address registration.

The automatic address registration is used to exchange addressing information between the end system and the switch at the UNI. This addressing information can be used as the Called Party Number and as the Calling Party Number in signalling messages. The addressing information must be known to both the end system and the switch in order to establish ATM connections via UNI signalling (see Chapter 3).

An ATM address at the UNI consists of three major parts: the End System Identifier (ESI), the Selector (SEL) and the rest (network prefix). The ESI and the SEL are determined by the end system. Because of the irrelevance of the SEL (the SEL part is only interpreted inside the end system) for the address registration, only a real ESI value is provided by the end system IME. Usually, the ESI is a unique value provided by the ATM interface card. The SEL field is always set to zero, even if other values are allowed. The network prefix is determined by the switch and therefore provided by the IME of the switch.

Both parts, the ESI and the network prefix, are exchanged by the ILMI and later used for UNI signalling. It is possible that the switch provides more than one network prefix,[8] e.g. in a transitional period to allow the switch and its end systems to use the old and the new network prefix value. An end system may support more than one ESI, e.g. to support multiple functions.

Not all combinations of network prefix and ESI are valid, e.g. in public networks. ATM addresses in public networks are usually E.164 addresses. In that case the network side (the switch) supplies the whole ATM address. The user side (the end system) does not provide any part of the address.

Three different procedures of automatic address registration exist. These procedures include the exchange of addressing information on new ILMI connections, the dynamic modification of addressing information and the de-registration on lost ILMI connections.

On a new ILMI connection, i.e. when the end system or switch are going up or when the end system is plugged into the switch, addressing information between the end system and the switch are exchanged. This information can then be used for UNI signalling.

[8] A switch may also associate different network prefixes with different ports, but that is irrelevant for ILMI, because the ILMI communication is always done in the context of one switch port and its MIB.

The addressing information may change during operation, e.g. the valid network prefix may change due to topology changes. That is why it is possible to update the exchanged addressing information during the operation. The list of network prefixes supported by the switch and the list of ESIs supported by the end system may be changed.

When an end system is unplugged from the UNI, i.e. the ILMI connectivity is lost, the address of the end system interface will be de-registered from the switch. This makes it possible to plug the same end system into another port of the same switch (which will usually have the same network prefix) without any problems. Of course, it is also possible to plug the end system into another switch. After plugging in the end system the automatic configuration procedure on the new ILMI connection will be executed.

7.4.2.1 Automatic Address Registration Details

As we saw in Section 7.3, two MIB tables are defined for the exchange of addressing information. The first table contains the network prefixes, one entry for each prefix. The second table contains the registered addresses, one entry for each address. We will now look at the procedures of information exchange.

The IME of the end system implements the Network Prefix MIB table so that the end system has local access to the current entries. These entries are supplied by the switch—the IME of the switch issues SetRequest messages to create or delete entries in the Network Prefix table in order to register or de-register network prefixes. The end system IME can accept or ignore the supplied entries.

The IME of the switch implements the Address MIB table. The entries are supplied by the end system. Each address entry represents an ATM address consisting of a network prefix, ESI and SEL=0. The address entries are supplied by the end system. Thus, the IME of the end system issues SetRequest messages to create or delete entries in the Address table in order to register or de-register addresses. The switch IME can accept or ignore the supplied entries.

In operation, the registration of network prefixes occurs first. Then, the end system combines its own ESI/SELs with one or more of the supplied network prefixes to create ATM addresses. The end system then registers these addresses, and the switch IME performs a final check to avoid the use of unsupplied network prefixes. In the case of E.164 addresses, the network prefixes represent the complete address. Thus, the end system can simply echo one or more of the supplied addresses.

At any time, one or both IMEs can check the consistency of a MIB. To do this, it can issue GetRequest and GetNextRequest messages.

7.5 ILMI Communication Examples

This section shows seven typical ILMI communication examples. All of these examples show the communication between the ATM interface of an end system and an ATM port of a switch as presented in Figure 7.1. ILMI is running on both systems.

The communication examples were recorded in the GMD Fokus laboratory. An optical splitting box was inserted into the physical link between both systems to allow recording of both communication directions by the use of two additional ATM interface cards (in our case devices "`/dev/ty0`" and "`/dev/ty1`") inserted into a third system. The recorded ATM communication was decoded by the `a5r` tool and the `ilmidump` tool of the TINA package of BEGEMOT [Beg].

To decode the ILMI messages of both directions (ILMI messages use VPI=0 and VCI=16) the commands

```
a5r -c 16 < /dev/ty0 | ilmidump
a5r -c 16 < /dev/ty1 | ilmidump
```

were executed on the third system. See Section 1.3 for details about capturing and decoding of ATM communication.

The following sections show the output of the Tina tools. The messages from the switch to the end system (e⇐s) and from the end system to the switch (e⇒s) were merged. Comments and time stamps (relative to the beginning of the experiment) were added. The messages contain the names of MIB entries. Internally, such a name is simply a sequence of numbers. The `ilmidump` tool tries to translate these numbers into the symbolic names, but not all translations are known to the tool. Therefore, some entries look more symbolic (e.g. `mib-2.system.sysUpTime.0`) than others (e.g. `enterprise.326.2.2.2.1.6.2.1.1.1.21.4294967295.0`).

7.5.1 An Unattached ATM End System Interface

The first example shows an ATM end system interface that is not attached to a switch port (or to anything else). It looks the same if an attached switch port is down. The end system's IME tries to establish ILMI communication, but, of course, it fails.

The following messages are transmitted every 5 s:

```
1  ILMI e⇒s  0.000   Trap    (45) enterprise.3.1.1 (0.0.0.0) coldStart 0 mib-2.
2                    system.sysUpTime.0=
3  ILMI e⇒s  0.000   GetNext(36) enterprise.atmForum.atmForumUni.6.1.1.3
```

7.5.2 An Unattached ATM Switch Port

This example shows an ATM switch port that is not attached to an end system interface (or to anything else). It looks the same if an attached end system interface is down. The switch port's IME tries to establish ILMI communication but, of course, it fails.

The following messages are transmitted every 4 s:

```
1  ILMI e⇐s  0.144   Trap    (45) enterprise.3.1.1 (0.0.0.0) coldStart 0 mib-2.
2                    system.sysUpTime.0=
3  ILMI e⇐s  0.145   GetNext(34) enterprise.atmForum.atmForumUni.7.1
4  ILMI e⇐s  1.884   GetReq (32) mib-2.system.sysUpTime.0
5  ILMI e⇐s  1.885   GetReq (35) enterprise.atmForum.atmForumUni.
6                    atmfPhysicalGroup.2.0
7  ILMI e⇐s  1.886   GetReq (37) enterprise.atmForum.atmForumUni.
8                    atmfPhysicalGroup.atmfPortTable.atmfPortEntry.atmfPortIndex.0
9  ILMI e⇐s  1.887   GetReq (37) enterprise.atmForum.atmForumUni.
10                   atmfAtmLayerGroup.atmfAtmLayerTable.atmfAtmLayerEntry.
11                   atmfAtmLayerMaxVCCs.0
12 ILMI e⇐s  1.888   GetReq (37) enterprise.atmForum.atmForumUni.
13                   atmfAtmLayerGroup.atmfAtmLayerTable.atmfAtmLayerEntry.9.0
```

7.5.3 An ATM Link is Going Up

This example shows what happens if an ATM end system interface and switch port are connected, and the automatic configuration procedure is executed (see also Section 7.4). The communication is very similar if both systems are plugged together or after a link failure.

Before the experiment, the ATM port of the switch is down and the end system and its ATM interface are already up (see also Section 7.5.1):

```
1  ILMI e=>s  0.000   Trap    (45) enterprise.3.1.1 (0.0.0.0) coldStart 0 mib-2.
2                     system.sysUpTime.0=
3  ILMI e=>s  0.000   GetNext(36) enterprise.atmForum.atmForumUni.6.1.1.3
```

These two messages are sent by the end system every 5 s to check whether there is some IME at the other end of the link. For the first 15 s the switch port was down:

```
1  ILMI e=>s  5.007   Trap    (45) enterprise.3.1.1 (0.0.0.0) coldStart 0 mib-2.
2                     system.sysUpTime.0=
3  ILMI e=>s  5.007   GetNext(36) enterprise.atmForum.atmForumUni.6.1.1.3
4  ILMI e=>s  10.008  Trap(45) enterprise.3.1.1 (0.0.0.0) coldStart 0 mib-2.
5                     system.sysUpTime.0=
6  ILMI e=>s  10.008  GetNext(36) enterprise.atmForum.atmForumUni.6.1.1.3
7  ILMI e=>s  15.009  Trap    (45) enterprise.3.1.1 (0.0.0.0) coldStart 0 mib-2.
8                     system.sysUpTime.0=
9  ILMI e=>s  15.009  GetNext(36) enterprise.atmForum.atmForumUni.6.1.1.3
```

Now the switch port is up (set by switch management):

```
1  ILMI e=>s  15.034  Trap    (45) enterprise.3.1.1 (0.0.0.0) coldStart 0 mib-2.
2                     system.sysUpTime.0=
3  ILMI e=>s  15.035  GetNext(34) enterprise.atmForum.atmForumUni.7.1
4  ILMI e=>s  15.036  GetResp(34) noSuchName@1 enterprise.atmForum.atmForumUni.
5                     7.1
```

The switch requests some information (physical port, max. VCCs, ATM layer):

```
1  ILMI e<=s  15.036  GetReq (35) enterprise.atmForum.atmForumUni.
2                     atmfPhysicalGroup.2.0
3  ILMI e<=s  15.037  GetReq (37) enterprise.atmForum.atmForumUni.
4                     atmfPhysicalGroup.atmfPortTable.atmfPortEntry.atmfPortIndex.0
5  ILMI e<=s  15.039  GetReq (37) enterprise.atmForum.atmForumUni.
6                     atmfAtmLayerGroup.atmfAtmLayerTable.atmfAtmLayerEntry.
7                     atmfAtmLayerMaxVCCs.0
8  ILMI e<=s  15.040  GetReq (37) enterprise.atmForum.atmForumUni.
9                     atmfAtmLayerGroup.atmfAtmLayerTable.atmfAtmLayerEntry.9.0
```

The switch gets responses from the end system to some of these requests (physical port):

```
1  ILMI e=>s  15.040  GetResp(35) noSuchName@1 enterprise.atmForum.
2                     atmForumUni.atmfPhysicalGroup.2.0
```

```
3  ILMI e⇒s  15.040  GetResp(36) enterprise.atmForum.atmForumUni
4                    .atmfPhysicalGroup.atmfPortTable.atmfPortEntry.atmfPortIndex.0=0
```

The switch sets the 13 byte ATM address prefix (57.39.111.49.0.1.239.0.0.4.1.112.6) in the end system's MIB:

```
1  ILMI e⇐s  15.050  SetReq (53) enterprise.atmForum.atmForumUni.7.1.1.3.0.
2                    13.57.39.111.49.0.1.239.0.0.4.1.112.6=1
```

The switch gets responses from the end system to earlier requests (max. VCCs, ATM layer):

```
1  ILMI e⇒s  15.053  GetResp(37) enterprise.atmForum.atmForumUni.
2                    atmfAtmLayerGroup.atmfAtmLayerTable.atmfAtmLayerEntry.
3                    atmfAtmLayerMaxVCCs.0=1024
4  ILMI e⇒s  15.054  GetResp(36) enterprise.atmForum.atmForumUni.
5                    atmfAtmLayerGroup.atmfAtmLayerTable.atmfAtmLayerEntry.9.0=3
```

The end system reports an error while setting the ATM address prefix (the end system is not yet ready to receive the prefix):

```
1  ILMI e⇒s  15.054  GetResp(53) badValue@1 enterprise.atmForum.atmForumUni.
2                    7.1.1.3.0.13.57.39.111.49.0.1.239.0.0.4.1.112.6(err objVal!=NULL)1
```

The switch requests some enterprise-specific information:

```
1  ILMI e⇐s  15.055  GetReq (78) enterprise.326.2.2.2.1.6.2.1.1.1.20.
2                    4294967295.0 enterprise.326.2.2.2.1.6.2.1.1.1.21.4294967295.0
```

The switch sets the ATM address prefix (again):

```
1  ILMI e⇐s  15.064  SetReq (53) enterprise.atmForum.atmForumUni.7.1.1.3.0.
2                    13.57.39.111.49.0.1.239.0.0.4.1.112.6=1
```

The end system reports an error while setting the ATM address prefix (again):

```
1  ILMI e⇒s  15.066  GetResp(53) badValue@1 enterprise.atmForum.atmForumUni.
2                    7.1.1.3.0.13.57.39.111.49.0.1.239.0.0.4.1.112.6(err objVal!=NULL)1
```

The switch requests further information:

```
1  ILMI e⇐s  15.067  GetReq (38) enterprise.326.2.2.2.1.1.2.0
```

The end system cannot supply the requested information:

```
1  ILMI e⇒s  15.067  GetResp(38) noSuchName@1 enterprise.326.2.2.2.1.1.2.0
```

The switch sets the ATM address prefix (again):

```
1  ILMI e⇐s  15.070  SetReq (53) enterprise.atmForum.atmForumUni.7.1.1.3.0.
2                    13.57.39.111.49.0.1.239.0.0.4.1.112.6=1
```

The end system reports an error while setting the ATM address prefix (again):

```
1  ILMI e⇒s  15.071  GetResp(53) badValue@1 enterprise.atmForum.atmForumUni.
2                    7.1.1.3.0.13.57.39.111.49.0.1.239.0.0.4.1.112.6(err objVal!=NULL)1
```

The switch requests further information (max. VCCs):

```
1  ILMI e⇐s  15.073  GetReq (37) enterprise.atmForum.atmForumUni.
2                    atmfAtmLayerGroup.atmfAtmLayerTable.atmfAtmLayerEntry.
3                    atmfAtmLayerMaxVCCs.0
```

The end system provides the requested information (max. VCCs):

```
1  ILMI e⇒s  15.074  GetResp(37) enterprise.atmForum.atmForumUni.
2                    atmfAtmLayerGroup.atmfAtmLayerTable.atmfAtmLayerEntry.
3                    atmfAtmLayerMaxVCCs.0=1024
```

The switch sets the ATM address prefix, and the end system reports an error while setting the prefix (not shown again). The switch requests the up-time of the end system, and the end systems answers:

```
1  ILMI e⇐s  18.953  GetReq (32) mib-2.system.sysUpTime.0
2  ILMI e⇒s  18.953  GetResp(34) mib-2.system.sysUpTime.0=491198586
```

The switch requests further information (ATM layer, physical layer) and the end system answers or reports an error:

```
1  ILMI e⇐s  18.954  GetReq (37) enterprise.atmForum.atmForumUni.
2                    atmfAtmLayerGroup.atmfAtmLayerTable.atmfAtmLayerEntry.9.0
3  ILMI e⇒s  18.955  GetResp(36) enterprise.atmForum.atmForumUni.
4                    atmfAtmLayerGroup.atmfAtmLayerTable.atmfAtmLayerEntry.9.0=3
5  ILMI e⇐s  18.964  GetReq (35) enterprise.atmForum.atmForumUni.
6                    atmfPhysicalGroup.2.0
7  ILMI e⇐s  18.965  GetReq (37) enterprise.atmForum.atmForumUni.
8                    atmfPhysicalGroup.atmfPortTable.atmfPortEntry.atmfPortIndex.0
9  ILMI e⇐s  18.967  GetReq (37) enterprise.atmForum.atmForumUni.
10                   atmfAtmLayerGroup.atmfAtmLayerTable.atmfAtmLayerEntry.9.0
11 ILMI e⇒s  18.967  GetResp(35) noSuchName@1 enterprise.atmForum.
```

```
12                          atmForumUni.atmfPhysicalGroup.2.0
13   ILMI e⇒s  18.967  GetResp(36) enterprise.atmForum.atmForumUni.
14                          atmfPhysicalGroup.atmfPortTable.atmfPortEntry.atmfPortIndex.0=0
15   ILMI e⇒s  18.968  GetResp(36) enterprise.atmForum.atmForumUni.
16                          atmfAtmLayerGroup.atmfAtmLayerTable.atmfAtmLayerEntry.9.0=3
```

At this point the switch requests enterprise specific information that cannot be supplied by the end system:

```
1   ILMI e⇐s  18.978  GetReq (78) enterprise.326.2.2.2.1.6.2.1.1.1.20.
2                          4294967295.0 enterprise.326.2.2.2.1.6.2.1.1.1.21.4294967295.0
3   ILMI e⇒s  18.979  GetResp(78) noSuchName@1 enterprise.326.2.2.2.1.6.2.1.1.
4                          1.20.4294967295.0
5   ILMI e⇐s  18.983  GetReq (38) enterprise.326.2.2.2.1.1.2.0
6   ILMI e⇒s  18.984  GetResp(38) noSuchName@1 enterprise.326.2.2.2.1.1.2.0
```

Now the switch requests other information (max. VCCs) and the end system answers:

```
1   ILMI e⇐s  18.986  GetReq (37) enterprise.atmForum.atmForumUni.
2                          atmfAtmLayerGroup.atmfAtmLayerTable.atmfAtmLayerEntry.
3                          atmfAtmLayerMaxVCCs.0
4   ILMI e⇒s  18.987  GetResp(37) enterprise.atmForum.atmForumUni.
5                          atmfAtmLayerGroup.atmfAtmLayerTable.atmfAtmLayerEntry.
6                          atmfAtmLayerMaxVCCs.0=1024
```

The end system still checks whether there is some IME at the other end of the link (this shows that the client and the server code work independent of each other in the same IME):

```
1   ILMI e⇒s  20.019  Trap   (45) enterprise.3.1.1 (0.0.0.0) coldStart 0 mib-2.
2                          system.sysUpTime.0=
3   ILMI e⇒s  20.019  GetNext(36) enterprise.atmForum.atmForumUni.6.1.1.3
```

The switch sets the ATM address prefix, and the end system reports an error while setting the prefix (not shown again).
The following shows a request for further information (ATM layer) by the switch and the answer from the end system:

```
1   ILMI e⇐s  20.023  GetReq (37) enterprise.atmForum.atmForumUni.
2                          atmfAtmLayerGroup.atmfAtmLayerTable.atmfAtmLayerEntry.9.0
3   ILMI e⇒s  20.024  GetResp(36) enterprise.atmForum.atmForumUni.
4                          atmfAtmLayerGroup.atmfAtmLayerTable.atmfAtmLayerEntry.9.0=3
```

The switch response to an earlier request:

```
1   ILMI e⇐s  20.026  GetResp(35) .iso.org.dod.internet.6.3.2.1.1.1.0=1
```

The end system requests some information (ATM layer):

```
1   ILMI e⇒s  20.027  GetReq (37) enterprise.atmForum.atmForumUni.
2                     atmfAtmLayerGroup.atmfAtmLayerTable.atmfAtmLayerEntry.9.0
```

The switch sets the ATM address prefix (again, but this time it will be successful):

```
1   ILMI e⇐s  20.029  SetReq (53) enterprise.atmForum.atmForumUni.7.1.1.3.0.
2                     13.57.39.111.49.0.1.239.0.0.4.1.112.6=1
```

The switch answers an end system request (ATM layer):

```
1   ILMI e⇐s  20.037  GetResp(38) enterprise.atmForum.atmForumUni.
2                     atmfAtmLayerGroup.atmfAtmLayerTable.atmfAtmLayerEntry.9.0=3
```

The end system acknowledges the setting of the ATM address prefix:

```
1   ILMI e⇒s  20.042  GetResp(53) enterprise.atmForum.atmForumUni.7.1.1.3.0.
2                     13.57.39.111.49.0.1.239.0.0.4.1.112.6=1
```

The end system sets the full ATM interface address (prefix + SEL + (ESI=0) = 57.39.111.49.0.1.239.0.0.4.1.112.6 + 0.32.72.16.4.205 + 0) in the switch's MIB. The switch acknowledges:

```
1   ILMI e⇒s  20.042  SetReq (61) enterprise.atmForum.atmForumUni.6.1.1.3.0.
2                     20.57.39.111.49.0.1.239.0.0.4.1.112.6.0.32.72.16.4.205.0=1
3   ILMI e⇐s  20.046  GetResp(61) enterprise.atmForum.atmForumUni.6.1.1.3.0.
4                     20.57.39.111.49.0.1.239.0.0.4.1.112.6.0.32.72.16.4.205.0=1
```

Now the address registration is complete. The end system knows the ATM address prefix, and the switch knows the full end system's ATM address.

The switch requests further information (up-time, ATM layer, physical layer) and the end system answers:

```
1   ILMI e⇐s  23.952  GetReq (32) mib-2.system.sysUpTime.0
2   ILMI e⇒s  23.953  GetResp(34) mib-2.system.sysUpTime.0=491199086
3   ILMI e⇐s  23.954  GetReq (37) enterprise.atmForum.atmForumUni.
4                     atmfAtmLayerGroup.atmfAtmLayerTable.atmfAtmLayerEntry.9.0
5   ILMI e⇒s  23.955  GetResp(36) enterprise.atmForum.atmForumUni.
6                     atmfAtmLayerGroup.atmfAtmLayerTable.atmfAtmLayerEntry.9.0=3
7   ILMI e⇐s  23.964  GetReq (35) enterprise.atmForum.atmForumUni.
8                     atmfPhysicalGroup.2.0
9   ILMI e⇒s  23.967  GetResp(35) noSuchName@1 enterprise.atmForum.
10                    atmForumUni.atmfPhysicalGroup.2.0
```

Now all needed information is exchanged. The automatic configuration is complete and the link is up (see Section 7.5.4).

7.5.4 An ATM Link is Up (Normal Operation)

When the ATM link is up the IMEs only need to track changes and to observe the link state. To check whether the link is still up and the partner at the other side is still there, the following messages are exchanged every 5 seconds (only the up-time of the end system changes from time to time):

```
1  ILMI e⇒s  25.929  GetNext(36) enterprise.atmForum.atmForumUni.6.1.1.3
2  ILMI e⇐s  25.932  GetResp(61) enterprise.atmForum.atmForumUni.6.1.1.3.0.
3                    20.57.39.111.49.0.1.239.0.0.4.1.112.6.0.32.72.16.4.205.0=1
4  ILMI e⇐s  28.952  GetReq (32) mib-2.system.sysUpTime.0
5  ILMI e⇒s  28.953  GetResp(34) mib-2.system.sysUpTime.0=491199586
6  ILMI e⇐s  28.954  GetReq (37) enterprise.atmForum.atmForumUni.
7                    atmfAtmLayerGroup.atmfAtmLayerTable.atmfAtmLayerEntry.9.0
8  ILMI e⇒s  28.955  GetResp(36) enterprise.atmForum.atmForumUni.
9                    atmfAtmLayerGroup.atmfAtmLayerTable.atmfAtmLayerEntry.9.0=3
```

The following messages are exchanged every 25 seconds (the switch tries to read a value which the end system cannot supply):

```
1  ILMI e⇐s  33.988  GetReq (35) enterprise.atmForum.atmForumUni.
2                    atmfPhysicalGroup.2.0
3  ILMI e⇒s  33.991  GetResp(35) noSuchName@1 enterprise.atmForum.atmForumUni.
4                    atmfPhysicalGroup.2.0
```

7.5.5 An ATM Link is Going Down

This example shows what happens when a link between an ATM end system interface and a switch port is up (see Section 7.5.4), but then the switch port is going down (set by switch management).

First, the communication is as shown in Section 7.5.4. Then the switch port is going down. Before the port is down it sends a trap and requests some information from the end system:

```
1  ILMI e⇐s  337.049 Trap   (45) enterprise.3.1.1 (0.0.0.0) coldStart 0 mib-2.
2                     system.sysUpTime.0=
3  ILMI e⇐s  337.050 GetNext(34) enterprise.atmForum.atmForumUni.7.1
4  ILMI e⇐s  337.051 GetReq (35) enterprise.atmForum.atmForumUni.
5                     atmfPhysicalGroup.2.0
6  ILMI e⇐s  337.052 GetReq (37) enterprise.atmForum.atmForumUni.
7                     atmfPhysicalGroup.atmfPortTable.atmfPortEntry.atmfPortIndex.0
8  ILMI e⇐s  337.053 GetReq (37) enterprise.atmForum.atmForumUni.
9                     atmfAtmLayerGroup.atmfAtmLayerTable.atmfAtmLayerEntry.
10                    atmfAtmLayerMaxVCCs.0
11 ILMI e⇒s  337.080 GetReq (37) enterprise.atmForum.atmForumUni.
12                    atmfAtmLayerGroup.atmfAtmLayerTable.atmfAtmLayerEntry.9.0
13 ILMI e⇒s  337.080 GetResp(34) noSuchName@1 enterprise.atmForum.atmForumUni.
14                    7.1
15 ILMI e⇒s  337.083 GetResp(35) noSuchName@1 enterprise.atmForum.atmForumUni.
16                    atmfPhysicalGroup.2.0
17 ILMI e⇒s  337.096 GetResp(37) enterprise.atmForum.atmForumUni.
```

```
18                        atmfAtmLayerGroup.atmfAtmLayerTable.atmfAtmLayerEntry.
19                        atmfAtmLayerMaxVCCs.0=1024
```

Now the switch port is down, but the end system still needs a few seconds to recognise the situation. It sends GetRequests, but they are not answered by the switch. The switch is quiet:

```
1  ILMI e⇒s  342.086 GetReq (37)   enterprise.atmForum.atmForumUni.
2                     atmfAtmLayerGroup.atmfAtmLayerTable.atmfAtmLayerEntry.9.0
3  ILMI e⇒s  347.087 GetReq (37) enterprise.atmForum.atmForumUni.
4                     atmfAtmLayerGroup.atmfAtmLayerTable.atmfAtmLayerEntry.9.0
```

Then the end system recognises that the switch port is down. It behaves as in Section 7.5.1.

7.5.6 Addition of an ATM Address Prefix

This example shows what happens when the switch adds an ATM address prefix associated with the port that is connected to the end system interface. The primary (new) prefix is the 13-byte prefix 57.39.111.49.0.1.239.0.0.4.1.112.119, and the secondary (old) prefix is the 13-byte prefix 57.39.111.49.0.1.239.0.0.4.1.112.6. Based on these two prefixes, the end system creates a primary and a secondary ATM end system address (prefix + SEL=0.32.72.16.4.205 + (ESI=0)).

We assume that the link is up as in Section 7.5.4. The new prefix is added to the ATM switch port by management. The switch sets both prefixes in the end system's MIB and the end system sets both ATM end system addresses in the switch's MIB:

```
1  ILMI e⇐s  145.364 SetReq (53) enterprise.atmForum.atmForumUni.7.1.1.3.0.
2                     13.57.39.111.49.0.1.239.0.0.4.1.112.6=2
3  ILMI e⇒s  145.365 GetResp(53) enterprise.atmForum.atmForumUni.7.1.1.3.0.
4                     13.57.39.111.49.0.1.239.0.0.4.1.112.6=2
5  ILMI e⇐s  145.365 SetReq (53) enterprise.atmForum.atmForumUni.7.1.1.3.0.
6                     13.57.39.111.49.0.1.239.0.0.4.1.112.119=1
7  ILMI e⇒s  145.365 SetReq (61) enterprise.atmForum.atmForumUni.6.1.1.3.0.
8                     20.57.39.111.49.0.1.239.0.0.4.1.112.6.0.32.72.16.4.205.0=2
9  ILMI e⇐s  145.371 GetResp(61) enterprise.atmForum.atmForumUni.6.1.1.3.0.
10                    20.57.39.111.49.0.1.239.0.0.4.1.112.6.0.32.72.16.4.205.0=2
11 ILMI e⇒s  145.378 GetResp(53) enterprise.atmForum.atmForumUni.7.1.1.3.0.
12                    13.57.39.111.49.0.1.239.0.0.4.1.112.119=1
13 ILMI e⇒s  145.379 SetReq (61) enterprise.atmForum.atmForumUni.6.1.1.3.0.
14                    20.57.39.111.49.0.1.239.0.0.4.1.112.119.0.32.72.16.4.205.0=1
15 ILMI e⇐s  145.388 GetResp(61) enterprise.atmForum.atmForumUni.6.1.1.3.0.
16                    20.57.39.111.49.0.1.239.0.0.4.1.112.119.0.32.72.16.4.205.0=1
```

The link is still up. The address information is updated, and from now the new primary address is used if nothing else is requested. Therefore, the further communication is as described in Section 7.5.4, except the fact that the old address (57.39.111.49.0.1.239.0.0.4.1.112.*6*.0.32.72.16.4.205.0) is replaced by the new primary address (57.39.111.49.0.1.239.0.0.4.1.112.*119*.0.32.72.16.4.205.0).

7.5.7 Removal of an ATM Address Prefix

This example shows what happens when the switch port has two ATM address prefixes, and the primary prefix is removed by management. At the beginning the situation is as the situation at the end of Section 7.5.6. After the removal, the 13-byte prefix 57.39.111.49.0.1.239.0.0.4.1.112.6 remains active. Based on the new primary prefix, the end system creates a new primary ATM end system address (prefix + SEL=0.32.72.16.4.205 + 0).

We assume that the link is up as at the end of Section 7.5.6. The primary prefix is removed from the ATM switch port by management. First, the switch changes the primary prefix (which will be removed soon) to a secondary. The end system changes the primary end system address (which will be removed soon) to a secondary:

```
1  ILMI e⇐s  222.497 SetReq (53) enterprise.atmForum.atmForumUni.7.1.1.3.0.
2                     13.57.39.111.49.0.1.239.0.0.4.1.112.119=2
3  ILMI e⇒s  222.499 GetResp(53) enterprise.atmForum.atmForumUni.7.1.1.3.0.
4                     13.57.39.111.49.0.1.239.0.0.4.1.112.119=2
5  ILMI e⇒s  222.499 SetReq (61) enterprise.atmForum.atmForumUni.6.1.1.3.0.
6                     20.57.39.111.49.0.1.239.0.0.4.1.112.119.0.32.72.16.4.205.0=2
7  ILMI e⇐s  222.503 GetResp(61) enterprise.atmForum.atmForumUni.6.1.1.3.0.
8                     20.57.39.111.49.0.1.239.0.0.4.1.112.119.0.32.72.16.4.205.0=2
```

Then the IME of the end system restarts. The switch sets the new prefix in the end system's MIB and the end system sets the new ATM end system addresses in the switch's MIB. Other information is exchanged too:

```
1  ILMI e⇒s  223.171 Trap   (45) enterprise.3.1.1 (0.0.0.0) coldStart 0 mib-2.
2                     system.sysUpTime.0=
3  ILMI e⇒s  223.171 GetNext(36) enterprise.atmForum.atmForumUni.6.1.1.3
4  ILMI e⇐s  223.173 SetReq (53) enterprise.atmForum.atmForumUni.7.1.1.3.0.
5                     13.57.39.111.49.0.1.239.0.0.4.1.112.6=1
6  ILMI e⇐s  223.175 GetReq (37) enterprise.atmForum.atmForumUni.
7                     atmfAtmLayerGroup.atmfAtmLayerTable.atmfAtmLayerEntry.9.0
8  ILMI e⇒s  223.177 GetResp(53) badValue@1 enterprise.atmForum.atmForumUni.
9                     7.1.1.3.0.13.57.39.111.49.0.1.239.0.0.4.1.112.6(err objVal!=NULL)1
10 ILMI e⇒s  223.177 GetResp(36) enterprise.atmForum.atmForumUni.
11                    atmfAtmLayerGroup.atmfAtmLayerTable.atmfAtmLayerEntry.9.0=3
12 ILMI e⇐s  223.179 GetResp(35) .iso.org.dod.internet.6.3.2.1.1.1.0=1
13 ILMI e⇒s  223.180 GetReq (37) enterprise.atmForum.atmForumUni.
14                    atmfAtmLayerGroup.atmfAtmLayerTable.atmfAtmLayerEntry.9.0
15 ILMI e⇐s  223.181 SetReq (53) enterprise.atmForum.atmForumUni.7.1.1.3.0.
16                    13.57.39.111.49.0.1.239.0.0.4.1.112.6=1
17 ILMI e⇐s  223.186 GetResp(38) enterprise.atmForum.atmForumUni.
18                    atmfAtmLayerGroup.atmfAtmLayerTable.atmfAtmLayerEntry.9.0=3
19 ILMI e⇒s  223.194 SetReq (61) enterprise.atmForum.atmForumUni.
20                    6.1.1.3.0.20.57.39.111.49.0.1.239.0.0.4.1.112.6.0.32.72.16.4.205.0=1
21 ILMI e⇒s  223.195 GetResp(53) enterprise.atmForum.atmForumUni.
22                    7.1.1.3.0.13.57.39.111.49.0.1.239.0.0.4.1.112.6=1
23 ILMI e⇐s  223.199 GetResp(61) enterprise.atmForum.atmForumUni.6.1.1.3.0.
24                    20.57.39.111.49.0.1.239.0.0.4.1.112.6.0.32.72.16.4.205.0=1
```

The link is still up. The address information is updated. Therefore further communication is as described in Section 7.5.4.

7.6 Summary

In this chapter we analysed the operation of the Integrated Local Management Interface (ILMI). This protocol is used to autoconfigure ATM links and to exchange address information between switches and end systems. As we have seen, the ILMI is based on SNMP which runs directly on an AAL5. Management information bases (MIBs) are used to hold and access the information. We have traced protocol operation in a number of everyday situations and have observed how the protocol recovers after link failures.

8

Protocols on Top of ATM Signalling

8.1 Introduction

In this chapter we introduce two protocol families that use ATM signalling: Classical IP over ATM (CLIP) and LAN Emulation over ATM (LANE). Both protocol families can be used to replace LANs such as Ethernet/IEEE 802.3[1] or Token Ring/IEEE 802.5 under some circumstances. This is not trivial because the services offered by LANs differ from those offered by ATM. The main differences are:

- LANs use connectionless message transport. ATM is connection oriented.
- Because of the shared media, LANs implement multicast or broadcast without extra effort. In ATM networks sets of point-to-point connections or point-to-multipoint connections are needed—both variants are not that simple in use.

CLIP and LANE differ in the supported features and in their complexity. These differences will be outlined in the following two sections. A good understanding of Local Area Network (LAN) protocols (especially Ethernet) and IP is assumed for these sections.

A third protocol family, Multi Protocol Over ATM (MPOA), also provides LAN services on top of ATM. This protocol family is an extension of LANE and allows shortcuts over subnet boundaries. MPOA is very complex—a thorough description would fill another book. Refer to [MPOA1.0] and [MPOA1.1] instead.

The final section of this chapter shows how to use ATM directly without emulating IP or LANs. This allows optimal usage of the ATM Quality of Service (QoS). The problem of this approach is that all communicating entities need direct access to an ATM network.

8.2 CLIP: Classical IP over ATM

8.2.1 Overview

Classical IP over ATM (CLIP) was developed as a direct replacement for LAN segments connecting IP end systems and IP routers. CLIP allows, as other "IP over ATM" technologies, the usage of existing IP applications in ATM networks and the interoperability of these

---

[1] Ethernet and IEEE 802.3 are slightly different in the frame encoding. IEEE 802.3 has an additional IEEE 802.2 Logical Link Control (LLC) header and an additional IEEE 802.2 SubNetwork Attachment Point (SNAP) header.

applications with the Internet. The usage of ATM with CLIP as a backbone technology between existing LANs and as a technology for dedicated circuits between IP routers is possible, too.

The basic idea of CLIP is to keep the deployment of an ATM network using CLIP similar to the deployment of a "classical" LAN based IP network. That means that an ATM host adapter is treated as an ordinary LAN network interface from the point of view of the IP protocol stack. The network configuration follows a similar model as classical IP networks including routers and firewalls. CLIP emulates the services of the IP network layer. IP means IP version 4 (IPv4) in the context of CLIP. But only a few changes would be need to apply CLIP to IP version 6 (IPv6).

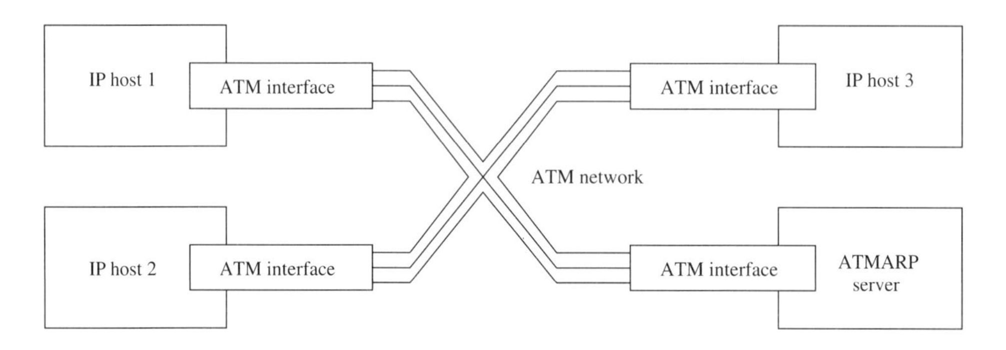

Figure 8.1: LIS scenario with three IP hosts, an ATMARP server and the maximum number of SVCs

The scope of CLIP is a Logical IP Subnetwork (LIS). An LIS is a direct replacement of a classical LAN segment. If networks of a larger scope are desired, then routers are required that are members of different LISs to connect these LISs. A router that is a member of a Logical IP Subnetwork (LIS) and of a classical LAN-based IP network is needed to connect this LIS with the classical LAN-based IP network.

Each LIS consists of its members, i.e. of its IP end systems and at least one IP router.[2] In addition an ATM Address Resolution Protocol (ATMARP) server is needed inside the LIS. Figure 8.1 shows an example. All members of an LIS need to follow some requirements:

- All members have the same IP subnet number and address mask.
- Each member knows its own IP address.
- All members are connected to the same signalled ATM network, i.e. they can communicate with each other using Switched Virtual Channels (SVCs). It is possible to use Permanent Virtual Connections (PVCs) instead of SVCs but then a fully mashed PVC network between all LIS members would be needed. In this book we only focus on CLIP using SVCs, which is the typical configuration.
- All hosts outside the LIS are accessed via an IP router.
- Each member knows its own ATM address.
- All members have a mechanism for resolving IP addresses to ATM addresses via ATMARP

[2] An LIS does not need an IP router if no host outside the LIS must be reached.

and vice versa via the Inverse ATM Address Resolution Protocol (InATMARP). These mechanisms are described later in this chapter.
- Each member knows the ATM address of the ATMARP server within the LIS.

The communication inside an LIS is quite simple. The basis is a virtual network consisting of bi-directional point-to-point SVCs that is fully meshed, i.e. each member is connected to each other member of the LIS and to the ATMARP server. Not all SVCs are established all the time. Instead, only the SVCs that are needed will be established using UNI signalling. Idle SVCs are released after a specific timeout. The virtual network allows the easy transport of each IP Protocol Data Unit (PDU).[3] The PDUs are simply encapsulated in a special frame and sent from the source to the destination or next router via the direct SVC. It is obvious that this mechanism cannot handle multicast or broadcast massages which is a known limitation of CLIP. Every member of an LIS is connected to the ATMARP server of the LIS. This connection is used to convert the IP address of the destination or next router into its corresponding ATM address to be able to establish an SVC to the destination or next router. If an SVC is already established, then this established SVC will be used to transport the IP PDUs.

Each LIS works independently of other LISs, even if they are using the same ATM network. The ATM network is only used to provide a point-to-point connection between any two members of a given LIS. It is important to mention that two IP hosts of different LISs must communicate via an intermediate IP router even though they are connected to the same ATM network and could therefore communicate directly at the ATM layer.

CLIP and its ATMARP are based on the Internet standards [RFC1577] and [RFC1483].

8.2.2 IP PDU Encapsulation

In CLIP, IP PDUs are directly transmitted to the destination or next router using a separate SVC. Each IP packet is encapsulated using LLC/SNAP. This means that an 8-byte header is added. The header plus the IP PDU are transmitted using ATM Adaptation Layer 5 (AAL5). This allows unassured transport of the IP PDU using an SVC. The Maximum Transport Unit (MTU) size for IP PDUs is 9180 bytes.

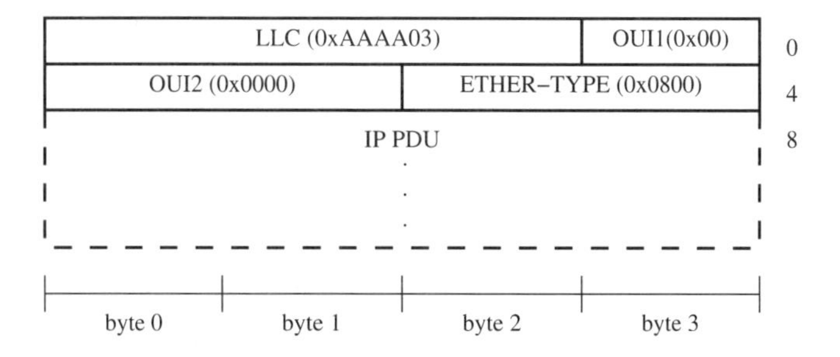

Figure 8.2: An AAL5 payload that contains an encapsulated IP PDU

[3] An IP PDU is also often called an IP datagram.

Figure 8.2 shows the encapsulated IP PDU that will be carried in the AAL5 payload. The three header fields LLC, Organisationally Unique Identifier (OUI) and ETHER-TYPE can be used to select different protocols. Because here we always have IP this feature is not needed and the fields are always set to the same values.[4]

8.2.3 ATMARP: ATM Address Resolution Protocol in CLIP

Address resolution plays a central role in CLIP. The main reason is that a member who wants to send an IP PDU to another member of the LIS needs to know the ATM address of the receiver to be able to establish an SVC, if not yet done. Because the IP address of the receiver is known[5] a mechanism is needed to translate the IP address into an ATM address. It is the same motivation as for the Address Resolution Protocol (ARP) service in LAN based IP networks.

The implementation of the ARP is specific to ATM/CLIP. The main functionality is performed by the ATMARP server. Each LIS contains one ATMARP server which is responsible for the authoritative resolving of all ATMARP requests of all members within the LIS. The ATMARP server maintains a database with all known IP address–ATM address assignments. This database is also called the ATMARP table cache. The term "cache" is used because table entries are usually not permanent, but are maintained dynamically. This is important to allow the LIS to adapt to new network configurations. Table entries are removed after 20 minutes if the entry cannot be validated.

But how does the ATMARP server get all the entries? Each member in the LIS plays the role of an ATMARP client, which means it establishes an SVC to the ATMARP server to allow the exchange of ATMARP and InATMARP requests and responses. After the establishment of the SVC the ATMARP server sends an InATMARP_REQUEST to the client, i.e. to the LIS member. The client answers with the InATMARP_REPLY which contains the IP and the ATM addresses of the client. Now the ATMARP server has its entry. This procedure is repeated regularly to validate the entries in the server ATMARP table cache.

The SVC between the ATMARP server and client will also be used by the client to perform the required translation of IP into ATM addresses. To do so the client sends an ATMARP_REQUEST inclusive IP address to the server. The server answers with the ATMARP_REPLY which contains the requested ATM address. If the server cannot resolve the request, then an ATMARP_NAK packet is sent back to the client to indicate the problem.

Each client implements a caching mechanism to reduce the number of ATMARP request to the server. Each entry in the client ATMARP table cache will be invalid after 15 minutes. After this time a new request needs to be sent to the ATMARP server if an address translation is needed.

It is possible to implement the ATMARP server in a robust way consisting of multiple servers that synchronise their databases. All these servers could be reached by the same ATM Anycast address. A protocol for the needed database synchronisation is not standardised.

ATMARP and InATMARP use five different messages, as we already have seen. Each message corresponds to a packet that will be transmitted in the payload over AAL5. All five

---

[4] The CLIP Standard requires the header fields. There is no technical reason why we need them.

[5] The IP address of the receiver of an IP PDU will be determined by ordinary IP algorithms. The receiver address is the destination IP address of the PDU or the IP address of the next router depending on the destination address and network mask configuration.

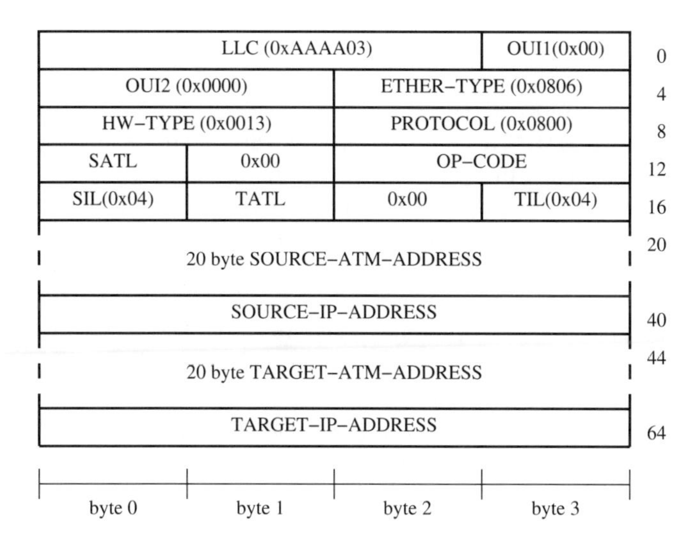

Figure 8.3: An AAL5 payload that contains an ATMARP or InATMARP packet

packets have the same structure as shown in Figure 8.3. The first 12 bytes specify the PDU type and are always the same for ATMARP and InATMARP PDUs. The SATL field stores the type and length of the source ATM address. The field has the value 0x14 for a 20-byte NSAP address and the value 0x54 for a 20-byte E.164 address. The OP-CODE field specifies the specific PDU type (see Table 8.1). The SIL field stores the length of the source IP address. The TATL field stores the type and length of the target ATM address and has the value 0x14 for a 20 byte-NSAP address and the value 0x54 for a 20-byte E.164 address. TIL is the length of the target IP address.

Table 8.1: ATMARP and InATMARP PDU types

PDU type	OP-CODE
ATMARP_REQUEST	0x0001
ATMARP_REPLY	0x0002
InATMARP_REQUEST	0x0008
InATMARP_REPLY	0x0009
InATMARP_NAK	0x000a

How are source and target address fields filled for the different PDUs? The source address fields are always set to the addresses of the sender of the PDU. The target addresses are set if they are known. All unknown address fields are set to zero. For example, an ATMARP_REQUEST PDU has the TARGET-ATM-ADDRESS set to zero, and the TARGET-IP-ADDRESS contains the IP address that should be resolved. The responding ATMARP_REPLY PDU contains the resolved ATM addressin the SOURCE-ATM-ADDRESS field, and the SOURCE-IP-ADDRESS contains the corresponding IP address (=TARGET-IP-ADDRESS field of the request PDU). This is the same mechanism as defined for the Ethernet ARP in [RFC826] and InARP in [RFC1293].

8.3 LANE: LAN Emulation over ATM

8.3.1 Overview

LAN Emulation over ATM (LANE) was developed as a direct replacement for LAN segments. In contrast to CLIP it emulates the MAC services including the encapsulation of MAC frames. Because of this, LANE is not limited to the support of IP. LANE emulates the LAN over the ATM network and is independent of the used network layer protocol, i.e. LANE can be used together with IP (IPv4), IPv6, NetBIOS, AppleTalk or any other network layer protocol. The different network layer protocols can even be mixed in a LANE environment. The second big difference to CLIP is the support of multicast and broadcast messages. This is important to support a maximum number of existing applications, and many of these applications cannot run on top of CLIP, but will run on top of LANE.

LANE was developed with the objective of keeping the deployment of an ATM network using LANE similar to the deployment of a “classical” LAN. This means that an ATM host adapter is treated as an ordinary LAN network interface from the point of view of the software. LANE is supported in the operation system through special network device drivers that emulate the LAN adapter functionality using ATM.

A LANE environment provides the service of user data transport among all members of the emulated LAN, similar to a physical LAN. Each emulated LAN works independently of other emulated LANs, even if they are using the same ATM network. The ATM network is only used to provide point-to-point or point-to-multipoint connections between members of the emulated LAN. It is important to mention that two hosts of different emulated LANs must communicate via an intermediate bridge or router even though they are connected to the same ATM network and can therefore communicate directly at the ATM layer.

LANE is able to emulate three different types of LANs:

- Ethernet;
- IEEE 802.3; and
- Token Ring/IEEE 802.5.

The handling of these types is quite similar. Some differences result from the additional LLC and SNAP headers in IEEE 802.3 and from different addressing mechanisms and the source routing feature of Token Ring/IEEE 802.5. Because of the widespread use of the Ethernet, we will focus on that technology. This allows an easier description of LANE.

Each emulated LAN consists of its LAN Emulation Clients (LECs), i.e. its end systems and at least one bridge or router.[6] In addition the LAN Emulation Service (LE Service) is needed inside the emulated LAN. The LE Service itself is implemented by at least one LAN Emulation Configuration Server (LECS), at least one LAN Emulation Server (LES) and at least one Broadcast and Unknown Server (BUS). The interface between an LAN Emulation Client (LEC) and the LE Service is called the LAN Emulation User–Network Interface (LUNI) and is specified in [LUNI2]. We will describe it in this section. The interface inside the LE Service, i.e. between LECS, LES and BUS, is called the LAN Emulation Network–Network Interface (LNNI). The LNNI is specified in [LNNI2] and will not be described in this book.

The LEC performs data transport, provides an MAC level emulated Ethernet service interface to the higher layer software and implements the client part of the address resolution

---

[6] An emulated LAN does not need a bridge or router if no host outside the emulated LAN needs to be reached.

mechanism. The data transport to other clients will be done by using a direct SVC to that client. The address resolution service of the LES is needed to get the destination ATM address to establish the SVC. If the destination ATM address cannot be resolved fast enough or a multicast or broadcast message is sent, then the BUS is used to distribute the message.

The LECS assigns each LEC to its emulated LAN. This means that configuration parameters (e.g. the ATM address of the LES) will be sent to a client which is just joining the emulated LAN. This configuration is done in the Configuration state. It is possible to use several LESs in an emulated LAN for scalability. In such case the LECS performs a load distribution among all LESs. If the LECS serves more than one emulated LAN, then the LECS decides, based on its configuration database and the ATM address of the LEC, which emulated LAN will be assigned to the LEC.

The LES implements the central control functions of the emulated LAN. It provides the service to register and resolve unicast and multicast MAC addresses to ATM addresses. Each LEC connects to an LES and registers its own unicast MAC address(es) together with its own ATM address. The LEC also registers the multicast MAC addresses where it wishes to receive messages. The BUS is then responsible for the distribution of the multicast messages to the registered LECs. The LES is also responsible for address resolution. This means that an LEC can use its LES to translate a unicast or multicast MAC address (if it was registered before) into an ATM address.

The BUS is responsible for the distribution of messages to the broadcast MAC address (0xffffffffffff) to all LECs and for the distribution of messages to a multicast MAC address to all LECs that are registered to receive messages to this specific multicast address. In addition, unicast messages are distributed to all LECs if the LEC did not yet resolve the ATM address and establish an SVC to the destination. In some networks a separate Selective Multicast Server (SMS) is used for load distribution. The SMS is then responsible for the distribution of multicast messages.

In LANE all messages are sent on top of AAL5.

Figure 8.4 shows an example scenario of an emulated LAN. LANE requires point-to-point and point-to-multipoint SVCs. The point-to-multipoint connections are used by the LES to distribute control information and by the BUS to distribute messages to multiple LECs. This can be a problem because some (public) ATM networks do not support multicast.

8.3.2 LANE Connections

LANE uses different connection types to transport user and control data. These connections are called flows. In LANE seven different flows are used. The Control Distribute flow and the Multicast Forward flow are carried on point-to-multipoint SVCs. The Configure Direct flow, the Control Direct flow, the Data Direct flow, the Default Multicast Send flow and the Selective Multicast Send flow are carried on point-to-point SVCs. In some situations the use of PVCs is possible, but in this book we focus on SVCs.

Since LANE version 2 it is possible to multiplex several Data Direct flows on one SVC, possibly of different emulated LANs. This multiplexing requires an additional LLC header for each message sent on the SVC to be able to distinguish messages of different flows. In this book we assume that all SVCs are non-multiplexed.

The Configure Direct flow is a bi-directional SVC set up by the LEC to the LECS in the LECS Connect state. This SVC is signalled using the B-LLI to indicate it carries “LE Control” messages. The Configure Direct flow is used to get configuration information, e.g. the ATM

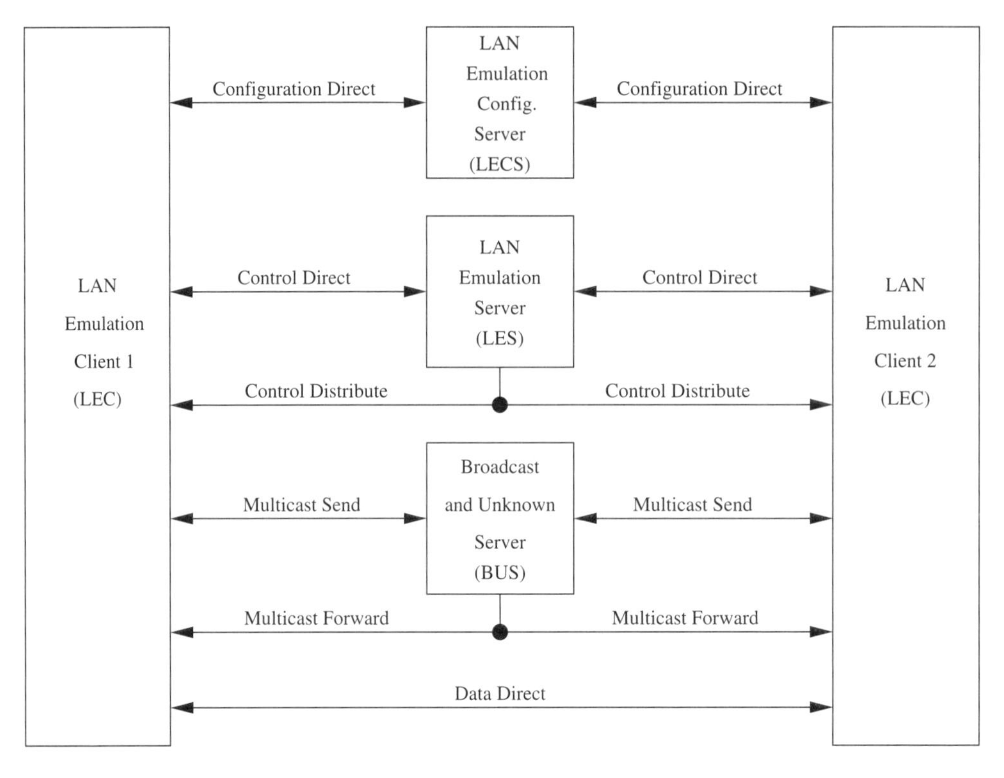

Figure 8.4: LANE scenario with two LAN Emulation Clients (LECs), one LAN Emulation Configuration Server (LECS), one LAN Emulation Server (LES) and one Broadcast and Unknown Server (BUS), with seven point-to-point and two point-to-multipoint SVCs

address of the LES.

The Control Direct flow is a bi-directional SVC established by the LEC to the LES. It is established in the Initialisation state. The SVC of the Control Direct flow will not be released while the LEC is participating in the emulated LAN.

The Control Distribute flow is a unidirectional point-to-multipoint SVC from the LES to one or more LEC. The LES has the option to establish this SVC. The Control Distribute flow is used to distribute control messages to a group of LECs more effectively. If a Control Distribute SVC is established, then it will not be released while the LEC is participating in the emulated LAN.

A Data Direct flow is a bi-directional SVC. It is establish from one LEC to another LEC to exchange unicast user data. The establishing LEC needs to know the ATM address of the second LEC. To find it out, the establishing LEC uses the address resolution service of LANE to translate the user data destination MAC address into an ATM address of an LEC. The SVCs of Data Direct flows will be released after a timeout.

An LEC sets up one bi-directional SVC to the BUS to transport the Default Multicast Send flow. An LEC sends all user data messages using this flow that cannot be sent over other flows. Such user data messages can be messages to the broadcast MAC address (0xffffffffffff), to an MAC address with unestablished Data Direct flow or to a multicast MAC address, where the selective multicast procedure has not provided an alternative path. The ATM address of the BUS to establish the SVC for the multicast flow will be determined by resolving the broadcast MAC address via the LANE address resolution service. The SVC of the Default Multicast Send flow will not be released while the LEC is participating in the emulated LAN.

An LEC can set up additional bi-directional SVCs to the BUS. They transport the Selective Multicast Send flows. The ATM address of the BUS will be determined by resolving a multicast MAC address using the LANE address resolution service. The resolved ATM address may vary to distinguish different Multicast Send flows. One BUS may have several ATM addresses.

The BUS is responsible for distributing multicast messages, broadcast messages and some unicast messages. To do so it establishes one or more unidirectional point-to-multipoint SVCs to each LEC to transport the Multicast Forward flows. The BUS uses these SVCs to forward user data messages received by its incoming Multicast Send and Default Multicast Send flows. The SVC of the Multicast Forward flow will not be released while the LEC is participating in the emulated LAN.

The BUS may also use a Multicast Send flow to forward user data messages. In such a case it is the responsibility of the BUS that an LEC does not receive duplicated messages.

8.3.3 LEC States

An LEC that wants to be a member of an emulated LAN can be in one of seven states. These states show more details of the LANE procedures. Even if we call the states "states", in the LANE standard documents they are called "phases".

An LEC starts with the Initial State. This state can be reached if the LEC is going up or after an LEC has left an emulated LAN. In this state the LEC needs to know several parameters. These parameters are, for example, its own ATM address(es), its own unicast MAC address(es), the LAN type (Ethernet, IEEE 802.3 or IEEE 802.5), the Maximum Transport Unit (MTU) size, the ELAN name (to distinguish multiple emulated LANs in one ATM network) or the ATM address of the LECS.

The LECS Connect state is the second state. The state is used by the LEC to establish an SVC that carries a Configuration Direct flow to the LECS. There are three mechanisms to determine the ATM address of the LECS. The first and preferred mechanism is to use a preconfigured ATM address of the LECS. If such an address is not configured, then an LECS ATM address is obtained via ILMI. A proper MIB definition is available in the LANE standard documents. If this fails too, then the LEC tries to establish an SVC of one of two well-known ATM addresses. These addresses are:

0xc50079000000000000000000000a03e00000100;
0x47007900000000000000000000000a03e00000100.

In this case these addresses must be configured to point to an LECS.

The Configuration state follows after the LECS Connect state. In this state the LEC obtains configuration parameters from the LECS. The most important parameter is the ATM address of the LES. To request a parameter an LE_CONFIGURE_REQUEST PDU is sent to the LECS. The LECS answers with an LE_CONFIGURE_RESPONSE PDU. Most of the parameters are encoded as a Type/Length/Value (TLV). Such a TLV is a pair formed by a 4-byte name (defined in the standard documents) and a value of variable length. After completion of the configuration the SVC will be released by the LEC.

The next state is the Join state. Here the LEC establishes an SVC to the LES that carries a Configuration Direct flow. After the connection establishment the LEC sends an LE_JOIN_REQUEST PDU to the LES. Now the LEC belongs to the emulated LAN. The PDU will also be used to register the first ATM address and the first unicast MAC address of the LEC that will be used for address resolution. The LES will add the LEC to the point-to-multipoint SVC that carries the Control Distribute flow. The request is answered by the LES with an LE_JOIN_RESPONSE PDU. This response contains the LAN Emulation Client ID (LECID) which is a unique ID assigned to the LEC in the emulated LAN.

The next state is the Initial Registration state. In this state the LEC can, but does not have to, register additional unicast and multicast addresses that will be handled by the LEC. The registration is used by the address resolution service. This registration(s) can be changed in the Operational state.

Now the LEC is in the BUS Connect state. The LEC connects the BUS to transport the Default Multicast Send flow. The LEC obtains the ATM address of the BUS by using the address resolution service and resolving the MAC broadcast address (0xffffffffffff) into an ATM address. The LEC may also establish additional SVCs to the BUS to carry Selective Multicast Send flows. Then the BUS adds the LEC to one or more point-to-point SVCs that carry Multicast Forward flows. The LEC receives multicast and broadcast messages over these SVCs.

At the end the LEC is in the Operational state. Now the LEC can deliver unicast and multicast MAC messages. In addition MAC addresses can be registered and unregistered and the address resolution service can be used.

What happens when an LEC wants to finish its membership in the emulated LAN or needs to handle the return to the Initial State because of an error? The answer is quite simple. The LEC simply releases all SVCs using UNI signalling. The LEC and BUS will not try to reestablish the released connections.

8.3.4 Address Registration

An LEC can register additional MAC address and ATM address pairs not registered in the Join state at the LES using the Data Direct flow. The LEC can also unregister address pairs. This procedure also includes the (un)registration of multicast MAC addresses where the LEC wants to receive messages. The broadcast MAC address cannot be registered.

The registration and unregistration uses four different PDU types:

- LE_REGISTER_REQUEST. This PDU is sent by the LEC to the LES to register a new address pair.
- LE_REGISTER_RESPONSE. This is the answer by an LES in response to a request from an LEC. This PDU confirms or rejects (e.g. if invalid addresses were used) the registration request.
- LE_UNREGISTER_REQUEST. This PDU type is sent by the LEC to the LES to unregister an address pair.
- LE_UNREGISTER_RESPONSE. This PDU is the answer of an LES to confirm or reject (e.g. if the address pair was not registered) the unregistration request.

Figure 8.5 shows one of the four PDUs just described. The PDU is sent on top of AAL5. The OP-CODE field is 0x0004 for LE_REGISTER_REQUEST, 0x0104 for

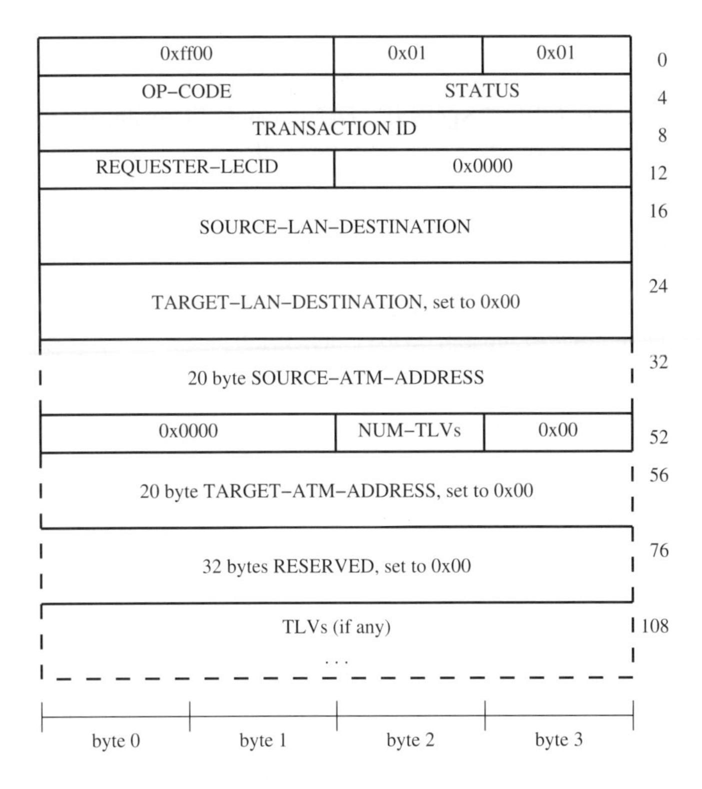

Figure 8.5: An AAL5 payload that contains an (un)registration PDU

LE_REGISTER_RESPONSE, 0x0005 for LE_UNREGISTER_REQUEST and 0x0105 for LE_UNREGISTER_RESPONSE. The STATUS field is 0x0000 for requests and successful responses. The field has a different value if the response indicates an error. The TRANSACTION ID is set by the LEC when sending a request to the LES. The response PDU contains the same TRANSACTION ID. The REQUESTER-LECID is a unique ID of the LEC and was assigned to the LEC while in the Join state. The SOURCE-LAN-DESTINATION field is the two-byte value 0x0001 plus the six-byte MAC address of the address pair. The SOURCE-ATM-ADDRESS is the ATM address of the address pair. TLVs may also be carried by a PDU to transmit additional parameters (e.g. QoS associated with the address pair).

8.3.5 Address Resolution

Address resolution is a service that translates unicast, multicast or broadcast MAC addresses into ATM addresses of an LEC or BUS and vice versa. The service is needed by an LEC to be able to transmit user data messages to the correct destination. The LEC transmits the user data messages over the SVCs that carry the Data Direct or Multicast Send flows.

To resolve an MAC address, an LEC sends an ARP request to the LEC over the Control Direct flow. The ARP request is encoded as LE_ARP_REQUEST PDU.

The LES answers the ARP request by sending an LE_ARP_RESPONSE PDU back to the requesting client on the same Control Direct flow. The answer is based on the address pairs registered and stored at the LES. If the LES cannot find an appropriate entry in its address registration cache, then one or more LECs are asked. This means that in such a case the LES forwards the original LE_ARP_RESPONSE PDU to many LECs using the Control Distribute flow or many Control Direct flows. The LES hopes that at least one LEC can resolve the address. This could be possible if the requested address was not registered by an LEC, e.g. because the LEC is a bridge that does not register all external MAC addresses. If an LEC receives a ARP request that it can resolve, then it sends an LE_ARP_RESPONSE PDU to the LES on the Control Direct flow. If the LEC cannot resolve the request, then it does nothing.

Each LEC caches address pairs resolved by ARP requests. An aging mechanism is responsible for the removal of old entries to be able to adapt to new topologies.

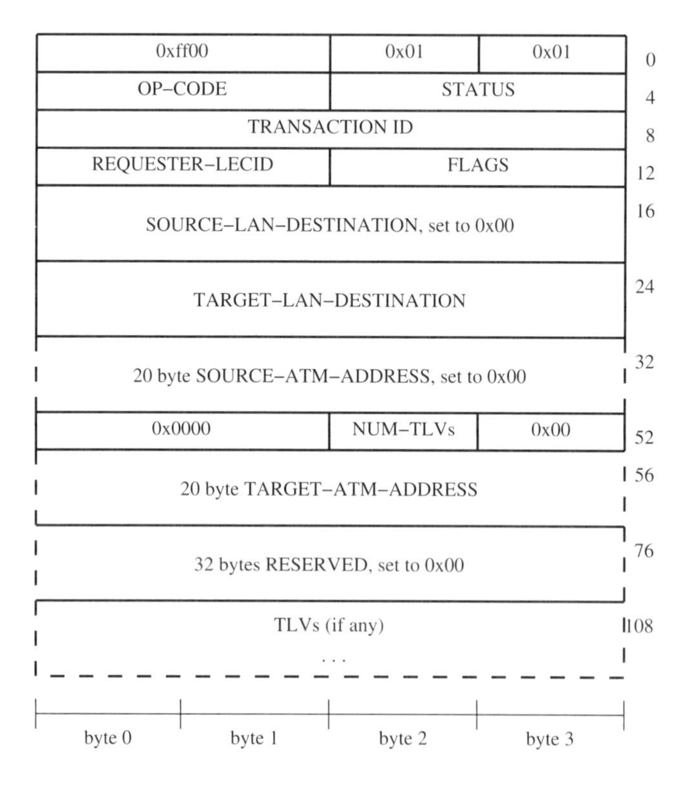

Figure 8.6: LANE ARP request/response PDU

Figure 8.6 shows the LE_ARP_REQUEST PDU which will be delivered on top of AAL5. The OP-CODE field is 0x0006 for LE_ARP_REQUEST and 0x0106 for LE_ARP_RESPONSE. The STATUS field is 0x0000 for requests and successful responses. The field has a different value if the response indicates an error. The TRANSACTION ID is set by the LEC when sending a request to the LES. The response PDU contains the same TRANSACTION ID. The REQUESTER-LECID is a unique ID of the LEC that was sending the request to the LES. The FLAGS field is usually set to 0x0000. It is set to 0x0001 in a response PDU if the TARGET-LAN-DESTINATION represents an address that was not registered at the LES (e.g. an external MAC address of a bridge). The TARGET-LAN-DESTINATION field is the two-byte value 0x0001 plus the six-byte MAC address of the address pair. The TARGET-ATM-ADDRESS field is 0x00...00 in request PDUs. In response PDUs it is the resolved ATM address that was assigned to the TARGET-LAN-DESTINATION. TLVs may also be carried with a PDU to transmit additional parameters (e.g. QoS associated with the address pair).

The LE_NARP_REQUEST PDU is an additional PDU type defined for address resolution. Such a PDU is issued by an LEC if a registered address pair is changed. The message is sent to the LES. If an LES receives such a PDU, then it forwards it to all LECs over the Control Distribute flow or all Control Direct flows. An LEC that receives the PDU removes the invalid address pair from, and adds the new address pair to, the address resolution table cache in the LEC. The invalid address pair and the new address pair are parameters of the LE_NARP_REQUEST PDU.

For the LANE address resolution mechanism an LE_TOPOLOGY_REQUEST PDU, is defined. Such a PDU is sent by an LEC to the LES if the LEC wants to advertise massive changes in address mappings. This is needed for LECs that work as transparent bridges and need to clear their internal address database. If an LES receives an LE_TOPOLOGY_REQUEST PDU then it forwards it to all LECs over the Control Distribute flow or all Control Direct flows. An LEC that receives such a PDU clears its address resolution table cache. This allows easy adaption to a new network topology.

8.3.6 User Data Transport

User data is sent from one LEC to another over a Data Direct flow SVC. The ATM address of the destination LEC is obtained by resolving the MAC address via LANE address resolution. If the destination address cannot be resolved into an ATM address, then the message is sent over the Default Multicast Send flow SVC. An example of a user data message is an IP PDU.

Multicast user data messages are sent over a Selective Multicast flow SVC to the BUS. The ATM address of the BUS is obtained by resolving the destination multicast MAC address. If the destination multicast address cannot be resolved or if it is the broadcast address, then the message is sent over the Default Multicast Send flow SVC.

Figure 8.7 shows an Ethernet user data message that will be delivered on top of AAL5. The two-byte LE-HEADER field contains the LAN Emulation Client ID (LECID) of the sending LEC. It is allowed to set this field to 0x0000 on Direct Data flows. DESTINATION ADDRESS and SOURCE ADDRESS are the MAC addresses. The TYPE field is 0x0600 or greater and contains the Ethernet payload type, e.g. 0x0800 for IP. The INFO field contains the Ethernet payload, e.g. the IP PDU. The payload is padded with zeros so that the payload is at least 46 bytes long.[7] A length field is not present. The length can be determined as the difference between the AAL5 payload length minus 16 bytes, but this calculation does not work if the Ethernet payload length is less than 46 bytes. A Cyclic Redundancy Check (CRC) or Frame Check Sequence (FCS) field is not needed because AAL5 already protects the message against bit modification errors.

An LEC can receive user data messages on different flows. It is responsible for discarding all user data messages that are not addressed to the higher layers of the LEC.

---

[7] A minimum payload length of 46 bytes is needed to ensure a minimum message size in Ethernet which is important for the Ethernet collision detection mechanism.

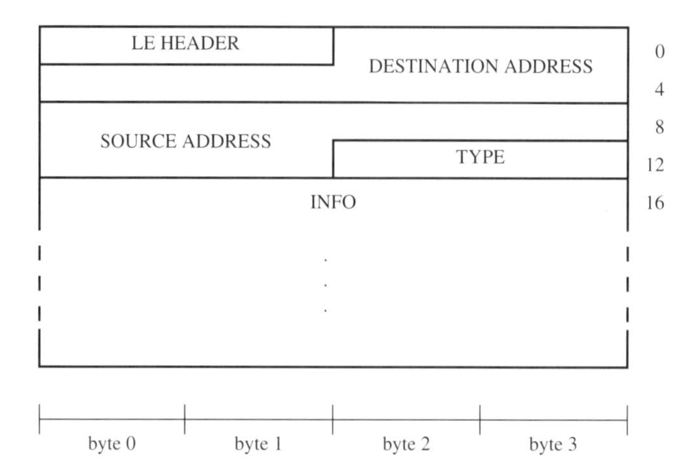

Figure 8.7: LANE Ethernet user data message

All LECs of an emulated LAN use the same MTU. This value is configurable. For Ethernet emulation the maximum user data size should be 1500 bytes.

LANE is prepared to support different Quality of Service (QoS) parameters on different flows. However, ordinary Ethernet applications do not expect any QoS support. Therefore, the LANE flows will usually be transported on Unspecified Bit Rate (UBR) connections. The QoS parameters are specified when establishing a new SVC.

8.3.7 Flush Message Protocol

When an LEC sends a user data message to a specific destination there may be two different flows that can be used: the Data Direct flow and the Default Multicast Send flow. It is also possible that an LEC sends some user data messages to one flow and then changes to the other flow. This can cause the problem that the messages are received in the wrong order because both flows transport the messages independently of the other flow. ATM only guarantees the correct message order on on VC, i.e. on one flow.

If an LEC receives user data messages then the original order cannot be recovered. This can cause problems in higher level applications because they know and expect that the Ethernet LAN transports all user data messages in the correct message order.

To avoid reordering of unicast messages the Flush Message Protocol was defined in LANE. It consists of the two PDU type LE_FLUSH_REQUEST and LE_FLUSH_RESPONSE.

Let us assume the a source LEC is sending user data messages to the destination LEC on an original flow. Now the LEC wants to change the flow. First, it will stop sending user data flows to the destination LEC. Then it sends an LE_FLUSH_REQUEST PDU to the destination LEC on the original flow. Then the source LEC waits for the LE_FLUSH_RESPONSE PDU from the destination LEC. The LE_FLUSH_RESPONSE PDU will be sent by the destination LEC using the Control Direct flow to the LES and received by the source LEC on the Control Distribute flow from the LES. After the source LEC has received the LE_FLUSH_RESPONSE PDU it is clear that there is no user data message on the way any more. Now the source LEC can choose to use a different flow to send unicast user data messages to the destination LEC.

This guarantees that the user messages are received in the order in which they were sent.

This mechanism cannot be used for multicast and broadcast messages because too many destinations are involved. Therefore, a source LEC waits for a configured path switching delay when it changes the flow.

8.3.8 Verify Protocol

LANE defines the optional Verify Protocol. It allows an LEC to check whether the calling party of a Multicast Forward flow SVC is a BUS of the emulated LAN. The Verify Protocol was defined to recognise misconfigurations in a LANE.

The Verify Protocol is quite unimportant and not needed for a working emulated LAN.

8.3.9 Interface to Higher Layer Services

One objective of LANE is the ability to work together with existing network layer protocol stacks like IP, NetBIOS, Apple Talk and others. These protocol stacks expect to communicate with an MAC device driver.[8] Some industry standards for MAC device drivers are in common use, e.g. Network Driver Interface Specification (NDIS) from Microsoft and Open Data-link Interface (ODI) from Novell. These industry standards specify the access to a MAC device driver. All these specifications use similar service primitives. These service primitives can be provided by the LANE service, i.e. LANE emulates the behaviour of existing MAC device drivers. This allows an easy migration from a LAN to LANE because the upper protocol layer does not need to be changed.

The following is a description of the two most important service primitives defined at the interface to the higher layer in the LEC. These primitives provide the capability to exchange user data (e.g. IP PDUs) over the LANE service.

The first important service primitive is LE_UNITDATA.request. This service primitive is generated by the higher layer whenever it has a PDU to be transferred to a peer entity or peer entities. Parameters of this service primitive are the unicast or multicast MAC address of the destination, the unicast MAC address of the source, the frame type to specify Ethernet or IEEE 802.3 and the user data payload. Beginning with LANE version 2, QoS parameters can be assigned to the LE_UNITDATA.request. Usually, this is not used because classical LAN applications do not know anything about QoS, i.e. the default QoS parameters will be used.

The second important service primitive is LE_UNITDATA.indication. It is passed from the LANE layer to the next higher layer to indicate the arrival of a PDU. Parameters of this service primitive are unicast or multicast MAC address of the destination, the unicast MAC address of the source, the frame type and the user data payload of the incoming PDU.

---

[8] More formally a network protocol layer stack communicates with the Logical Link Control (LLC) which is the upper sublayer of the data link layer, and the LLC communicates with the Multiple Access Control (MAC) layer which is the lower sublayer of the data link layer. In the case of Ethernet the LLC has no functionality. Therefore we can say that the network layer communicates with the MAC layer. But this is not true for IEEE 802.3 and Token Ring/IEEE 802.5.

8.3.10 Management of a LEC

The Simple Network Management Protocol (SNMP) is widely used to allow easy management of network attached devices. It can also be used for devices of an emulated LAN. In [LECM2] algorithms and MIBs are defined for configuration management, performance management and fault management of LECs.

8.4 Sylvia: A Native ATM Multimedia Application

The previous sections described how to use classical IP/LAN applications on top of ATM. The problem of this applications is that they cannot use the advantages of QoS that can be provided by ATM networks. Therefore it is an interesting alternative approach to develop native ATM applications. In this section we want to give a short introduction to such a native ATM application, namely Sylvia.

Sylvia is an extensible, high-end video conference software. It offers excellent audio and video quality for different kinds of multimedia services. Because it is based directly on ATM, Sylvia is able to use the QoS guarantees offered by the ATM network. The modular structure of Sylvia enables easy installation of new services. The security mechanisms of Sylvia ensure privacy of communication.

Services currently provided by Sylvia include a multipoint video conference, audio and video telephony, Audio on Demand (AoD) and Video on Demand (VoD), and a shared white board. Other possible services are an audio and video mail box, teleteaching, and TV and radio distribution. Sylvia runs on multimedia workstations connected together by an ATM network.

SVCs or PVCs are used to deliver audio and video streams as well as control information from and to the users. The audio and video streams especially profit from the QoS of the ATM network. Different service categories and traffic parameter sets can be used. Sylvia was tested with UBR connections in a local environment and with Constant Bit Rate (CBR) connections[9] in international sessions using a public ATM network.

Depending on the available bandwidth Sylvia can offer a broad range of audio and video quality. The audio bandwidth can be from 36 kbit/s for telephone like quality up to 1.6 Mbit/s for CD-quality in stereo. The video bandwidth ranges from below 1 MBit/s for small pictures up to 6 Mbit/s for 786×576 frames.

Sylvia supports different audio (8/16-bit PCM, G711 A-law and μ-Law, G.721/G.723 ADPCM, MPEG bit stream) and video (Motion-JPEG, MPEG) compression standards.

To establish switched connections through private and public ATM networks different signalling protocols can be selected: UNI 3.1, Q.2931, Q.2932.1 and Fore SPANS. NSAP as well as E.164 address formats are supported. Multicast connections are not used because of the availability problems in public networks.

Sylvia is implemented as a collection of interacting modules written in Java and C for high portability and performance. Sylvia currently runs on Sun and Linux workstations with different audio and video equipment (e.g. Parallax, SunVideo).

A multimedia server can be used as an optional component of Sylvia. It runs centralised services like AoD and VoD, or TV distribution. The server is designed for scalability and extensibility—services can be added and removed without interrupting server operation. It

[9] The use of CBR requires shaping.

is possible to run the server distributed on several machines for maximum performance and reliability. The multimedia server is implemented in C for high portability and maximum performance. It currently runs on Sun workstations under the Solaris operating system.

The flexible Sylvia and multimedia server architecture supports the easy introduction of new services. In the example configuration the video telephony (VT), and the AoD and VoD services are installed. The AoD and VoD service needs a module running in the multimedia server because the server stores video and audio clips and handles user navigation and accounting.

Sylvia was developed in the context of research activities at the Research Institute for Open Communication (GMD Fokus). Further information is available from the Sylvia Internet home page [BTT+98].

8.5 Summary

In this chapter we looked briefly at some protocols that are used on top of ATM signalling. In today's world the most important protocols are the Internet protocols. A major effort of the standardisation bodies was the definition of protocols to support the Internet in ATM. Two of these protocols are: CLIP (Classical IP over ATM) which, as it turns out, is very easy to implement, but suffers from security problems, bad scaleability and missing multicast, and LANE (LAN emulation) which emulates traditional local area networks like Ethernet on top of ATM. The chapter closed with an example of an application that builds directly on ATM and thus can employ the Quality of Service features of native ATM.

Appendix A

ITU-T Standards

The following Standard are available from the International Telecommunication Union (ITU-T).

- [Q.2010] Broadband integrated services digital network overview Signalling capability set 1, release 1
- [Q.2100] B-ISDN signalling ATM adaptation layer (SAAL) overview description
- [Q.2110] B-ISDN ATM adaptation layer Service specific connection oriented protocol (SSCOP)
- [Q.2119] B-ISDN ATM adaptation layer Convergence function for SSCOP above the frame relay core service
- [Q.2120] B-ISDN meta-signalling protocol
- [Q.2130] B-ISDN signalling ATM adaptation layer Service specific coordination function for support of signalling at the user–network interface (SSCF at the UNI)
- [Q.2140] B-ISDN ATM adaptation layer Service specific coordination function for signalling at the network node interface (SSCF at the NNI)
- [Q.2144] B-ISDN signalling ATM adaptation layer (SAAL) Layer management for the SAAL at the network node interface (NNI)
- [Q.2210] Message transfer part level 3 functions and messages using the services of ITU-T Recommendation
- [Q.2610] Usage of cause and location in B-ISDN user part and DSS 2
- [Q.2650] Interworking between Signalling System No. 7 broadband ISDN User Part (B-ISUP) and digital subscriber Signalling System No. 2 (DSS 2)
- [Q.2660] Interworking between Signalling System No. 7 Broadband ISDN User Part (B-ISUP) and Narrow-band ISDN User Part (N-ISUP)
- [Q.2721.1] B-ISDN user part Overview of the B-ISDN Network Node Interface Signalling Capability Set 2, Step 1
- [Q.2722.1] B-ISDN User Part Network Node Interface specification for point-to-multipoint call/connection control
- [Q.2723.1] B-ISDN User Part Support of additional traffic parameters for Sustainable Cell Rate and Quality of Service
- [Q.2723.2] Extensions to the B-ISDN User Part Support of ATM transfer capability in the broadband bearer capability parameter
- [Q.2723.3] Extensions to the B-ISDN User Part Signalling capabilities to support traffic parameters for the Available Bit Rate (ABR) ATM transfer capability

- [Q.2723.4] Extensions to the B-ISDN User Part Signalling capabilities to support traffic parameters for the ATM Block Transfer (ABT) ATM transfer capability
- [Q.2723.6] Extensions to the Signalling System No. 7 B-ISDN User Part Signalling capabilities to support the indication of the Statistical Bit Rate configuration 2 (SBR 2) and 3 (SBR 3) ATM transfer capabilities
- [Q.2724.1] B-ISDN User Part Look-ahead without state change for the Network Node Interface (NNI)
- [Q.2725.1] B-ISDN User Part Support of negotiation during connection setup
- [Q.2725.2] B ISDN User Part Modification procedures
- [Q.2725.3] Extensions to the B-ISDN User Part Modification procedures for sustainable cell rate parameters
- [Q.2725.4] Extensions to the Signalling System No. 7 B-ISDN User Part Modification procedures with negotiation
- [Q.2726.1] B-ISDN user part ATM end system address
- [Q.2726.2] B-ISDN user part Call priority
- [Q.2726.3] B-ISDN user part Network generated session identifier
- [Q.2726.4] Extensions to the B-ISDN user part Application generated identifiers
- [Q.2727] B-ISDN user part Support of frame relay
- [Q.2730] Signalling System No. 7 B-ISDN User Part (B-ISUP) Supplementary services
- [Q.2735.1] Stage 3 description for community of interest supplementary services for B-ISDN using SS No.7: Closed User Group (CUG)
- [Q.2751.1] Extension of Q.751.1 for SAAL signalling links
- [Q.2761] Functional description of the B-ISDN user part (B-ISUP) of signalling system No. 7
- [Q.2762] General Functions of messages and signals of the B-ISDN user part (B-ISUP) of Signalling System No. 7
- [Q.2763] Signalling System No. 7 B-ISDN User Part (B-ISUP) Formats and codes
- [Q.2764] Signalling System No. 7 B-ISDN User Part (B-ISUP) Basic call procedures
- [Q.2766.1] Switched virtual path capability
- [Q.2767.1] Soft PVC capability
- [Q.2931] Digital Subscriber Signalling System No. 2 (DSS 2) User–Network Interface (UNI) layer 3 specification for basic call/connection control
- [Q.2931a1] Digital Subscriber Signalling System No. 2 User–Network Interface (UNI) layer 3 specification for basic call/connection control
- [Q.2932.1] Digital subscriber signalling system No. 2 Generic functional protocol: Core functions
- [Q.2933] Digital subscriber Signalling System No. 2 (DSS 2) Signalling specification for Frame Relay service
- [Q.2934] Digital subscriber signalling system No. 2 Switched virtual path capability
- [Q.2939.1] Digital Subscriber Signalling System No. 2 Application of DSS 2 service-related information elements by equipment supporting B-ISDN services
- [Q.2941.1] Digital subscriber Signalling System No. 2 Generic identifier transport
- [Q.2951] Stage 3 description for number identification supplementary services using B-ISDN digital subscriber Signalling System No. 2 (DSS 2) Basic Call
- [Q.2955.1] Stage 3 description for community of interest supplementary services using B-ISDN Digital Subscriber Signalling System No. 2 (DSS 2): Closed User Group (CUG)
- [Q.2957.1] User-to-user signalling (UUS)

- [Q.2959] Digital subscriber signalling system No. 2 Call priority
- [Q.2961] Digital subscriber signalling system No. 2 Additional traffic parameters
- [Q.2961.2] Support of ATM transfer capability in the broadband bearer capability information element
- [Q.2961.3] Digital Subscriber Signalling System No. 2 Additional traffic parameters: Signalling capabilities to support traffic parameters for the available bit rate (ABR) ATM transfer capability
- [Q.2961.4] Digital Subscriber Signalling System No. 2 Additional traffic parameters: Signalling capabilities to support traffic parameters for the ATM Block Transfer (ABT) ATM transfer capability
- [Q.2961.6] Digital Subscriber Signalling System No. 2 Additional traffic parameters: Additional signalling procedures for the support of the SBR2 and SBR3 ATM transfer capabilities
- [Q.2962] Digital Subscriber Signalling System No. 2 Connection characteristics negotiation during call/connection establishment phase
- [Q.2963.1] Digital Subscriber Signalling System No. 2 Connection modification: Peak cell rate modification by the connection owner
- [Q.2963.2] Digital Subscriber Signalling System No. 2 Connection modification: Modification procedures for sustainable cell rate parameters
- [Q.2963.3] Digital Subscriber Signalling System No. 2 Connection modification: ATM traffic descriptor modification with negotiation by the connection owner
- [Q.2964.1] Digital Subscriber Signalling System No. 2: Basic Look-Ahead
- [Q.2971] Broadband integrated services digital network (B-ISDN) Digital subscriber signalling system No. 2 (DSS 2) User–network interface layer 3 specification for point-to-multipoint call/connection control

Appendix B

Source Code Availability

In this book we use and refer to many publicly available software programs and documents. This appendix provides information on how to obtain these software programs and documents

B.1 Standards

B.1.1 ATM Forum Standards

The standards and specifications of the ATM Forum (ATMF) are freely available via the Internet. Starting from the home page `http://www.atmforum.com`, go to "Technical Specifications" and then to "Approved Specifications". Here the downloadable documents are listed. The documents are available in one or more of the following formats: PDF, PostScript, MS Word. If the file names are known, then the documents can be directly downloaded via anonymous FTP from `ftp://ftp.atmforum.com/pub/approved-specs/`.

B.1.2 ITU-T Standards

The Q.2xxx standards/recommendations of the International Telecommunication Union (ITU-T) are available via the Internet or from CD-ROM. In both cases they must be purchased. The home page is `http://www.itu.int`. To download documents via the online shop go to "Standardisation", then to "ITU-T Publications" and then to "Recommendations in force". The documents are available in one or more of the following formats: PDF, PostScript, RTF, MS Word.

B.1.3 RFCs

Every Internet standard Request for Comments (RFC) can be freely downloaded from many sites in the Internet. One URL is `http://www.ietf.org/rfc.html`. The documents are plain ASCII.

B.2 Protocol Tracing Tools

The Tina framework can be freely downloaded from the software section of the home page of Begemot Computer Associates. The URL is `http://www.begemot.org`. Distributions for different platforms (Solaris, FreeBSD and Linux) are available.

B.3 ATM Protocol Software

The FreeUNI ATM protocol software can be freely downloaded from the software section of the home page of Begemot Computer Associates, `http://www.begemot.org`. It is a source code distribution which must be built on the destination platform.

References

[ADDR1.0] The ATM Forum, Technical Committee (January 1999) *ATM Forum Addressing: User Guide Version 1.0*. af-ra-0105.000.

[ADDRR] The ATM Forum, Technical Committee (February 1999) *ATM Forum Addressing: Reference Guide*. af-ra-0106.000.

[ADDRUNI] The ATM Forum, Technical Committee (February 1999) *Addressing Addendum for UNI Signalling 4.0*. af-ra-0107.000.

[AINI1.0] The ATM Forum, Technical Committee (July 1999) *ATM Inter-Network Interface (AINI) Specification*. af-cs-0125.000.

[Beg] Begemot Computer Associates *ATM Software*. `http://www.begemot.org`.

[BICI2.0] The ATM Forum, Technical Committee (December 1995) *B-ISDN Inter Carrier Interface (B-ICI) Specification Version 2.0 (Integrated)*. af-bici-0013.003.

[BICI2.1] The ATM Forum, Technical Committee (November 1996) *Addendum to B-ISDN Inter Carrier Interface (B-ICI) Specification Version 2.0 (B-ICI Specification Version 2.1)*. af-bici-0068.000.

[BTT+98] Brandt H., Todorova P., Tittel C., Welk M., Tchouto J.-J., and Hapke C. (1998) *Sylvia Homepage*. GMD Fokus, `http://www.fokus.gmd.de/research/cc/cats/products/sylvia/`.

[E.164] International Telecommunication Union (ITU-T) (May 1997) *ITU-T Recommendation E.164: The international public telecommunication numbering plan*.

[E.191] International Telecommunication Union (ITU-T) (October 1996) *ITU-T Recommendation E.191: B-ISDN numbering and addressing*.

[GMD] GMD Fokus *Tanya ATM network interface*. `http://www.fokus.gmd.de/research/cc/tip/products`.

[ILMI4] The ATM Forum, Technical Committee (September 1996) *Integrated Link Management Interface (ILMI) Specification—Version 4*. af-ilmi-0065.000.

[ISO3166] International Standardisation Organisation (December 1993) *ISO 3166: Codes for the representation of the names of countries*.

[ISO8348] International Standardisation Organisation (September 1996) *ISO/IEC 8348: Information Techology—Open Systems Interconnection—Network Service Definition*.

[LECM2] The ATM Forum, Technical Committee (October 1998) *LAN Emulation Client Management Specification Version 2*. af-lane-0093.000.

[LNNI2] The ATM Forum, Technical Committee (February 1999) *LAN Emulation Over ATM Version 2—LNNI Specification*. af-lane-0112.000.

[LUNI2] The ATM Forum, Technical Committee (July 1997) *LAN Emulation Over ATM Version 2—LUNI Specification*. af-lane-0084.000.

[MPOA1.0] The ATM Forum, Technical Committee (July 1997) *Multi-Protocol over ATM Version 1.0*. af-mpoa-0087.000.

[MPOA1.1] The ATM Forum, Technical Committee (May 1999) *Multi-Protocol over ATM Version 1.1*. af-mpoa-0114.000.

[PNNI] The ATM Forum, Technical Committee (March 1996) *Private Network–Network Interface Specification Version 1.0 (PNNI 1.0)*. af-pnni-0055.000.

[Q.2010] International Telecommunication Union (ITU-T) (February 1995) *ITU-T Recommendation Q.2010: Broadband integrated services digital network overview Signalling capability set 1, release 1*.

[Q.2100] International Telecommunication Union (ITU-T) (July 1994) *ITU-T Recommendation Q.2100:*

B-ISDN Signalling ATM Adaptation Layer (SAAL) overview description.
[Q.2110] International Telecommunication Union (ITU-T) (July 1994) *ITU-T Recommendation Q.2110: B-ISDN ATM Adaptation Layer—Service Specific Connection Oriented Protocol (SSCOP).*
[Q.2119] International Telecommunication Union (ITU-T) (July 1996) *ITU-T Recommendation Q.2119: B-ISDN ATM Adaptation Layer—Convergence Function for SSCOP above the Frame Relay Core Service.*
[Q.2120] International Telecommunication Union (ITU-T) (February 1995) *ITU-T Recommendation Q.2120: B-ISDN Meta-Signalling Protocol.*
[Q.2130] International Telecommunication Union (ITU-T) (July 1994) *ITU-T Recommendation Q.2130: B-ISDN Signalling ATM Adaptation Layer—Service Specific Coordination Function for Support of Signalling at the User–Network Interface (SSCF at the UNI).*
[Q.2140] International Telecommunication Union (ITU-T) (February 1995) *ITU-T Recommendation Q.2140: B-ISDN ATM Adaptation Layer—Service Specific Coordination Function for Signalling at the Network Node Interface (SSCF at the NNI).*
[Q.2144] International Telecommunication Union (ITU-T) (October 1995) *ITU-T Recommendation Q.2144: B-ISDN Signalling ATM Adaptation Layer (SAAL)—Layer Management for the SAAL at the Network Node Interface (NNI).*
[Q.2210] International Telecommunication Union (ITU-T) (July 1996) *ITU-T Recommendation Q.2210: Message Transfer Part Level 3 Functions and Messages using the Services of ITU-T Recommendation Q.2140.*
[Q.2610] International Telecommunication Union (ITU-T) (February 1995) *ITU-T Recommendation Q.2610: Usage of Cause and Location in B-ISDN User Part and DSS 2.*
[Q.2650] International Telecommunication Union (ITU-T) (February 1995) *ITU-T Recommendation Q.2650: Interworking between Signalling System No. 7—Broadband ISDN User Part (B-ISUP) and Digital Subscriber Signalling System No. 2 (DSS 2).*
[Q.2660] International Telecommunication Union (ITU-T) (February 1995) *ITU-T Recommendation Q.2660: Interworking between Signalling System No. 7—Broadband ISDN User Part (B-ISUP) and Narrow-band ISDN User Part (N-ISUP).*
[Q.2721.1] International Telecommunication Union (ITU-T) (July 1996) *ITU-T Recommendation Q.2721.1: B-ISDN User Part—Overview of the B-ISDN Network Node Interface Signalling Capability Set 2, Step 1.*
[Q.2722.1] International Telecommunication Union (ITU-T) (July 1996) *ITU-T Recommendation Q.2722.1: B-ISDN User Part—Network Node Interface specification for point-to-multipoint Call/Connection Control.*
[Q.2723.1] International Telecommunication Union (ITU-T) (July 1996) *ITU-T Recommendation Q.2723.1: B-ISDN User Part—Support of Additional Traffic Parameters for Sustainable Cell Rate and Quality of Service.*
[Q.2723.2] International Telecommunication Union (ITU-T) (September 1997) *ITU-T Recommendation Q.2723.2: Extensions to the B-ISDN User Part—Support of ATM Transfer Capability in the Broadband Bearer Capability Parameter.*
[Q.2723.3] International Telecommunication Union (ITU-T) (September 1997) *ITU-T Recommendation Q.2723.3: Extensions to the B-ISDN User Part—Signalling Capabilities to Support Traffic Parameters for the Available Bit Rate (ABR) ATM Transfer Capability.*
[Q.2723.4] International Telecommunication Union (ITU-T) (September 1997) *ITU-T Recommendation Q.2723.4: Extensions to the B-ISDN User Part—Signalling Capabilities to Support Traffic Parameters for the ATM Block Transfer (ABT) ATM Transfer Capability.*
[Q.2723.6] International Telecommunication Union (ITU-T) (May 1998) *ITU-T Recommendation Q.2723.6: Extensions to the Signalling System No. 7—B-ISDN User Part—Signalling Capabilities to Support the Indication of the Statistical Bit Rate Configuration 2 (SBR 2) and 3 (SBR 3) ATM Transfer Capabilities.*
[Q.2724.1] International Telecommunication Union (ITU-T) (July 1996) *ITU-T Recommendation Q.2724.1: B-ISDN User Part—Look-ahead without State Change for the Network Node Interface (NNI).*

[Q.2725.1] International Telecommunication Union (ITU-T) (May 1998) *ITU-T Recommendation Q.2725.1: B-ISDN User Part—Support of Negotiation during Connection Setup.*

[Q.2725.2] International Telecommunication Union (ITU-T) (July 1996) *ITU-T Recommendation Q.2725.2: B ISDN User Part—Modification Procedures.*

[Q.2725.3] International Telecommunication Union (ITU-T) (September 1997) *ITU-T Recommendation Q.2725.3: Extensions to the B-ISDN User Part—Modification Procedures for Sustainable Cell Rate Parameters.*

[Q.2725.4] International Telecommunication Union (ITU-T) (May 1998) *ITU-T Recommendation Q.2725.4: Extensions to the Signalling System No. 7 B-ISDN User Part—Modification Procedures with Negotiation.*

[Q.2726.1] International Telecommunication Union (ITU-T) (July 1996) *ITU-T Recommendation Q.2726.1: B-ISDN User Part—ATM End System Address.*

[Q.2726.2] International Telecommunication Union (ITU-T) (July 1996) *ITU-T Recommendation Q.2726.2: B-ISDN User Part—Call Priority.*

[Q.2726.3] International Telecommunication Union (ITU-T) (July 1996) *ITU-T Recommendation Q.2726.3: B-ISDN User Part—Network Generated Session Identifier.*

[Q.2726.4] International Telecommunication Union (ITU-T) (September 1997) *ITU-T Recommendation Q.2726.4: Extensions to the B-ISDN User Part—Application Generated Identifiers.*

[Q.2727] International Telecommunication Union (ITU-T) (September 1996) *ITU-T Recommendation Q.2727: B-ISDN User Part—Support of Frame Relay.*

[Q.2730] International Telecommunication Union (ITU-T) (February 1995) *ITU-T Recommendation Q.2730: Signalling System No. 7—B-ISDN User Part (B-ISUP)—Supplementary services.*

[Q.2735.1] International Telecommunication Union (ITU-T) (June 1997) *ITU-T Recommendation Q.2735.1: Stage 3 Description for Community of Interest Supplementary Services for B-ISDN using SS No. 7: Closed User Group (CUG).*

[Q.2751.1] International Telecommunication Union (ITU-T) (September 1997) *ITU-T Recommendation Q.2751.1: Extension of Q.751.1 for SAAL Signalling Links.*

[Q.2761] International Telecommunication Union (ITU-T) (February 1995) *ITU-T Recommendation Q.2761: Functional Description of the B-ISDN User Part (B-ISUP) of Signalling System No. 7.*

[Q.2762] International Telecommunication Union (ITU-T) (February 1995) *ITU-T Recommendation Q.2762: General Functions of Messages and Signals of the B-ISDN User Part (B-ISUP) of Signalling System No. 7.*

[Q.2763] International Telecommunication Union (ITU-T) (February 1995) *ITU-T Recommendation Q.2763: Signalling System No. 7 B-ISDN User Part (B-ISUP)—Formats and Codes.*

[Q.2764] International Telecommunication Union (ITU-T) (February 1995) *ITU-T Recommendation Q.2764: Signalling System No. 7 B-ISDN User Part (B-ISUP)—Basic Call Procedures.*

[Q.2766.1] International Telecommunication Union (ITU-T) (May 1998) *ITU-T Recommendation Q.2766.1: Switched Virtual Path Capability.*

[Q.2767.1] International Telecommunication Union (ITU-T) (May 1998) *ITU-T Recommendation Q.2767.1: Soft PVC Capability.*

[Q.2931] International Telecommunication Union (ITU-T) (February 1995) *ITU-T Recommendation Q.2931: Digital Subscriber Signalling System No. 2 (DSS 2)—User–Network Interface (UNI) layer 3 Specification for Basic Call/Connection Control.*

[Q.2931a1] International Telecommunication Union (ITU-T) (June 1997) *ITU-T Recommendation Q.2931 Amd 1: Digital Subscriber Signalling System No. 2—User-Network Interface (UNI) Layer 3 Specification for Basic Call/Connection Control.*

[Q.2932.1] International Telecommunication Union (ITU-T) (July 1996) *ITU-T Recommendation Q.2932.1: Digital Subscriber Signalling System No. 2 —Generic Functional Protocol: Core Functions.*

[Q.2933] International Telecommunication Union (ITU-T) (July 1996) *ITU-T Recommendation Q.2933: Digital Subscriber Signalling System No. 2 (DSS 2)—Signalling Specification for Frame Relay Service.*

[Q.2934] International Telecommunication Union (ITU-T) (May 1998) *ITU-T Recommendation*

Q.2934: Digital Subscriber Signalling System No. 2 —Switched Virtual Path Capability.

[Q.2939.1] International Telecommunication Union (ITU-T) (September 1997) *ITU-T Recommendation Q.2939.1: Digital Subscriber Signalling System No. 2 —Application of DSS 2 Service-related Information Elements by Equipment Supporting B-ISDN Services.*

[Q.2941.1] International Telecommunication Union (ITU-T) (September 1997) *ITU-T Recommendation Q.2941.1: Digital Subscriber Signalling System No. 2 —Generic Identifier Transport.*

[Q.2951] International Telecommunication Union (ITU-T) (May 1998) *ITU-T Recommendation Q.2951: Stage 3 Description for Number Identification Supplementary Services using B-ISDN Digital Subscriber Signalling System No. 2 (DSS 2)—Basic Call.*

[Q.2955.1] International Telecommunication Union (ITU-T) (June 1997) *ITU-T Recommendation Q.2955.1: Stage 3 Description for Community of Interest Supplementary Services using B-ISDN Digital Subscriber Signalling System No. 2 (DSS 2): Closed User Group (CUG).*

[Q.2957.1] International Telecommunication Union (ITU-T) (February 1995) *ITU-T Recommendation Q.2957.1: User-to-user Signalling (UUS).*

[Q.2959] International Telecommunication Union (ITU-T) (July 1996) *ITU-T Recommendation Q.2959: Digital Subscriber Signalling System No. 2 —Call Priority.*

[Q.2961] International Telecommunication Union (ITU-T) (June 1997) *ITU-T Recommendation Q.2961: Digital Subscriber Signalling System No. 2 —Additional Traffic Parameters.*

[Q.2961.2] International Telecommunication Union (ITU-T) (June 1997) *ITU-T Recommendation Q.2961.2: Support of ATM Transfer Capability in the Broadband Bearer Capability Information Element.*

[Q.2961.3] International Telecommunication Union (ITU-T) (September 1997) *ITU-T Recommendation Q.2961.3: Digital Subscriber Signalling System No. 2 —Additional Traffic Parameters: Signalling capabilities to support Traffic Parameters for the Available Bit Rate (ABR) ATM transfer capability.*

[Q.2961.4] International Telecommunication Union (ITU-T) (September 1997) *ITU-T Recommendation Q.2961.4: Digital Subscriber Signalling System No. 2 —Additional Traffic Parameters: Signalling Capabilities to support Traffic Parameters for the ATM Block Transfer (ABT) ATM Transfer Capability.*

[Q.2961.6] International Telecommunication Union (ITU-T) (May 1998) *ITU-T Recommendation Q.2961.6: Digital Subscriber Signalling System No. 2 —Additional Traffic Parameters: Additional Signalling Procedures for the support of the SBR2 and SBR3 ATM Transfer Capabilities.*

[Q.2962] International Telecommunication Union (ITU-T) (May 1998) *ITU-T Recommendation Q.2962: Digital Subscriber Signalling System No. 2 —Connection Characteristics Negotiation during Call/Connection Establishment Phase.*

[Q.2963.1] International Telecommunication Union (ITU-T) (July 1996) *ITU-T Recommendation Q.2963.1: Digital Subscriber Signalling System No. 2 —Connection Modification: Peak Cell Rate Modification by the Connection Owner.*

[Q.2963.2] International Telecommunication Union (ITU-T) (September 1997) *ITU-T Recommendation Q.2963.2: Digital Subscriber Signalling System No. 2 —Connection Modification: Modification Procedures for Sustainable Cell Rate Parameters.*

[Q.2963.3] International Telecommunication Union (ITU-T) (May 1998) *ITU-T Recommendation Q.2963.3: Digital Subscriber Signalling System No. 2 —Connection Modification: ATM Traffic Descriptor Modification with Negotiation by the Connection Owner.*

[Q.2964.1] International Telecommunication Union (ITU-T) (July 1996) *ITU-T Recommendation Q.2964.1: Digital Subscriber Signalling System No. 2: Basic Look-Ahead.*

[Q.2971] International Telecommunication Union (ITU-T) (October 1995) *ITU-T Recommendation Q.2971: Broadband Integrated Services Digital Network (B-ISDN)—Digital Subscriber Signalling System No. 2 (DSS 2)—User–network Interface Layer 3 Specification for point-to-multipoint Call/Connection Control.*

[Q.704] International Telecommunication Union (ITU-T) (July 1996) *ITU-T Recommendation Q.704: Signalling network functions and messages.*

[Q.850] International Telecommunication Union (ITU-T) (May 1998) *ITU-T Recommendation Q.850: Usage of cause and location in the Digital Subscriber Signalling System No. 1 and the Signalling*

System No. 7 ISDN User Part.

[RFC1155] Network Working Group (May 1990) *RFC 1155: Structure and Identification of Management Information for TCP/IP-based Internets.*

[RFC1157] Network Working Group (May 1990) *RFC 1157: Simple Network Management Protocol.*

[RFC1213] Network Working Group (March 1991) *RFC 1216: Management Information Base for Network Management of TCP/IP-based Internets: MIB-II.*

[RFC1293] Bradly T. and Brown C. (January 1992) *RFC 1293: Inverse Address Resolution Protocol.* Wellfleet Communications.

[RFC1483] Heinanen J. (July 1993) *RFC 1483: Multiprotocol Encapsulation over ATM adaptation Layer 5.* Telecom Finland.

[RFC1577] Laubach M. (January 1994) *RFC 1577: Classical IP and ARP over ATM.* Hewlett-Packard Laboratories.

[RFC826] Plummer D. C. (November 1982) *RFC 826: An Ethernet Address Resolution Protocol.* MIT.

[UNI3.1] The ATM Forum, Technical Committee (September 1994) *ATM User–Network Interface Specification; Version 3.1.* af-uni-0010.002.

[UNI4.0] The ATM Forum, Technical Committee (July 1996) *ATM User–Network Interface (UNI) Signalling Specification; Version 4.0.* af-sig-0061.000.

[VH98] Venieris I. and Hussmann H. (eds) (1998) *Intelligent Broadband Networks.* John Wiley & Sons, Chichester, New York, Weinheim, Brisbrane, Singapore, Toronto.

[WATM] The ATM Forum, Technical Committee (October 1998) *Wireless ATM Capability Set 1 Specification—Draft.* btd-watm-01.09.

[X.213] International Telecommunication Union (ITU-T) (November 1995) *ITU-T Recommendation X.213: Information Techology—Open Systems Interconnection—Network Service Definition.*

Index